Clinical Optics
and Refraction

A Guide for Optometrists, Contact Lens Opticians and Dispensing Opticians

For Elsevier:

Commissioning Editor: Robert Edwards
Development Editor: Veronika Krcilova
Project Manager: Anne Dickie
Design Direction: George Ajayi
Illustrator: Oxford Illustrators
Illustration Buyer: Gillian Richards

Clinical Optics and Refraction

A Guide for Optometrists, Contact Lens Opticians and Dispensing Opticians

Andrew Keirl BOptom (Hons), MCOptom, FBDO

Optometrist, Dispensing Optician and Contact Lens Practitioner, Noakes and Habermehl, Cornwall, UK
Low Vision Practitioner, Royal Eye Infirmary, Plymouth
Optometric Educator and CET Provider
Formerly Senior Lecturer and Field Leader, Department of Optometry and Ophthalmic Dispensing, Anglia Ruskin University, Cambridge

Caroline Christie BSc (Hons), FCOptom, DipCLP

Optometrist and Contact Lens Practitioner
Optometric Educator and CET Provider
Consultant to the Contact Lens Industry

BAILLIÈRE TINDALL

ELSEVIER

EDINBURGH LONDON NEW YORK OXFORD PHILADELPHIA ST LOUIS SYDNEY TORONTO 2007

BAILLIÈRE
TINDALL
ELSEVIER

An imprint of Elsevier Limited

First published 2007

© 2007, Elsevier Ltd

ISBN-13 978 0 7506 8889 5

British Library Cataloguing in Publication Data
A catalogue record for this book is available from the British Library.

Library of Congress Cataloging in Publication Data
A catalog record for this book is available from the Library of Congress.

Note
Neither the Publisher nor the Authors assume any responsibility for any loss or injury and/or damage to persons or property arising out of or related to any use of the material contained in this book. It is the responsibility of the treating practitioner, relying on independent expertise and knowledge of the patient, to determine the best treatment and method of application for the patient.

The
publisher's
policy is to use
**paper manufactured
from sustainable forests**

Printed in China

Contents

Contents

Contributors

Bruce Evans, BSc, PhD, FCOptom, DipCLP, DipOrth, FAAO
Director of Research, Institute of Optometry, London, UK; Visiting Professor, City University, London, UK

Andrew Franklin, BSc, FBCO, DOrth, DCLP
Professional Programme Tutor, Boots Opticians Examiner, College of Optometrists, UK; Optometrist in private practice, Gloucestershire, UK

William Harvey, MCOptom
Visiting Clinician and Boots Opticians' Tutor Practitioner, Fight for Sight Optometry Clinic, City University, London, UK; Clinical Editor, Optician, Reed Business Information, Sutton, UK; Examiner, College of Optometrists, UK

Preface

Visual optics must be one of the widest ranging topics in clinical optometry, spanning as it does ocular anatomy, geometrical and physical optics, physiological optics, ophthalmic lens theory, refraction, instrumentation and many other aspects. There are of course many substantial textbooks on the subject of visual optics available to students and practitioners, the two most popular and also most respected being *Introduction to Visual Optics* by Alan Tunnacliffe and *Bennett & Rabbetts Clinical Visual Optics* by Ronald Rabbetts. These major works have stood the test of time and it is no coincidence that this book will bear some similarities to the aforementioned texts. However, this work should be thought of as a 'how-to-do-it' book as opposed to a rigorous theoretical textbook.

The authors believe that in order to fully understand any optics-based subject, a student must invest a considerable amount of time mastering the various calculations that are so important in illustrating and underpinning optical theory. It is for this reason that the majority of the chapters within this book contain worked examples to allow the student to solve and explore problems relevant to the subject being discussed. Wherever possible, the worked examples employ first principles and thus avoid the need for lengthy and tedious derivations. The interested reader can of course source such derivations from other readily available texts.

The book has been split into two parts. Part 1 discusses the basic visual optics of the eye along with emmetropia, ametropia and the correction of ametropia with spectacle lenses. Clinical refraction and instrumentation relevant to refraction have also been included in Part 1 along with visual acuity and presbyopia. Part 1 is particularly suitable for undergraduate optometrists student dispensing opticians and ophthalmologists in training.

Part 2 discusses the optics of contact lenses and the use of contact lenses in vision correction. To the best of our knowledge, Part 2 contains subject areas that have not been covered, to any great extent, in previous contact-lens-related texts. These topics are over-refraction techniques in contact lens practice and binocular vision considerations in contact lens practice. It is interesting to speculate as to why these subject areas have been given little attention in other contact-lens-related texts. However, in the authors' opinion, an over-refraction should be included as part of all contact lens fitting and aftercare appointments. In addition, as most patients seen in clinical practice have two eyes, a basic knowledge of binocular vision and its assessment is essential. Part 2 is particularly suitable for undergraduate optometrists and student contact lens opticians.

This book is designed for both undergraduate (student) and newly qualified optometrists, contact lens opticians and dispensing opticians. The content of this guide has been derived from the visual optics programmes used by some of the UK optometric teaching establishments and also the Association of British Dispensing Opticians. Although the assessment of students is moving away from the traditional examination setting and towards competency-based schemes, the authors hope that this book will be of use to all the varieties of current and future students – optometric, contact lens and dispensing. The authors sincerely hope that the content reflects the title and the book does what it says on the cover!

Andrew Keirl
Callington

Caroline Christie
London

Acknowledgements

A substantial amount of material within the early chapters of this book was originally written as part of a distance learning course in visual optics for Anglia Distance Learning Ltd. The authors are grateful to Anglia Distance Learning Ltd for giving kind permission for some of this earlier material to be revised, developed and included in this text.

The majority of the calculations used in this book are based on past examination questions produced by the Association of British Dispensing Opticians. The authors are grateful to Mark Chandler of the Association of British Dispensing Opticians for providing relevant past examination papers.

Thanks are also due to Andrew Franklin, Bill Harvey and Professor Bruce Evans for their contributions to this book and also to Ron Beerten for providing several of the images used in Part 2 of this text.

Many colleagues have been involved in reading and commenting on earlier versions of the chapters. Such individuals include Dr David Adams, Ron Beerton, Esther Hobbs, Richard Payne and Eleanor Parke. The authors are grateful for the advice and candour given.

Finally, thanks are given to the many contact lens and ophthalmic instrument manufacturing companies who readily provided technical advice and information to support the development of this text.

Dedications

To Alison, Ray and Sarah for two years of suffering, enforced patience and late dinners!

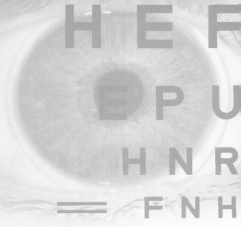

Part 1

Optics of the Eye, Ametropia and its Correction

Background Optical Principles

Andrew Keirl

Introduction

This chapter is designed to provide a brief and general revision of the optical principles and skills that are used throughout this book. For many readers, there is nothing new here because much of the material will have been considered in greater detail in other texts relating to the subjects of geometrical and visual optics. However, if knowledge of these basic topics and more importantly, an ability to apply these principles is deficient, then the inclusion of this chapter is justified.

Chapter contents

Refractive index

Much of the theory relating to optics, ophthalmic lenses and visual optics is based on the laws of refraction and reflection. Basically, light is considered to travel in straight lines, called **rays**, until it meets a polished surface separating two transparent media, e.g. a spectacle lens. When light is incident at such a surface, some is reflected but most is refracted at the surface and passes into the new medium. The percentage of light that undergoes reflection depends on the refractive index of the medium and the light that is refracted undergoes a change in velocity. This change in velocity is important because the ratios of the velocities in the first and second media are used to define **refractive index**. As is often the case, there are several variations to understand. **Absolute refractive index** is given the symbol n and is defined as the ratio of the velocity of light in a vacuum to the velocity of light in the medium.

$$n = \frac{\text{Velocity of light in vacuum}}{\text{Velocity of light in the medium}}$$

The velocity of light in a vacuum is internationally recognised as $299792.458\,\mathrm{km\,s^{-1}}$. As refractive index is a ratio, it is not given a unit. The refractive index of a particular material, e.g. glass, is given by:

$$n_\mathrm{g} = \frac{\text{Velocity of light in vacuum}}{\text{Velocity of light in glass}}$$

The refractive index of the glass would be given the symbol n_g, the subscript g denoting glass. Again, as refractive index is a ratio, it is not given a unit.

Although the absolute refractive index of air is 1.00029, it is generally acceptable to state that the refractive index of air is 1.0. When light passes from one optical medium to another, e.g. from air to glass, glass to water, the refractive effect is the result of the ratio of the velocity of light in one medium divided by the velocity of light in the other. This ratio defines the term **relative refractive index** (Figure 1.1). Typical refractive index values are given in Table 1.1.

So what is the importance of refractive index in optometric practice? Essentially, the refractive index of a lens material has a major influence on the appearance and thickness of a lens. In general terms, the higher the refractive index, the thinner the lens. This is achieved by the fact that a lens made using a higher refractive index material requires flatter curves to produce the required power when compared with a lower refractive index material. Flatter curves mean smaller sags and ultimately a thinner lens.

The definitions of refractive index given above were, in fact, somewhat simplified. Strictly speaking, refractive index is defined as the ratio of the velocity of light of a given frequency in air to the velocity of the same frequency in the refracting medium. The use of the word frequency means that this definition includes a reference to the **wavelength** of the light used to determine the refractive index. It is therefore important to note that the refractive index of an optical material varies with the wavelength of light used. This is illustrated in Figure 1.2.

When stating the refractive index of a lens material, ophthalmic lens manufacturers commonly use two wavelengths: The helium d-line (wavelength 587.6 nm) is used in the UK and the USA whereas the mercury e-line (wavelength 546.1 nm) is used in continental Europe. For crown glass, the d-line produces a refractive index of 1.523, whereas the e-line gives a refractive index of 1.525 (Table 1.2). The materials are of course identical. When working in optometric practice it is therefore important to know to which wavelength a particular manufacturer is

Table 1.1 Typical refractive index values

Material	Refractive index
Air at STP	1.00029
Water at 25°C	1.3334
Spectacle crown glass	1.523
Optical glass	1.523–1.885
CR39	1.498
Optical plastics	1.498–1.74
Polycarbonate	1.586
Diamond	2.4173

STP = standard temperature and pressure.

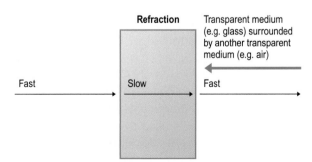

Figure 1.1 A ray of light is incident at the surface of a glass block (fast), passes through (slow) and emerges (fast) from the block. The ratio of the velocities determines the refractive index of the glass. The incident light is assumed to be travelling from left to right.

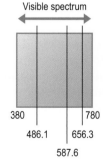

Yellow light (587.6 nm) n_d = 1.500
Red light (656.3 nm) n_c = 1.496
Blue light (486.1 nm) n_F = 1.505

Figure 1.2 The refractive index of a material varies with the wavelength of the incident light. A material will display many refractive indices, depending on the wavelength used.

3

Table 1.2 Refractive index of various glasses

Material	n_d	n_e
15 white	1.523	1.525
16 white	1.600	1.604
17 white	1.700	1.705
18 white	1.802	1.807
19 white	1.885	1.892

The refractive index of a material varies with the wavelength of the incident light.

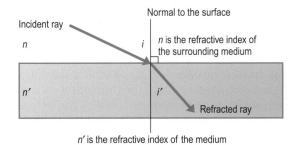

Figure 1.3 Refraction at a plane surface.

Table 1.3 The classification of refractive index

Normal index	1.48 but < 1.54
Mid-index	1.54 but < 1.64
High index	1.64 but < 1.74
Very high index	1.74 and above

From BS 7394 Part 2: Specification for prescription spectacles.

working. Strictly speaking, all the values given in Table 1.1 should have a stated wavelength, e.g. $n_d = 1.523$. The classification of materials in terms of refractive index is given in Table 1.3.

Refraction

When light is incident obliquely at the boundary of a plane surface separating two media with different refractive indices, its direction changes on passing from the first medium into the second. This change in direction is caused by the change in the velocity of the refracted light as it enters the new medium and is known as **refraction**; the surface being referred to as a **refracting surface**. In Figure 1.3 the angle i is known as the angle of incidence and the angle i' is the angle of refraction. The angle of incidence i of the refracting surface is the angle contained between the ray of light at the point of incidence and the normal (or perpendicular) to the surface at that point. The angle of refraction i' is the angle contained between the ray of light in the second medium at the point where the light enters this medium, and the normal to the refracting surface at that point. **Snell's law**

is a statement relating the angles of incidence and refraction, and the refractive indices of the two media, and may be stated as follows:

> The incident and refracted rays and the normal to the surface at the point of incidence lie in the one plane on opposite sides of the normal and the ratio of the sine of the angle of incidence to the sine of the angle of refraction is a constant for any two media for light of any one wavelength.

In other words:

$$n \sin i = n' \sin i'.$$

Snell's law is one of the most important laws in geometrical optics and although Figure 1.3 illustrates refraction at a plane surface, Snell's law also holds good for curved surfaces because the point of incidence by light on a curved surface may be considered as a minute plane surface. By means of Snell's law and the application of geometry, the effect of lenses and prisms can be determined.

Vergence

The term **vergence**, symbol L (or, more correctly, reduced vergence), is used to describe the curvature of a wavefront incident at a refracting surface. A spectacle lens is often described as a device for altering the vergence of light transmitted by the lens to the eye. When discussing the concept of vergence, three common terms are used: parallel light, divergence and convergence (Figures 1.4–1.6). Figure 1.4 shows a diverging pencil of light. The group of rays (or **pencil** of light) arising from a point source becomes larger as it gets further away from the source. Light in this pencil is **divergent**. Figure 1.6 shows **convergent** light as the pencil illustrated in the diagram is getting smaller along the direction of the light up to point F'. Beyond F' the pencil becomes divergent. Parallel light is illustrated in Figure 1.5.

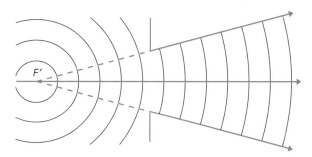

Figure 1.4 Divergent (from F'). (After Freeman and Hull 2003, with permission of Elsevier Ltd.)

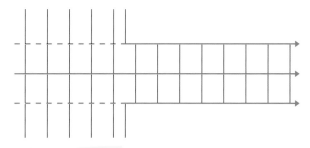

Figure 1.5 Parallel. (After Freeman and Hull 2003, with permission of Elsevier Ltd.)

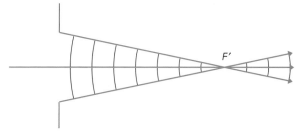

Figure 1.6 Convergent (up to F'). (After Freeman and Hull 2003, with permission of Elsevier Ltd.)

Sign convention

To calculate the positions of objects and images, it is necessary to adopt a standard way of measuring distances and angles. Objects and images that are in front of a lens or mirror have to be distinguished from those that are behind. This can be achieved by the use of positive and negative signs. In ophthalmic optics all measurements follow the Cartesian sign convention; the following are the main points:

- Distances are measured *from* the reference point or plane (lens, refracting surface, mirror etc.) *to* the point in question.
- Light is assumed to travel from left to right.
- All measurements are taken *from* the lens, mirror or surface. Measurements to the *left* of the lens, mirror or surface are *negative*, and those to the *right* of the lens, mirror or surface are *positive*.
- All measurements are taken from the axis. Measurements taken to objects *below* the axis are *negative* and those taken to objects *above* the axis are positive.
- Angle measurements are taken *from* the ray *to* the axis where *clockwise* angles are *negative* and anticlockwise angles *positive*.

The optical sign convention is illustrated in Figure 1.7 in which a spherical refracting surface is receiving light from an object O. In this illustration, the object distance l is measured *from* the surface to O and has a *negative* value because O is to the left of the surface, whereas the radius of curvature r has a positive value because C is to the right of the surface.

Surface curvature and surface power

A plane or optically flat surface may be described as a surface where normals at all points on it lie parallel to each other, whereas a spherical surface is one where all normals pass through a single point. Most optical elements (including the eye) have curved surfaces that can be

5

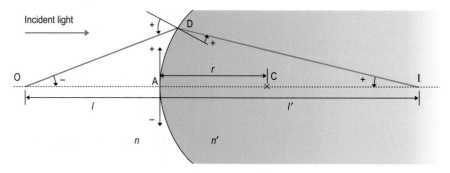

Figure 1.7 Sign convention. (After Freeman and Hull 2003, with permission of Elsevier Ltd.)

described as either convex or concave. The optical effect of a curved surface is to alter the vergence of incident light. The optical effects of curved surfaces are described in terms of their **surface power** and the power of a curved surface can be described as 'the vergence that the surface impresses on incident light', or 'how the surface changes the vergence of incident light'. It is important to note that these descriptions apply to both lenses and the eye. Ophthalmic prisms as commonly found in an optometrist's trial case have plane surfaces. Such devices alter the direction but not the vergence of incident light.

The power of any curved surface depends on two factors:

1. The refractive index of the material used to form the curved surface
2. The radius of curvature of the curved surface.

Once again, the above applies to *both* lenses *and* the eye. In their simplest from, curved surfaces are considered to be part of a spherical surface (Figure 1.8). Spherical surfaces are either **convex** or **concave** to incident light. A convex surface causes parallel incident light to become **convergent** (Figure 1.9), whereas a concave surface causes parallel incident light to become **divergent** (Figure 1.10). It is important to note that a spherical surface has the same curvature (and therefore power) in all meridians (Figure 1.11).

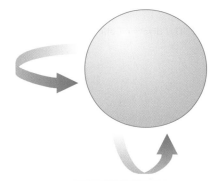

Figure 1.8 The sphere: a spherical surface has the same curvature (shape) along all meridians.

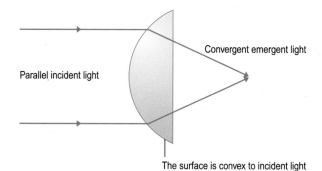

Convergent emergent light

Parallel incident light

The surface is convex to incident light

Figure 1.9 The converging effect of a convex refracting surface.

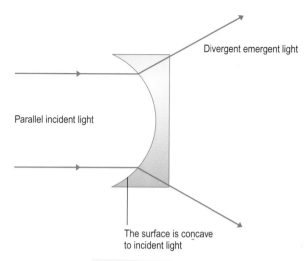

Figure 1.10 The diverging effect of a concave refracting surface.

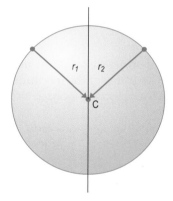

Figure 1.11 A spherical surface has the same curvature in all meridians: r_1 and r_2 have the same length but, according to the optical sign convention, r_1 is positive (because C is to the right of the surface and r_2 is negative because C is to the left of the surface).

The calculation of surface power

A spectacle lens always consists of two surfaces, one of which must be curved. Light is assumed to be incident at the **front** surface of the lens (the surface away from the eye) and leaves the **back** surface (the surface facing the eye). Light is assumed to travel from left to right.

Front surface power

The power of the front surface of a spectacle lens can be calculated using the equation:

$$F_1 = \frac{n' - n}{r_1}$$

where:

- F_1 is the power of the front surface of the lens in dioptres (D)
- n' is the refractive index of the lens material
- n is the refractive index of the surrounding medium
- r_1 is the radius of curvature of the front surface of the lens in metres.

Back surface power

The power of the back surface of a spectacle lens can be calculated using the equation:

$$F_2 = \frac{n - n'}{r_2}$$

where:

- F_2 is the power of the back surface of the lens in dioptres
- n' is the refractive index of the lens material
- n is the refractive index of the surrounding medium
- r_2 is the radius of curvature of the back surface of the lens in metres.

When dealing with ophthalmic lenses, n is usually taken to be the refractive index of air and is given a value of 1.00. However, this is often not the case when dealing with the eye and care must be taken when substituting values into equations. When using the surface power equations given above, the value for r must be substituted in metres and given the appropriate sign (+ or −). Always ensure that n and n' are in the correct order for the required surface power. Remember that n' is the refractive index that the light is *about to enter* and n the refractive index that the light is *about to leave*. The above surface power equations can be applied to the surfaces of spectacle lenses and the various refracting surfaces of the eye. The SI unit for the surface power of a spectacle lens is the dioptre, which is also the unit used when describing the power of the eye.

7

Thin lens power

A spectacle lens has two surfaces, the front surface, F_1 (the most positive value), which is taken to be the surface facing away from the eye, and the back surface, F_2 (the most negative value), which is taken to be the surface next to the eye. If the two surfaces are assumed to be in close contact, the thickness of the lens can be ignored and the spectacle lens is referred to as a **thin** lens. The power of a thin lens, F, is simply given by the sum of its surface powers, i.e.:

$$F = F_1 + F_2$$

The unit for both surface and lens power is again the dioptre (D).

Thin lens form

If the thickness of a lens is ignored, the total thin lens power of the lens is simply the **sum** of its surface powers. As an example, if a lens has a front surface power of +6.00 D and a back surface power of −2.00 D, then the sum of the surface powers and therefore its **thin lens power** is +4.00 D. These stated surface powers determine the **form** of the lens. In theory, this +4.00 D lens could be made in any form.

The fundamental paraxial equation

Thin lenses, thick lenses, lens systems, mirrors and single refracting surfaces all, in some way, form an image of an object. In the study of visual optics it is often necessary to determine the position of an image formed by such optical devices. This section discusses the use of the **fundamental paraxial equation**, which can be applied to thin lenses (and mirrors) as well as a single refracting surface. Thick lenses and complex systems require a different approach and are considered later. The word **paraxial** refers to a restricted region of a curved surface where only rays near the axis form the image. This 'restriction' of the aperture of a curved surface, so that only rays near the axis form the image, is a simple 'cure' for a defect in the imaging properties of the surface known as spherical aberration. This restricted region is known as the **paraxial region** and rays within this region are known as **paraxial rays**. For such rays, the angles of incidence and refraction are small and it is possible to calculate the positions of images using approximate expressions without incurring major errors. The fundamental paraxial equation is:

$$L' = L + F$$

where:

- L is the incident vergence arriving at the surface from an object (object vergence)
- F is the power of the thin lens or single refracting surface
- L' is the vergence leaving the surface, which goes on to form the image (image vergence).

L is given by:

$$L = \frac{n}{l}$$

where l is the object distance (in metres) and n is the refractive index of the medium containing the object. If the object is placed in air the expression becomes:

$$L = \frac{1}{l}$$

If the object distance is required and the object vergence known, the above can be rewritten as:

$$l = \frac{n}{L}$$

and if the object is in air the above expression becomes:

$$l = \frac{1}{L}$$

The process of changing from an object distance to an object vergence is therefore very straightforward. If L and F are known, the image vergence L' can be found using the fundamental paraxial equation given above. The image position l' can then be found using the expression:

$$l' = \frac{n'}{L'}$$

where n' is the refractive index of the medium containing the image or:

$$l' = \frac{1}{L'}$$

if the image is formed in air.

Example 1.1
An on-axis object is placed 50 cm to the left of a thin +5.00 D lens in air. Find the position of the image formed.

When computing with distances and vergences, it is advisable to work in two columns, one headed 'vergences' and the other 'distances'. As the object is positioned to the left of the lens, sign convention dictates that it is a negative distance and must be given a minus sign. It is advisable to work in metres at all times (50 cm $\equiv$ 0.50 m). The starting point in the computation is the given object distance l.

Vergences (D)   Distances (m)

$$L_1 = \frac{1}{-0.50} = -2.00\,D \quad \leftarrow \quad l = -0.50\,m$$

$$F = +5.00\,D$$

$$L' = L + F$$

$$L' = -2.00 + (+5.00) = +3.00\,D \rightarrow l' = \frac{n'}{L'}$$

$$l' = \frac{1}{+3.00} = +0.33\,m$$

The positive sign means that the image is formed 0.33 m (33 cm) to the right of the lens. Such an image is referred to as a **real image**.

Example 1.2
An on-axis object is placed 100 cm to the left of a convex refracting surface of radius of curvature +0.125 m. The surface has air to its left and a refractive index of 1.5 to its right. Find the position of the image formed.

The radius of curvature has been given a positive sign because the centre of curvature of the surface is to the right. First of all we need to find the power of the convex refracting surface using:

$$F = \frac{n' - n}{r}$$

We will call the refractive index to the left of the lens (air) n and the refractive index to the right (1.5) n'. This statement is important at the end of the computation. The radius of curvature must be entered in metres:

$$F = \frac{1.50 - 1.00}{+0.125} = +4.00\,D$$

And now the computation! The starting point is the object distance l.

Vergences (D)   Distances (m)

$$L_1 = \frac{1}{-1.00} = -1.00\,D \quad \leftarrow \quad l = -1.00\,m$$

$$F = +4.00\,D$$

$$L' = L + F$$

$$L' = -1.00 + (+4.00) = +3.00\,D \rightarrow l' = \frac{n'}{L'}$$

$$l' = \frac{1.5}{+3.00} = +0.50\,m$$

Please note that the refractive index of the medium to the right of the surface in Example 1.2 is not air! The correct refractive index (1.5) has to be used in the final step of the computation. Be careful not to fall into the trap of using 'one-over' every time! The positive sign means that the image is therefore formed 0.50 m (50 cm) to the right of the single refracting surface.

Lens systems and thick lenses

A 'real' lens cannot be considered to be thin and, by definition, a lens that cannot be considered 'thin' must be 'thick'. The term 'thick lens' relates to lenses with a centre thickness that cannot be ignored. In other words, the centre thickness, refractive index and surface powers all play a part in the overall power of the lens. This means that the total power of a thick lens does *not* equal the sum of the surface powers. For thick lenses:

$$F \neq F_1 + F_2$$

9

All lenses used in optometric practice are of course 'thick'. Thick lenses and systems of more than one thin lens can be treated in the same way because there are several similarities between the two, e.g. both have two powers and can be replaced by a single thin lens, and simple ray tracing methods can be applied to thick lenses and systems of thin lenses.

A thick lens (or lens system) actually has two powers. These are known collectively as the **vertex powers** and individually as the **back vertex power** and **front vertex power**. The back vertex power (BVP) is defined as the 'vergence of light leaving the back surface of the lens', whereas the front vertex power (FVP) is defined as the 'vergence of light leaving the front surface of the lens'. When describing the vertex powers of a spectacle lens, light is assumed (as always) to be travelling from left to right from a distant object, i.e. the incident light is parallel and the incident vergence is zero. As far as spectacle lenses are concerned, we are usually interested in the BVP of the lens. In fact every time we order a spectacle lens for a patient, we are specifying its BVP. To calculate the BVP of a thin lens system or a thick lens, step-along ray-tracing should be used. To calculate the FVP of a thin lens system or a thick lens, *reverse* the lens (or system) so that light is incident on the back surface, therefore leaving the front surface, and then use the step-along method again. Remember that light must always travel from left to right.

Ray-tracing

Ray-tracing is simply a method of getting from one side of a thick lens (or lens system) to the other. There are two ray-tracing procedures: step-along and step-back. Step-along takes us from left to right whereas step-back takes us from right to left. Remember that in both procedures light is always assumed to travel from left to right and we also assume that light is incident at the front surface (F_1) of a lens. When ray-tracing to find an FVP or BVP, the incident light must always be parallel with an incident vergence of zero. A reminder of the procedure for step-along ray tracing is given below. In some ways, step-along is a development of

computations using the fundamental paraxial equation. When ray-tracing, always work using two columns, one for vergences (in dioptres) and one for distances (in metres).

An example of the layout for step-along ray tracing is:

Vergences (D)	Distances (m)

$L_1 =$

$F_1 =$

$L_1' = L_1 + F_1$ $\rightarrow$ $l_1' = \dfrac{n'}{L_1'}$

$L_2 = \dfrac{n'}{l_2}$ $\leftarrow$ $l_2 = l_1' - d$ (step-along)

$L_2' = L_2 + F_2$

$L_2' =$

The terms used in the step-along procedure are given below:

- L_1 is the vergence of light (D) incident at the first (front) surface of a lens
- L_1' is the vergence of light (D) leaving the first (front) surface of a lens
- d is the distance between the surfaces of a lens system (m) and is always positive; in the case of a thick lens t is the centre or axial thickness (m)
- n is the refractive index of the medium separating the two surfaces; in the case of a thick lens, n' is the refractive index of the lens material
- L_2 is the vergence of light (D) incident at the second (back) surface of a lens
- L_2' is the vergence of light (D) leaving the second (back) surface of a lens.

If parallel light ($L_1 = 0$) is incident at the first surface of a thick lens or lens system, then the vergence leaving its second surface (L_2') is equivalent to the BVP of the thick lens or lens system. If the system or thick lens is reversed so that parallel light ($L_1 = 0$) is incident at the second surface of the lens (F_2), then the vergence leaving its second surface (L_2') is equivalent to the FVP of the thick lens or lens system.

Example 1.3

Two thin lenses in air are used to form a lens system. The two thin lenses are in air ($n = 1$) and

are separated by a distance of 20 mm. $F_1 = +8.00\,D$ and $F_2 = -5.00\,D$. Assuming a distant object, calculate the back and front vertex powers of the thin lens system. Refer to **Figure 1.12.** *[handwritten: $L'_2 = BVP = ?$]*

Both vertex powers can be calculated using a step-along ray-trace:

Vergences (D)   Distances (m)

$L_1 = 0.00$

$F_2 = +8.00\,D$

$L'_1 = L_1 + F_1 = +8.00\,D \quad \rightarrow \quad l'_1 = \dfrac{n}{L'_1}$

[handwritten: $0 + 8$]

$l'_1 = \dfrac{1}{+8.00} = +0.125\,m$ *[handwritten: l'_1]*

[handwritten: new step $+0.125 - 0.002\,m$ $l_2 = l'_1 - d$]

$l_2 = +0.125 - 0.020 = +0.105\,m$ *[handwritten: l_2]*

$L_2 = \dfrac{n}{l_2} \quad \leftarrow$

$L_2 = \dfrac{1}{+0.105} = +9.52\,D$

$L'_2 = L_2 + F_2$

$L'_2 = +9.52 + (-5.00) = +4.52\,D$

As $L_1 = 0.00$, $L'_2 = BVP = +4.52\,D$

The BVP of this thin lens system in air is therefore +4.52 D.

The usual symbol for BVP is F'_v and n/F'_v gives a distance known as the back vertex focal length (symbol f'_v). This distance is measured from the final lens (or back surface of a thick lens) to the image formed by the lens or lens system.

The FVP is also found using a step-along ray-trace, but this time with the system **reversed**. Parallel light is now incident at the second surface (F_2) and leaves the first surface. Refer to **Figure 1.13.**

Vergences (D)   Distances (m)

$L_1 = 0.00$

$F_2 = -5.00\,D$

$L'_1 = L_1 + F_2 = -5.00\,D \quad \rightarrow \quad l'_1 = \dfrac{n}{L'_1}$

$l'_1 = \dfrac{1}{-5.00} = -0.20\,m$

$l_2 = l'_1 - d$

$L_2 = \dfrac{n}{l_2} \quad \leftarrow \quad l_2 = -0.200 - 0.020 = -0.220\,m$

$L_2 = \dfrac{1}{-0.220} = -4.54\,D$

$L'_2 = L_2 + F_1$

$L'_2 = -4.54 + (+8.00) = +3.45\,D$

As $L_1 = 0.00$, $L'_2 = FVP = +3.45\,D$

The FVP of this thin lens system in air is therefore +3.45 D.

The usual symbol for FVP is F_v and $-n/F_v$ gives a distance known as the front vertex focal length (symbol f_v). The minus sign is necessary

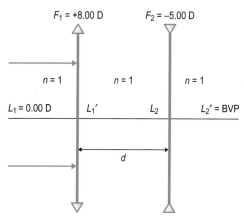

Figure 1.12 Diagram for the back vertex power (BVP).

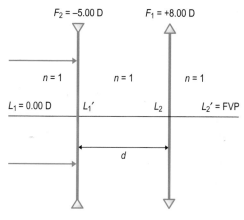

Figure 1.13 Diagram for the front vertex power (FVP): system reversed.

because the ray trace was performed with the lenses reversed. This distance is measured from the first lens (or front surface) to the image formed by the lens or lens system. As always, if a vertex power or vertex focal length is given, parallel incident light is assumed.

Please note that the only time a lens or a lens system is reversed in this way is when the FVP is required. You are not required to calculate a FVP very often in visual optics. Step-along ray-tracing can also be used to find the back surface power of a thick lens or lens system.

Example 1.4

A thin lens system consists of two thin lenses in air separated by a distance of 8 mm. The BVP of the system is +5.00 D and the power of the first thin lens (F_1) is +8.00 D. Find the power of the second thin lens (F_2).

As always, if a **vertex power** is specified a **distance object is assumed**, and if parallel light ($L_1 = 0$) is incident at the first surface of the lens the vergence leaving its second surface (L_2') is equivalent to its BVP. Ray-tracing takes us from one side of a lens (or a lens system) to the other and step-along takes us from left to right. In this example, the unknown value (F_2) is to the *left* of the system so a step-along ray-trace is required. As always, the computation takes place in two columns:

Vergences (D)   Distances (m)

$L_1 = 0.00$

$F_1 = +8.00\,D$

$L_1' = L_1 + F_1 = +8.00\,D \;\rightarrow\; l_1' = \dfrac{n}{L_1'}$

$$l_1' = \dfrac{1}{+8.00} = +0.125\,m$$

$$l_2 = l_1' - d$$

$L_2 = \dfrac{n}{l_2} \qquad \leftarrow \qquad l_2 = +0.125 - 0.008 = +0.117\,m$

$L_2 = \dfrac{1}{+0.117} = +8.55\,D$

$L_2' = L_2 + F_2 \qquad$ and $\qquad F_2 = L_2' - L_2$

$F_2 = +5.00 - (+8.55) = -3.55\,D$

The back surface power of this thin lens system in air is therefore –3.55 D.

Step-back ray-tracing

The technique of step-back ray-tracing takes us from right to left and is used when the unknown value is situated to the left of a lens or lens system. It does *not* require the system to be reversed. A revision of the procedure for step-back ray tracing is given below. As before, the calculation takes place using two columns, one for vergences (D) and one for distances (m):

Vergences (D)   Distances (m)

$L_2' =$

$L_2 = L_2' - F_2$

$L_2 = \qquad\qquad\qquad\rightarrow\qquad l_2 = \dfrac{n'}{L_2}$

$L_1' = \dfrac{n'}{l_1'} \qquad\qquad\qquad \leftarrow \qquad l_1' = l_2 + d$ (step-back)

As $L_1 = 0.00,\; L_1' = F_1$

Example 1.5

A lens system consists of two thin lenses in air separated by a distance of 10 mm. The BVP of the system is –6.00 D and the power of the second thin lens (F_2) is –10.00 D. Find the power of the first thin lens, F_1. Assume a distance object.

As with all lens system and thick lens problems, if parallel light ($L_1 = 0$) is incident at the first surface of the lens, the vergence leaving its second surface (L_2') is equivalent to its BVP. Remember that ray-tracing is simply a method of getting from one side of a lens (or a lens system) to the other and that there are two ray-tracing procedures: step-along and step-back. Step-along takes us from left to right, whereas step-back takes us from right to left. In this problem the unknown value (F_1) is to the *left* of the system so the 'tool' to solve the problem is therefore step-back. As usual, the calculation takes place in two columns:

Vergences (D)   Distances (m)

$L_2' = -6.00\,D$

$L_2 = L_2' - F_2$

$L_2 = -6.00 - (-10.00) = +4.00\,D$

$L_2 = +4.00\,D \qquad\rightarrow\qquad l_2 = \dfrac{1}{+4.00} = +0.250\,m$

$l_1' = l_2 + d$ (step-back)

$l'_1 = +0.250 + 0.010 = +0.260\,\text{m}$

$$\rightarrow L'_1 = \frac{1}{+0.260} = +3.85\,\text{D}$$

As $L_1 = 0.00$, $L'_1 = F_1$

$F_1 = +3.85\,\text{D}$

The front surface power of the lens system is therefore +3.85 D.

The above examples have both used a system of separated thin lenses in air. However, if the vertex powers or surface powers of a thick lens are required, the ray tracing procedure is almost identical. The only difference is that for a thick lens, the equivalent air distance (EAD) t/n is used for the separation or distance between the two surfaces instead of d. The EAD is also known as the reduced distance and is as usual stated in metres. The equivalent air distance is simply given by:

$$\text{EAD} = \frac{t}{n}$$

If the EAD is used the refractive index between the two surfaces can be assumed to be the same as air ($n = 1$). This approach can simplify ray-tracing with a thick lens and is especially useful when ray-tracing through multiple surfaces.

The equivalent thin lens

The optical effect of any thick lens or lens system can be replaced by an **equivalent thin lens,**

which is an imaginary lens and does not really exist! **Figure 1.14** shows parallel light being refracted by a thick lens. As the lens is thick there are two distinct refractions: D_1 and D_2, for the light coming from the distant object before it intersects with the axis at point F', the second principal focus. By extending the ray paths as shown in Figure 1.14, it can be seen that these two refractions are equivalent to a single refraction, taking place at H'. This suggests the idea that a lens, placed in the plane of H', performs the same function as the thick lens, i.e. mimics the BVP and produces a focus at F'. This equivalent thin lens can be seen in Figure 1.14 to have an **equivalent focal length** f'_E and therefore has an equivalent power F_E. H'P' is known as the **second principal plane** and P' the **second principal point**. The distance A_2P' is usually given the symbol e'. In Figure 1.14 f'_v is the **back vertex focal length**.

Figure 1.15 shows the same thick lens, this time with an object placed at the first principal focus F. As before, light arising from the object has two refractions at the actual lens surfaces. Once again, the ray paths can be extended to show the position of the equivalent refraction H. A thin lens placed in the in plane of H duplicates the effect of the FVP. The position HP of the equivalent thin lens is called the **first principal plane** and P is the **first principal point**. As long as the refractive indices in front of and behind the thick lens or system are the same, the equivalent focal length f_E is the same value as f'_E but of *opposite* sign. The equivalent power

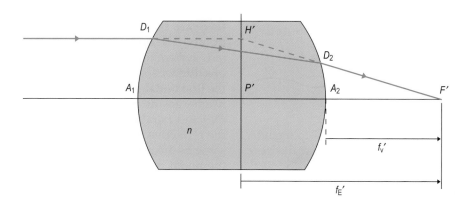

Figure 1.14 Thick lens: distant object. (After Freeman and Hull 2003, with permission of Elsevier Ltd.)

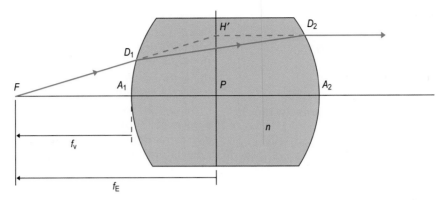

Figure 1.15 Thick lens: distant image. (After Freeman and Hull 2003, with permission of Elsevier Ltd.)

F_E is also the same as before. The distance A_1P is usually given the symbol e. In Figure 1.15 f_v is the front vertex focal length.

There are several methods of calculating the power of the equivalent thin lens.

For a two-lens system

$$F_E = F_1 \times \frac{L_2'}{L_2}$$

or:

$$F_E = F_1 + F_2 - dF_1F_2$$

For a thick lens

$$F_E = F_1 \times \frac{L_2'}{L_2}$$

or:

$$F_E = F_1 + F_2 - (t/n)F_1F_2$$

For a multi-lens system

$$F_E = F_1 \times \frac{L_2'}{L_2} \times \frac{L_3'}{L_3} \times \ldots\ldots\ldots$$

The equation above can be extended to include further vergences. It is suggested that the vergences in the equations be taken from the BVP step-along computation.

The second equivalent focal length f_E' is given by:

$$f_E' = \frac{n}{F_E}$$

The first equivalent focal length f_E is given by:

$$f_E = \frac{-n'}{F_E}$$

The position of the second principal point A_2P' is given by:

$$A_2P' = f_v' - f_E' = e'$$

The position of the first principal point A_1P is given by:

$$A_1P = f_v - f_E = e$$

The nodal points

When dealing with a thick lens or lens system there are two final items to consider: the **nodal points**. The nodal points of a thick lens or thin lens system are equivalent to the optical centre of a thin lens. If a ray of light is directed towards the **first nodal point** (N) of a thick lens or system, it emerges as if from the **second nodal point** (N') and parallel to the original incident ray. If the refractive index in front of the system is the same as the refractive index behind the system, the nodal points and the principal points coincide, i.e. N coincides with P and N' with P' if $n = n'$. If the refractive indices on either side of a thick lens or system are not equal, the two equivalent thin lens focal lengths f'_E and f_E are of different magnitudes. This leads to the nodal points and the principal points being separated. If $n \neq n'$ the nodal points are always displaced towards the medium with the higher refractive index. The displacement of the nodal points is given by:

$$PN = P'N' = f'_E + f_E$$

To summarise the preceding material on thick lenses and systems of separated thin lenses, we have essentially reviewed the calculation of six individual points. Collectively, these are known as the **cardinal points** and are listed in Table 1.4.

Example 1.6

A system of two thin lenses in air is specified as follows:

$$F_1 = +12.5\,D, F_2 = -28.00\,D, d = 5\,cm.$$

Find:

The back and front vertex powers
The equivalent thin lens power
The positions of the principal and nodal points.

This is a relatively straightforward example because the refractive indices in front, behind and in between the lenses are all the same as the system in air ($n = 1.00$). Both vertex powers can be calculated using a step-along ray trace. Refer to Figure 1.12.

The first step is to find the BVP:

Vergences (D)	Distances (m)

$L_1 = 0.00$

$F_1 = +12.50\,D$

$L'_1 = L_1 + F_1 = +12.50\,D \;\rightarrow\; l'_1 = \dfrac{n}{L'_1}$

$$l'_1 = \dfrac{1}{+12.50} = +0.08\,m$$

$$l_2 = l'_1 - d$$

$L_2 = \dfrac{n}{l_2} \qquad \leftarrow \qquad l_2 = +0.08 - 0.05 = +0.03\,m$

$$L_2 = \dfrac{1}{+0.03} = +33.33\,D$$

$L'_2 = L_2 + F_2$

$L'_2 = +33.33 + (-28.00) = +5.33\,D$

As $L_1 = 0.00$, $L'_2 = $ BVP $\qquad F'_v = +5.33\,D$

The BVP of this thin lens system in air is therefore **+5.33 D.**

The back vertex focal length of the system is given by:

$$f'_v = \dfrac{n}{F'_v} = \dfrac{1}{+5.33} = +0.1875\,m$$

The FVP is also found by using a step-along ray-trace but with the system **reversed**. Parallel light is now incident at the second surface (F_2) and leaves the first surface. Refer to Figure 1.13.

Vergences (D)	Distances (m)

$L_1 = 0.00$

$F_2 = -28.00\,D$

$L'_1 = L_1 + F_2 = -28.00\,D \;\rightarrow\; l'_1 = \dfrac{n}{L'_1}$

Table 1.4 The cardinal points

	Name	Symbol	Comment
1	Front vertex focal point	F'_v	Position is specified by stating the distance f_v
2	Back vertex focal point	F'_v	Position is specified by stating the distance f'_v
3	First principal point	P	Position is specified by stating the distance e
4	Second principal point	P'	Position is specified by stating the distance e'
5	First nodal point	N	Will coincide with P and P' if $n = n'$ If $n \neq n'$, N and N' are displaced towards the higher refractive index

15

$$l_1' = \frac{1}{-28.00} = -0.0357\,\text{m}$$

$$l_2 = l_1' - d$$

$$L_2 = \frac{n}{l_2} \quad \leftarrow \quad l_2 = -0.0357 - 0.0500 = -0.0857\,\text{m}$$

$$L_2 = \frac{1}{-0.0857} = -11.67\,\text{D}$$

$$L_2' = L_2 + F_1$$

$$L_2' = -11.67 + (+12.50) = +0.83\,\text{D}$$

As $L_1 = 0.00$, $L_2' = \text{FVP}$ $F_\text{v} = +0.83\,\text{D}$

The FVP of this thin lens system in air is therefore +0.83 D.

The front vertex focal length of the system is given by:

$$f_\text{v} = \frac{-n}{F_\text{v}} = \frac{-1}{+0.83} = -1.20\,\text{m}$$

(The minus sign is used because the system was reversed.)

The power of the equivalent thin lens can be found using the expression:

$$F_\text{E} = F_1 \times \frac{L_2'}{L_2}$$

Taking values from the BVP ray-trace:

$$F_\text{E} = +12.50 \times \frac{+5.33}{+33.33} = +2.00\,\text{D}$$

The second equivalent focal length f_E' is given by:

$$f_\text{E}' = \frac{n}{F_\text{E}}$$

As the system is in air:

$$f_\text{E}' = \frac{1}{+2.00} = +0.50\,\text{m}$$

The first equivalent focal length f_E is given by:

$$f_\text{E} = \frac{-n'}{F_\text{E}}$$

And again as the system is in air:

$$f_\text{E} = \frac{-1.00}{+2.00} = -0.50\,\text{m}$$

The position of the second principal point A_2P' is given by:

$$A_2P' = f_\text{v}' - f_\text{E}' = e'$$

$$e' = +0.1875 - (+0.5000) = -0.3125\,\text{m}$$

The position of the first principal point A_1P is given by:

$$A_1P = f_\text{v} - f_\text{E} = e$$

$$e = -1.20 - (+0.50) = -0.70\,\text{m}$$

As the refractive indices in front of and behind the system are the same, the nodal points coincide with the principal points, i.e. N coincides with P and N' with P'.

The cardinal points for this system are illustrated in Figure 1.16.

Example 1.7

A thick biconvex lens has a front surface radius of +32.500 mm and a back surface radius of −21.133 mm. The front surface is bounded by air and the back surface by water with a refractive index of 1.333. The refractive index of the lens is 1.650 and its centre thickness is 6.6 mm.

Find:

The powers of the two surfaces
The equivalent thin lens power
The positions of the cardinal points

This is a more complicated example because we have to deal with three refractive indices. Also, as a thick lens is given as opposed to a system of thin lenses, t/n must be used instead of d when ray-tracing. Both vertex powers can be calculated using a step-along ray-trace.

The first task is to calculate the two surface powers. The radius of curvature r must be substituted in metres and the appropriate sign used. Care must be exercised when using the given refractive indices! To find surface powers we must use:

$$F_1 = \frac{n'-n}{r_1} \quad \text{and} \quad F_2 = \frac{n-n'}{r_2}$$

$$F_1 = \frac{1.650 - 1.000}{+0.032500} = +20.00\,\text{D}$$

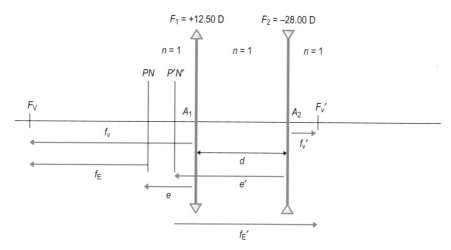

Figure 1.16 Diagram for Example 1.6 showing the relative positions of the cardinal points (not to scale).

$$F_2 = \frac{1.333 - 1.650}{-0.021133} = +15.00\,\text{D}$$

The equivalent air distance (with t substituted in metres) is given by:

$$\text{EAD} = \frac{t}{n} = \frac{0.0066}{1.650} = 0.004\,\text{m}$$

Now that the EAD has been determined, the refractive index between the two surfaces can be assumed to be **equal to that of air** ($n = 1$). This value is used in the following ray-traces.

The first step-along ray-trace will determine the BVP:

Vergences (D)   Distances (m)

$L_1 = 0.00$

$F_1 = +20.00\,\text{D}$

$L_1' = L_1 + F_1 = +20.00\,\text{D} \rightarrow l_1' = \dfrac{n}{L_1'}$

$$l_1' = \frac{1}{+20.00} = +0.050\,\text{m}$$

$$l_2 = l_1' - (t/n)$$

$L_2 = \dfrac{n}{l_2}$ $\leftarrow$ $l_2 = +0.050 - 0.004 = +0.046\,\text{m}$

$$L_2 = \frac{1}{+0.046} = +21.739\,\text{D}$$

$L_2' = L_2 + F_2$

$L_2' = +21.739 + (+15.00) = +36.739\,\text{D}$

As $L_1 = 0.00$, $L_2' = \text{BVP}$ $F_v' = +36.739\,\text{D}$

The BVP of this thick lens system is therefore +36.739 D.

As the image is formed in **water**, the back vertex focal length of the system is given by:

$$f_v' = \frac{n'}{F_v'} = \frac{1.333}{+36.739} = +0.03628\,\text{m}$$

The FVP is also found by using a step-along ray-trace but with the system **reversed**. Parallel light is now incident at the second surface (F_2) and leaves the first surface:

Vergences (D)   Distances (m)

$L_1 = 0.00$

$F_2 = +15.00\,\text{D}$

$L_1' = L_1 + F_2 = +15.00\,\text{D} \rightarrow l_1' = \dfrac{n}{L_1'}$

$$l_1' = \frac{1}{+15.00} = +0.06667\,\text{m}$$

$$l_2 = l_1' - (t/n)$$

$L_2 = \dfrac{n}{l_2}$ $\leftarrow$ $l_2 = +0.06667 - 0.00400 = +0.06267\,\text{m}$

$$L_2 = \frac{1}{+0.06267} = +15.957\,\text{D}$$

$L_2' = L_2 + F_1$

$L_2' = +15.957 + (+20.00) = +35.957\,\text{D}$

As $L_1 = 0.00$, $L_2' = \text{FVP}$ $F_v = +35.957\,\text{D}$

The FVP of this thick lens system is therefore +35.957 D.

As the image is formed in **air** the front vertex focal length is given by:

$$f_v = \frac{-n}{F_v} = \frac{-1}{+35.957} = -0.0278\,\text{m}$$

The minus sign is used because the system was reversed. The power of the equivalent thin lens can be found using the expression:

$$F_E = F_1 \times \frac{L_2'}{L_2}$$

Taking values from the BVP ray-trace:

$$F_E = +20.000 \times \frac{+36.739}{+21.739} = +33.800\,\text{D}$$

The second equivalent focal length f_E' is given by:

$$f_E' = \frac{n'}{F_E}$$

As **water** is to the **right** of the lens:

$$f_E' = \frac{1.3330}{+33.800} = +0.03944\,\text{m}$$

The first equivalent focal length f_E is given by:

$$f_E = \frac{-n'}{F_E}$$

As **air** is to the **left** of the lens:

$$f_E = \frac{-1.00}{+33.80} = -0.02959\,\text{m}$$

The position of the second principal point A_2P' is given by:

$$A_2P' = f_v' - f_E' = e'$$

$$e' = +0.03628 - (+0.03944) = -0.00316\,\text{m}$$

The position of the first principal point A_1P is given by:

$$A_1P = f_v - f_E = e$$

$$e = -0.0278 - (-0.02959) = +0.00179\,\text{m}$$

As the refractive index in front of the lens is air ($n = 1$) and that behind the lens is water ($n = 1.333$), the nodal points do *not* coincide with the principal points. The displacement of the nodal points is given by:

$$PN = P'N' = f_E' + f_E = +0.03944 + (-0.02959)$$
$$= +0.00985\,\text{m}$$

The nodal points are displaced towards the higher refractive index, which in this case is to the right. The cardinal points for this system are illustrated in Figure 1.17.

Concluding points for Chapter 1

To avoid the common slip-ups:

- always use the correct sign for the surface in question
- always use the correct SI unit
- always choose the correct 'tool' to solve the problem, either step-along or step-back

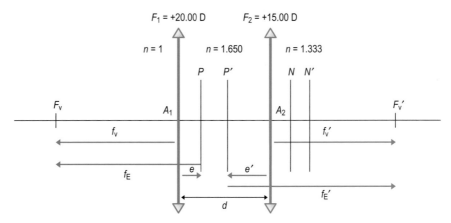

Figure 1.17 Diagram for Example 1.7 showing the relative positions of the cardinal points (not to scale).

- when performing ray tracing, distances should be entered given in metres
- with all lens system and thick lens problems, if parallel light ($L_1 = 0$) is incident at the first surface of the lens, the vergence leaving its second surface (L_2') is always equivalent to its BVP
- if parallel light ($L_1 = 0$) is incident at the second surface of the lens (the lens has been reversed) the vergence leaving its first surface (L_2') is always equivalent to its FVP

- a lens or lens system is reversed only if the FVP is required.

Further recommended reading

Freeman M H, Hull C C (2003) *Optics*. Butterworth-Heinemann, Oxford

Jalie M (2003) *Ophthalmic Lenses and Dispensing*. Butterworth-Heinemann, Oxford

Tunnacliffe A H, Hurst J G (1996) *Optics*. Association of British Dispensing Opticians, London

Schematic Eyes, Emmetropia and Ametropia

Andrew Keirl

Introduction

This short and relatively straightforward chapter is designed to provide a brief overview of **model eyes** encountered in visual optics. The model eye that is of most interest is the **standard reduced eye** because this 'eye' forms the basis for the vast majority of definitions and calculations that need to be grasped for the successful study of visual optics.

Chapter content

- Basic structure and optical components of the human eye
- Gullstrand's schematic eyes
- The standard +60.00 D reduced eye
- The parameters of the standard reduced emmetropic eye

Basic structure and optical components of the human eye

A detailed anatomical discussion of the human eye is beyond the scope of this book; however, the gross anatomical structure of the human eye is illustrated in **Figure 2.1**.

The components of the human eye that can be considered to have an 'optical function' are the cornea, aqueous, crystalline lens and the vitreous. These structures are considered to perform an optical function because they are transparent, have refractive indices > 1 and are bounded by curved surfaces. In Chapter 1 we were reminded that the power of any curved surface depends on two factors:

1. The refractive index of the material used to form the curved surface

2. The radius of curvature of the curved surface.

We were also reminded that these two points applied to both lenses and the eye. In simple terms, the optical system of the eye is nothing more than a strong plus lens! The concepts of curvature, transparency, refractive index and the separation of surfaces form the basis for the construction of model eyes.

Gullstrand's schematic eyes

Allvar Gullstrand (1862–1930) was a nineteenth-century ophthalmologist and professor of ophthalmology, whose work laid the foundations for our knowledge of the eye and vision. On the basis of measurements taken, he drew up a model eye consisting of a centred optical system with spherical surfaces. This likeness of the human eye was used to study its optical working and its performance together with spectacle

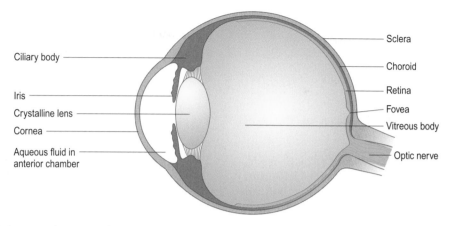

Figure 2.1 A horizontal section of the human eye.

Ciliary body

Iris

Crystalline lens

Cornea

Aqueous fluid in anterior chamber

Sclera

Choroid

Retina

Fovea

Vitreous body

Optic nerve

lenses and optical instruments. The importance of Gullstrand's eye lies in the fact that it gives a good idea of the average dimensions of the human eye. It must be stressed, however, that these are only *average* values arrived at either on a mathematical basis or by way of reason. They were not measurements taken from an existing eye. The purpose of a schematic eye is to provide a basis for theoretical studies of the eye as an optical instrument. Gullstrand provided data for two schematic eyes: the No. 1 version had six refracting surfaces, whereas the No. 2 eye consisted of a single surface cornea and a 'thin' crystalline lens.

Gullstrand also provided a separate version of each of his models to represent the eye when strongly accommodated. It should be appreciated that many other researchers and authors have published their own variations of a schematic eye. Any proposed model must include data relating to the radii of curvature of the refracting surfaces, the distance between each refracting surface and the refractive indices of the optical media between each refracting surface.

Gullstrand's No. 1 (exact) schematic eye

This model consists of six refracting surfaces as follows:

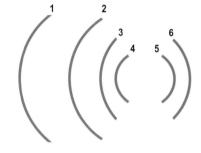

Figure 2.2 The six-surface eye (not to scale). Surfaces 1 and 2 represent the front and back surfaces of the cornea. Surfaces 3 and 6 represent the front and back surfaces of the crystalline lens. Surfaces 4 and 5 represent the front and back surfaces of the lens nucleus.

1. Anterior cornea
2. Posterior cornea
3. Anterior lens
4. Anterior nucleus
5. Posterior nucleus
6. Posterior lens.

The basic outline of this model eye is illustrated in Figure 2.2.

The surface radii of the No. 1 (exact) schematic eye are as follows:

21

- Anterior cornea +7.700 mm
- Posterior cornea +6.800 mm
- Anterior lens +10.00 mm
- Anterior nucleus +7.911 mm
- Posterior nucleus −5.760 mm
- Posterior lens −6.000 mm

Distances between the surfaces of the exact schematic eye are:

- Corneal thickness: 0.500 mm
- Posterior cornea to anterior lens: 3.100 mm
- Anterior lens to anterior nucleus: 0.546 mm
- Thickness of nucleus: 2.419 mm
- Posterior nucleus to posterior lens: 0.635 mm
- Lens thickness: 3.600 mm

The following are the refractive indices of the No. 1 (exact) schematic eye:

- Cornea: 1.376
- Aqueous: 1.336
- Vitreous: 1.336
- Crystalline lens (cortex): 1.386
- Crystalline lens (nucleus): 1.406

The axial length of the exact schematic eye is represented by the distance from the anterior cornea to the macula and is 24.385 mm. The powers of the eye in its relaxed and fully accommodated states are +58.64 D and +70.57 D, respectively.

Gullstrand's No. 2 (simplified) schematic eye

This model eye consists of three surfaces:

1. A single refracting surface (the reduced surface)
2. Anterior crystalline lens
3. Posterior crystalline lens

The basic outline of this model eye is illustrated in **Figure 2.3**.

The surface radii of the simplified schematic eye are as follows:

- Reduced surface: +7.800 mm
- Anterior lens: +10.00 mm
- Posterior lens: −6.000 mm

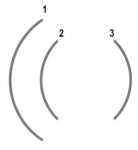

Figure 2.3 The three-surface eye (not to scale). Surface 1 represent the single refracting surface. Surfaces 2 and 3 represent the front and back surfaces of the single index crystalline lens.

The following are the refractive indices of the simplified schematic eye:

- Aqueous: 1.336
- Vitreous: 1.336
- Crystalline lens: 1.413

The following are the distances between the surfaces of the simplified schematic eye:

- Reduced surface to anterior lens: 3.600 mm
- Lens thickness: 3.600 mm
- Posterior lens to macula: 16.970 mm

The axial length of the simplified schematic eye is represented by the distance from the reduced surface to the macula and is 24.17 mm. The powers of the eye in its relaxed and fully accommodated states are +59.74 D and +70.54 D, respectively.

The standard +60.00 D reduced eye

The standard +60.00 D reduced eye is a single surface model and is generally used as the model for most definitions and calculations in visual optics. This simple model eye has a single refracting surface, providing all of the eye's power, a single refractive index and an axial length. There is a recognised set of symbols used with this eye. These are:

- F_e: the power of the single refracting reduced surface
- k': the distance from the reduced surface to the macula (the axial length)
- n_e: the symbol used for the single refractive index of the reduced eye

The standard reduced +60.00 D eye is illustrated in Figure 2.4.

The parameters of the standard reduced emmetropic eye

The standard reduced emmetropic eye is illustrated in Figure 2.5. For now, the term emmetropic is used to simply describe an eye that does not need a correction for a distance refractive error (it is explained fully in Chapter 3). Inspection of Figure 2.5 shows that the position of the second principal focus of the eye F_e' coincides with the macula, M'. This is necessary for emmetropia to occur. Note also that, as the eye is viewing a distant object (parallel light), it is assumed to be **unaccommodated**. Accommodation is discussed in Chapter 13.

The parameters of the standard reduced emmetropic eye are:

- Reduced surface power $F_e = +60.00\,\text{D}$
- Axial length $k' = 22.22\,\text{mm}$
- Refractive index $n_e = 4/3$

Concluding points for Chapter 2

This relatively short chapter has described:

- the optical components of the human eye
- the concept of model eyes
- the parameters of Gullstrand's schematic eyes

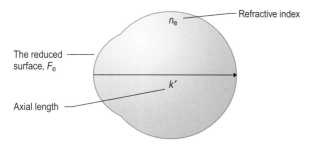

Figure 2.4 The standard reduced +60.00 D model eye.

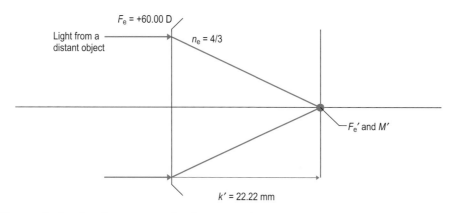

Figure 2.5 The standard reduced unaccommodated emmetropic eye.

- the parameters of the standard reduced emmetropic eye

The standard reduced +60.00 D model eye will be used extensively in future chapters.

Further recommended reading

Tunnacliffe A H (1993) *Introduction to Visual Optics*. Association of the British Dispensing Opticians, London

Rabbetts R B B (1998) *Bennett & Rabbetts' Clinical Visual Optics*. Butterworth-Heinemann, Oxford

Emmetropia and Ametropia

Andrew Keirl

Introduction

This chapter discusses and illustrates the definitions and concepts of emmetropia and ametropia. An understanding of this material is necessary before calculations involving the reduced eye are attempted. Indeed, a thorough understanding of this material makes such calculations easier to master!

Chapter content

- Emmetropia
- Ametropia
- The classification of spherical ametropia

Emmetropia

In Chapter 2, the term **emmetropia** was explained by stating that it described an eye that does not need a correction for a distance refractive error. This term is now defined and explained in a more robust way. There are two definitions of emmetropia that are commonly used in visual optics:

1. An unaccommodated eye is emmetropic if the second focal point F'_e coincides with the centre of the macula M' (**Figure 3.1**).
2. An unaccommodated eye is emmetropic if the far point (M_R) is at infinity.

Figure 3.1 shows parallel light from a distant object incident at the reduced surface of an emmetropic eye. The incident light is refracted by the reduced surface and is converged (the reduced surface is positive in power) towards a point on the retina called the **macula** (M'). If we think of the reduced surface of the eye as a simple thin lens, the point image formed by the reduced surface can be referred to as the **second principal focus** F'_e (the subscript 'e' simply stands for eye).

Emmetropia is said to exist if the second principal focus of the reduced surface F'_e coincides with the centre of the macula M'. If an image is formed on the macula, distance vision is clear *without* the need for a correcting lens. It should be remembered that the distance from any lens to its focus is called the focal length (f'). As the reduced surface of the eye is just a lens, it also has a focal length. Strictly speaking, this is the second principal focal length and is the distance from the reduced surface F_e to the macula M'. This distance is given the symbol f'_e (again, e is for eye). Inspection of Figure 3.1 shows that the axial length k' of the eye is the same as the focal length f'_e. This is another way of stating that the second principal focus of the reduced surface F'_e coincides with the centre of the macula M' and can also be used to define emmetropia. **Figure 3.2** also shows the standard reduced emmetropic eye, this time with the parameters added.

The second definition of emmetropia refers to something called the *far point*. In very simple terms, the far point is the **furthest** distance that a person can see **without** the help of spectacles or contact lenses. The far point of an eye can be in one of three places: at infinity, at some distance in front of the eye and at some distance behind the eye. As an emmetropic eye should be able to see at long distances without the need for a spectacle or contact lens correction, the far point of an emmetropic eye is at **infinity**. The actual definition of far point is 'that point conjugate with the centre of the macula by refraction at the unaccommodated and uncorrected eye'. The far point is also known as the **far point of distinct vision,** the **true far point** and the **far point of accommodation**. The far point is given the symbol M_R.

It is important to note that as the term 'emmetropia' refers to **distance vision** and the eye is always assumed to be **unaccommodated**. **Accommodation** is defined as the 'eye's ability to change in power in order to focus on objects at different distances'. The term 'unaccommodated' refers to the power of the eye in its weakest dioptric state. The term 'fully accom-

modated' refers to the power of the eye in its strongest dioptric state. If the word unaccommodated is added to a diagram or a particular situation, we can assume that the eye is in its weakest dioptric state and focused for distance vision. If the word accommodated is used, we can assume that the eye is focused for near vision. The **amplitude of accommodation** is the difference between maximum and minimum powers of a particular eye and therefore gives the total amount of accommodation available. Accommodation is discussed in Chapter 13.

Ametropia

Ametropia is simply a deviation from emmetropia. If an eye is not emmetropic it must be ametropic! The term **ametropia** refers to distance vision and once again, the eye is always assumed to be unaccommodated. The classification of ametropia is given in **Figure 3.3**, and essentially is a process of elimination because ametropia is either **spherical** or **astigmatic**. If spherical, the ametropia is either **myopia** or **hypermetropia**. Myopia and hypermetropia can be either **axial** or **refractive** in origin. Refractive ametropia can be classified as either **curvature** or **index**. **Astigmatic** ametropia is either **corneal** in origin (the astigmatic error originates at the corneal surface and is usually caused by the cornea being toroidal in shape) or **lenticular** (the astigmatic error is the result of toroidal lens surfaces or a tilted crystalline lens). Astigmatism is discussed in Chapter 7.

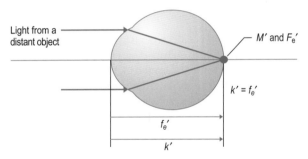

Figure 3.1 The standard reduced unaccommodated emmetropic eye.

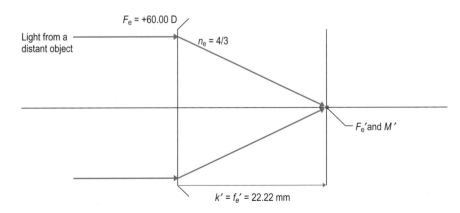

Figure 3.2 The standard reduced emmetropic eye (unaccommodated).

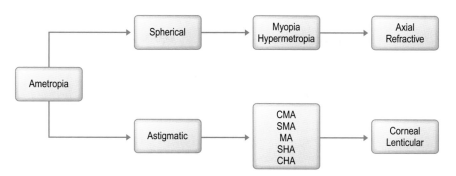

Figure 3.3 The classification of ametropia. CHA, compound hypermetropic astigmatism; CMA, compound myopic astigmatism; MA, mixed astigmatism; SHA, simple hypermetropic astigmatism; SMA, simple myopic astigmatism.

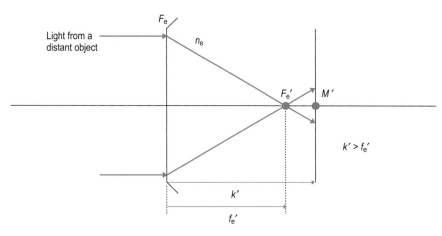

Figure 3.4 The unaccommodated myopic eye.

Myopia

An unaccommodated eye is myopic if the second focal point F'_e lies in **front** of the centre of the macula. This is commonly called short sight because the focus formed by the eye falls 'short' of the reduced surface. In myopia the eye is long compared with its focal length, because the axial length k' is always greater than the focal length f'_e (**Figure** 3.4). As with emmetropia, myopia can also be defined with reference to the far point. An unaccommodated eye is myopic if the far point lies at a finite distance in front of the eye. This is why short-sighted patients can see at near without their spectacles! The far point is simply the position at which an object has to be placed for an eye to see it clearly. The distance from the reduced surface F_e to the far point is called the **far point distance** and is given the symbol k. The far point in the myopic eye is illustrated in **Figure** 3.5. Myopia is corrected using negative spectacle (or contact) lenses.

Hypermetropia

An unaccommodated eye is hypermetropic if the second focal point F'_e lies behind the centre of the macula. This is commonly called long sight because the focus formed by the eye is 'long' compared with the position of the macula. In hypermetropia, the eye is short compared with its focal length, because the axial length k' is always less than the focal length f'_e (**Figure** 3.6). As with emmetropia and myopia,

27

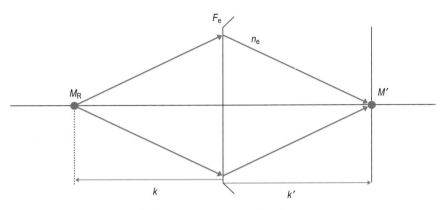

Figure 3.5 The position of the far point (M_R) in the unaccommodated myopic eye.

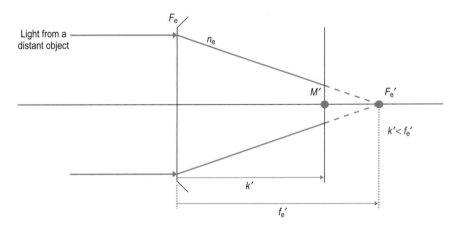

Figure 3.6 The unaccommodated hypermetropic eye.

hypermetropia can also be defined with reference to the far point. An unaccommodated eye is hypermetropic if the far point lies at a finite distance behind the eye. In the case of hypermetropia an object has to be placed behind the eye for it to be seen clearly. This, of course, cannot actually happen! The distance from the reduced surface F_e to the far point is again called the far point distance and is also given the symbol k. The far point in the hypermetropic eye is illustrated in **Figure 3.7**. Hypermetropia is corrected using positive spectacle (or contact) lenses.

The classification of spherical ametropia

As outlined in Figure 3.3, spherical ametropia can have several causes: essentially, the eye is either too long or too short (axial ametropia) or too strong or too weak (refractive ametropia). All classifications and comparisons are made with reference to the standard reduced emmetropic eye discussed in Chapter 2. In **axial ametropia**, the axial length of the reduced eye is either longer or shorter than the axial length of

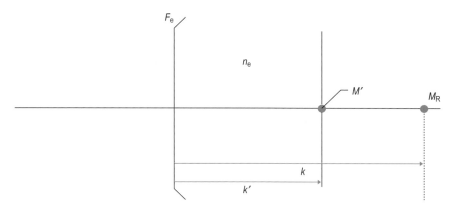

Figure 3.7 The position of the far point (M_R) in the unaccommodated hypermetropic eye.

the standard reduced emmetropic eye, which is 22.22 mm. In **refractive ametropia**, the power of the reduced surface is either greater or less than the power of the reduced surface in the standard reduced emmetropic eye, which is +60.00 D. As the power of any curved surface depends on its material and shape, refractive ametropia has two possible causes: **curvature ametropia** is when the radius of curvature of the reduced surface is either longer or shorter than the radius of curvature of the reduced surface in the standard reduced eye; and **index ametropia** is when the refractive index of eye is either higher or lower than the refractive index of the standard reduced eye. As a reminder, the parameters of the standard reduced emmetropic eye are:

Power of the reduced surface: F_e = +60.00 D
Radius of curvature of the reduced surface:
  r_e = +5.556 mm
Refractive index: n_e = 4/3
Axial length: k' = 22.22 mm

Axial myopia

Axial myopia occurs when the length of the eye is greater than 22.22 mm. This causes the second principal focus to form in front of the macula. In axial myopia, the parameters of the reduced eye are:

F_e = +60.00 D and r_e = +5.556 mm
n_e = 4/3
Axial length: k' > 22.22 mm

The eye is myopic because the axial length of the eye is *longer* than the axial length of the standard reduced eye.

Curvature myopia

This is a type of refractive myopia that means that the power of the eye is greater than +60.00 D. Curvature myopia occurs when the radius of curvature of the reduced surface is shorter (and therefore dioptrically stronger) than the radius of curvature of the standard reduced emmetropic eye. As the power of the eye is greater than +60.00 D, the second principal focus is again formed in front of the macula. In curvature myopia, the parameters of the reduced eye are:

Axial length: k' = 22.22 mm
F_e > +60.00 D because r_e < +5.556 mm
n_e = 4/3

Index myopia

This is also a type of refractive myopia (power of the eye > +60.00 D). However, this time the cause is not the shape but the material. In the case of index myopia, the eye is myopic because the refractive index of the eye is *higher* than the refractive index of the standard reduced eye. As long as the radius of curvature of the reduced surface is the standard +5.556 mm, the higher refractive index results in the power of the eye

29

being greater than +60.00 D, causing the second principal focus to form in front of the macula. Index myopia does exist in real eyes and is often seen in older patients with age-related media changes. An increase in myopia is often a sign of nuclear sclerotic lens opacities in an older patient. In index myopia, the parameters of the reduced eye are:

Axial length: $k' = 22.22$ mm
$F_e > +60.00$ D and $r_e = +5.556$ mm
$n_e > 4/3$

When attempting to get your head around curvature and index myopia, remember that:

$$F_e = \frac{n_e - 1}{r_e}$$

The value of F_e can be increased by increasing the value of n_e or reducing the value of r_e.

Axial hypermetropia

Axial hypermetropia occurs when the length of the eye is *less* than 22.22 mm. This causes the second principal focus to form behind the macula. In axial hypermetropia, the parameters of the reduced eye are:

Axial length: $k' < 22.22$ mm
$F_e = +60.00$ D and $r_e = +5.556$ mm
$n_e = 4/3$

The eye is hypermetropic because the axial length of the eye is shorter than the axial length of the standard reduced eye.

Curvature hypermetropia

Once again, this is a type of refractive hypermetropia that means that the power of the eye is less than +60.00 D. Curvature hypermetropia occurs when the radius of curvature of the reduced surface is longer (and therefore dioptrically weaker) than the radius of curvature of the standard reduced emmetropic eye. As the power

of the eye is less than +60.00 D, the second principal focus is formed behind the macula. In curvature hypermetropia, the parameters of the reduced eye are:

Axial length: $k' = 22.22$ mm
$F_e < +60.00$ D because $r_e > +5.556$ mm
$n_e = 4/3$

Index hypermetropia

This is also a type of refractive hypermetropia (the power of the eye is < +60.00 D). Once more, the cause is not the radius of curvature (shape) but the material. In this case the eye is hypermetropic because the refractive index of the eye is *lower* than the refractive index of the standard reduced eye. As long as the radius of curvature of the reduced surface is a standard +5.556 mm, the lower refractive index results in the power of the eye being less than +60.00 D, causing the second principal focus to form behind the macula. However, unlike its myopic counterpart, index hypermetropia does not exist in real eyes. In index hypermetropia, the parameters of the reduced eye are:

Axial length: $k' = 22.22$ mm
$F_e < +60.00$ D and $r_e = +5.556$ mm
$n_e < 4/3$

Concluding points for Chapter 3

This chapter has:

- defined emmetropia and ametropia
- classified spherical ametropia.

Further recommended reading

Tunnacliffe A H (1993) *Introduction to Visual Optics.* Association of the British Dispensing Opticians, London
Rabbetts R B B (1998) *Bennett & Rabbetts' Clinical Visual Optics.* Butterworth-Heinemann, Oxford

The Correction of Spherical Ametropia

Andrew Keirl

Introduction

Calculations involving the spectacle and ocular refractions frequently occur as part of a calculation within the subject of visual optics. The reader therefore needs to be able to solve such problems first time and every time. Other topics within this chapter, e.g. the vertex distance and effectivity, are a daily occurrence in clinical practice, so the material in this chapter has both theoretical and practical applications.

Chapter content

Ocular and spectacle refractions

Ocular refraction is the necessary vergence measured at the reduced surface so that a clear image is formed at the macula in the unaccommodated eye. The symbol used for ocular refraction is K and the unit for ocular refraction is the dioptre (D). Ocular refraction is often likened (although this is not strictly true) to the power of a contact lens because this particular optical device is of course placed in contact with the eye.

Spectacle refraction is the vergence necessary measured in the plane of the spectacle lens so that a clear image is formed at the macula in the unaccommodated eye. Common symbols used for spectacle refraction are F, F_{sp} and F_v'. The unit for spectacle refraction is also the dioptre. Unlike a contact lens, a spectacle lens is placed at some distance from the eye. This distance is known as the **vertex distance** and is defined as 'the distance from the corneal apex to the visual point on the lens'. A vertex distance should accompany a spectacle prescription if the power of the lens in any meridian is ±5.00 D and above, e.g. $-5.00/-1.00 \times 130$ @ 12 mm where '@ 12 mm' refers to the vertex distance of the optometrist's trial frame or refractor head. The value of an ocular refraction for a particular individual at a particular time is constant, but the corresponding spectacle refraction varies with the vertex distance. The vertex distance is discussed in detail at the end of this chapter.

It is important to be able to convert from a spectacle refraction, F_{sp}, to an ocular refraction, K, and vice versa. This can be easily performed using the following equations:

$$F_{sp} = \frac{K}{1+(dK)}$$

and:

$$K = \frac{F_{sp}}{1-(dF_{sp})}$$

or by using step-along and step-back ray-tracing. In the above equations, K is the ocular refraction in dioptres, F_{sp} is the spectacle refraction in dioptres and d is the vertex distance of the correcting spectacle lens in metres. The calculation of K and F_{sp} from first principles is shown in **Figures 4.1** and **4.2**. Note that the reciprocal of the far point distance, k, gives the ocular refraction, K, and vice versa.

The spectacle correction of ametropia

When working through the following section, please pay particular and careful attention to **Figures 4.3** and **4.4**. The following statement is the *key* to the understanding of the correction of ametropia using a spectacle lens and applies to both myopic and hypermetropic corrections.

The second principal focus of a correcting spectacle lens must coincide with the far point of the ametropic eye

Figure 4.3 illustrates the spectacle correct of hypermetropia: inspection shows that the parallel light incident at the spectacle lens is converged. If the eye were removed from Figure 4.3, the converging light would come to at focus at the second principal focus F'_{sp}. If the lens is to correct the hypermetropia shown in Figure 4.3, F'_{sp} of the correcting spectacle lens must coincide with the eye's far point, M_R, which is, of course, behind the eye in hypermetropia. The convergent light leaving the lens undergoes further refraction (convergence) when it arrives at the reduced surface and goes on to form an

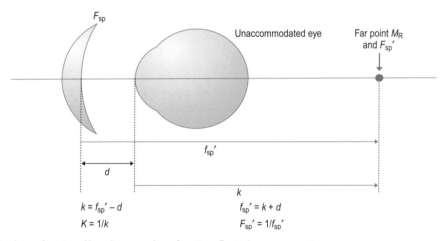

Figure 4.1 Ocular refraction, K, and spectacle refraction, F_{sp}, in hypermetropia.

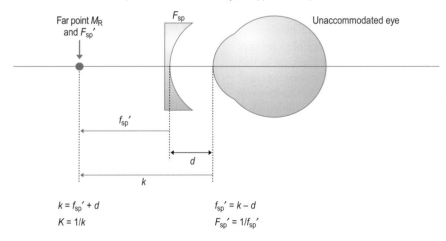

Figure 4.2 Ocular refraction, K, and spectacle refraction, F_{sp}, in myopia.

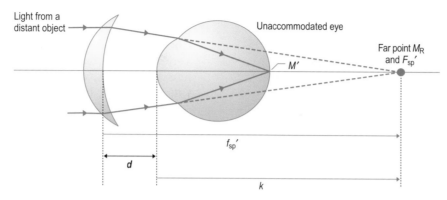

Figure 4.3 The spectacle correction of hypermetropia.

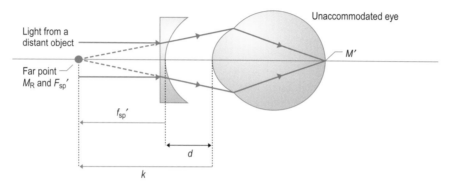

Figure 4.4 The spectacle correction of myopia.

image at the macula M'. The hypermetropia is therefore corrected.

Figure 4.4 illustrates the spectacle correct of myopia: inspection shows that the parallel light incident at the spectacle lens is diverged. If the eye were removed from the diagram, the divergent light would produce a virtual focus (image) to the left of the lens. This is a good time to remember some basic geometrical optics and the formation of the second principal focus, F'_{sp}, with minus lenses! The divergent rays must be traced backwards to find the position of the second principal focus. If the lens is to correct the myopia shown in Figure 4.4, F'_{sp} of the correcting spectacle lens must coincide with the eye's far point, which is in front of the eye in myopia. The divergent light leaving the lens undergoes further refraction (convergence) when it arrives at the reduced surface and goes on the form an image at M'. The myopia is

therefore corrected. Students of visual optics should be able to explain and illustrate the spectacle correction of myopia and hypermetropia. Table 4.1 summarises the terms used in the reduced eye.

Please note that, when performing calculations involving the dioptre, any distance must be entered in metres. Often, however, it is necessary to convert back to millimetres afterwards.

The **dioptric length** of the reduced eye is the value of the vergence immediately to the right of the reduced surface and is given the symbol K'.

The vertex distance

In the preceding section the term 'vertex distance' was introduced and defined. The vertex distance of a spectacle lens plays an important part in the spectacle correction of ametropia

Table 4.1 Summary of the terms used with the reduced eye

Term	Symbol
Spectacle refraction	F, F_{sp} (thin lens) or F'_v (thick lens)
Ocular refraction	K
Far point distance	k
Vertex distance	d
Reduced surface power	F_e
Focal length of the reduced surface	f'_e
Dioptric length	K'
Refractive index of the reduced eye	n_e
Axial length	k'

because the effective power of a spectacle lens used for distance vision depends on its position relative to the eye. It is important to realise that *any* lens moved away from the eye becomes more **positive**. This means that plus spectacle lenses become **stronger** (more plus) and minus spectacle lenses become **weaker** (less minus or more plus) if positioned **away** from the eye. These effects are often used by people with presbyopia, in particular aphakia (a patient without a crystalline or intraocular lens) to enhance near vision. As mentioned above, a vertex distance should accompany a spectacle prescription if the power of the lens in any meridian is ±5.00 D and above. So why is a vertex distance necessary for medium and high prescriptions? Well, as mentioned above, the effective power of a lens changes as it is moved away from or closer to the eye, and *all* lenses become more *positive* if they are moved away from the eye. This means that a minus lens becomes weaker (and therefore a stronger lens needs to be ordered to compensate for the change caused by effectivity) and a plus lens becomes stronger (and weaker lens is required) when moved away from the eye. The opposite occurs when the lens is moved towards the eye.

If a practitioner is presented with a prescription of ±5.00 D and above, there are three options:

1. Ensure that the chosen frame sits at the prescribed vertex distance.
2. Choose a frame that sits at a different vertex distance, but alter the power of the lenses accordingly.
3. Choose another frame, which fits at the required vertex distance.

The second point above can be achieved by the use of:

- knowledge of focal length and focal power
- appropriate conversion tables and charts
- appropriate formulae
- step-along

How to compensate for a change in vertex distance using first principles

For a spectacle lens to correct an eye, its second principal focus, F'_{sp}, must coincide with the eye's far point. The distance from the lens to F'_{sp} is the focal length of the lens, f'_{sp}, which is made up from the sum (in the case of hypermetropia) of the vertex distance, d, and the far point distance, k. In myopia, f'_{sp} is the difference between k and d. If the lens is moved to a position other than its prescribed vertex distance, F'_{sp} no longer coincides with M_R. For F'_{sp} to be returned to the position occupied by M_R, the focal length of the lens needs to be adjusted. In the case of a positive lens (**Figure 4.5**), if the vertex distance is increased, F'_{sp} lies to the left of M_R and, to compensate for the change in vertex distance, the focal length of the correcting lens must be increased to 'push' F'_{sp} back so that once again it coincides with M_R. This results in a weaker lens being required. If the vertex distance is reduced, the necessary shorter focal length results in a stronger lens being required.

As far as the correction of myopia is concerned (**Figure 4.6**), even though the far point lies in front of the eye, F'_{sp}, and M_R must again coincide. Once more, any movement in the position of the lens from its prescribed vertex distance must be allowed for by adjusting its focal length. In the case of a negative lens, if the vertex distance is increased, F'_{sp} lies to the right of M_R and, to compensate for the change in vertex distance, the focal length of the correcting lens must be decreased to 'pull' F'_{sp} back to

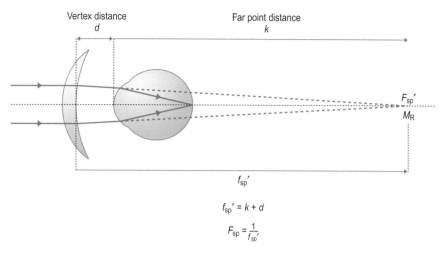

Figure 4.5 The significance of the vertex distance in the correction of hypermetropia.

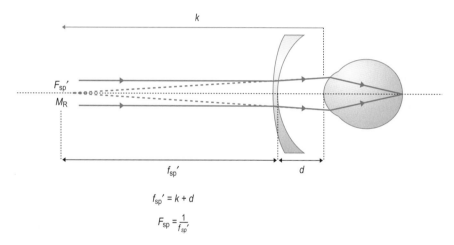

$$f_{sp}' = k + d$$

$$F_{sp} = \frac{1}{f_{sp}'}$$

Figure 4.6 The significance of the vertex distance in the correction of myopia.

coincide with M_R. This results in a stronger lens being required. If the vertex distance is reduced, the necessary longer focal length results in a weaker lens being required.

For both plus and minus lenses, the adjustment in focal length is always equal to the *change* in vertex distance. When dealing with minus lenses, it is a good idea to do the sums without using minus signs. A minus sign can be added once the sums are completed. As with all calculations in optics, always ask yourself the question: 'Does my answer make sense?'

Changes in power resulting from no compensation for alterations in vertex distance may affect visual acuity. Patients who wear strong plus lenses sometimes pull their spectacles down towards the end of their nose to read. The increase in vertex distance is effectively making the lenses stronger and thereby producing a reading addition! Another common example of the significance of the vertex distance is to be found in contact lens work because the power of a patient's contact lenses is often different to the power of the spectacles. This is a result of the fact that a lens in contact with the eye has an effective vertex distance of zero, whereas the spectacle prescription has a vertex distance in the region of 12 mm. Examples of lens powers

35

with various vertex distances are given in Table 4.2. In the calculation of the contact lens power, a zero vertex distance was assumed.

Inspection of Table 4.2 reveals three important points:

1. If the vertex distance is increased, a weaker plus lens or a stronger minus lens is required.
2. If the vertex distance is decreased, a stronger plus lens or a weaker minus lens is required.
3. A change in vertex distance has the greatest effect on higher powers.

The vertex distance can be measured using a vertex distance measuring gauge or simply by using a millimetre ruler (Figures 4.7 and 4.8a). When compared with traditional steep lenses, the flatter form lenses in current use actually make this procedure easier and more accurate because the plane of the frame is closer to the back vertex of the lens. Most trial frames employ a millimetre scale position along each side, which can be used to measure the vertex distance assigned to a given spectacle refraction (Figure 4.8b). Occasionally, a spectacle frame fitting at some vertex distance other than the prescribed one causes non-tolerance to a new pair of spectacles. When solving any problem (and this is no exception), it is always a good idea to be aware of clues based on clinical experience and also verbal clues given by the patient. Possible clues to this particular problem are the following:

- The patient has medium/high myopia or hypermetropia (this situation is unlikely to occur with lower powered lenses).
- The patient is generally unhappy with vision in primary gaze.
- The patient may state: 'My new spectacles are not as good as my old ones.'
- The patient may state: 'I can see better when the spectacles are placed closer/further away from my face.'

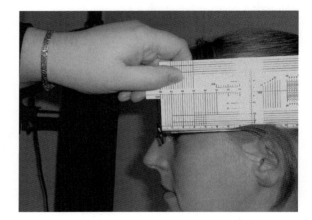

Figure 4.7 The vertex distance measured using a millimeter ruler.

Table 4.2 Spectacle versus contact lens prescriptions

Prescription	Vertex distance 10 mm (D)	Vertex distance 14 mm (D)	Contact lens (D)
+4.00 D at 12	+4.03	+3.97	+4.20
+6.00 D at 12	+6.07	+5.93	+6.47
+8.00 D at 12	+8.13	+7.87	+8.85
+10.00 D at 12	+10.20	+9.80	+11.36
−4.00 D at 12	−3.97	−4.03	−3.82
−6.00 D at 12	−5.93	−6.07	−5.60
−8.00 D at 12	−7.87	−8.13	−7.30
−10.00 D at 12	−9.80	−10.20	−8.93

The values are the actual powers required when compared with a vertex distance of 12 mm.

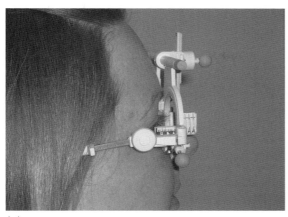

(a)

(b)

Figure 4.8 (a) The vertex distance measured scale on a trial frame. (b) The measurement of vertex distance using a vertex distance measuring gauge.

Good communication is vital. One may think a frame is very suitable and is sitting at the same vertex distance as the trial frame, but the patient may like to wear the spectacles slightly further down the nose. It is always, therefore, sensible to confirm with patients that the frame is comfortable and sitting where they like to wear their spectacles. It is a good idea to compare the position of the new frame with their old spectacles. The necessary equations to compensate for a change in vertex distance are given below.

If the vertex distance is decreased:

$$F_{new} = \frac{F_{old}}{1-(dF_{old})}$$

If the vertex distance is increased:

$$F_{new} = \frac{F_{old}}{1+(dF_{old})}$$

In both expressions, d (in metres) represents the *change* in vertex distance and not the vertex distance itself.

Astigmatism is dealt with in Chapter 7. However, it is necessary to mention here that, when dealing with an astigmatic prescription, it is incorrect to compensate for the sphere and the cylinder. Each principal power must be compensated in turn. The final result can then be re-written in the usual sph–cyl (spherical–cylindrical) form.

Summary of equations and expressions

Ocular and spectacle refraction

$$F_{sp} = \frac{K}{1+(dK)} \qquad K = \frac{F_{sp}}{1-(dF_{sp})}$$

d *must* be entered in metres

Ocular refraction and far point distance

$$K = \frac{1}{k} \qquad k = \frac{1}{K}$$

Axial length and dioptric length

$$k' = \frac{n_e}{K'} \quad \text{and} \quad K' = \frac{n_e}{k'}$$

Ocular refraction, reduced surface power and dioptric length

$$K' = K + F_e$$

Focal length and power of the reduced surface

$$f_e' = \frac{n_e}{F_e} \quad \text{and} \quad F_e = \frac{n_e}{f_e'}$$

f_e' should be entered in metres

Equations for vertex distance compensation

If the vertex distance is decreased:

$$F_{new} = \frac{F_{old}}{1 - (dF_{old})}$$

If the vertex distance is increased:

$$F_{new} = \frac{F_{old}}{1 + (dF_{old})}$$

d must be entered in metres

Elementary reduced eye calculations

Example 4.1

For each of the reduced eyes given below find the ocular refraction (K) and state the type of ametropia present. Assume that the refractive index of all six eyes is 4/3:

1. **Far point distance (k) +25 cm, axial length (k′) 22.22 mm**
2. **Power of the reduced surface (Fₑ) +64.00 D, axial length (k′) 20.833 mm**
3. **Dioptric length (K′) +60.00 D, power of the reduced surface (Fₑ) +66.00 D**
4. **Dioptric length (K′) +58.00 D, power of the reduced surface (Fₑ) +62.00 D**
5. **Radius of curvature of the reduced surface (rₑ) +6.410 mm, axial length (k′) 22.22 mm**
6. **Radius of curvature of the reduced surface (rₑ) +5.556 mm, far point distance (k) −12.50 cm**

Reduced eye (1)

The ocular refraction K can be calculated easily using the far point distance k. The required expression (with k in metres) is:

$$K = \frac{1}{k} = \frac{1}{+0.25} = +4.00\,D$$

As the given axial length is the standard value, this eye must have **refractive hypermetropia**.

Reduced eye (2)

Inspection of the data given in the question shows that both the power of the reduced surface and the axial length are non-standard

values. As the axial length has been given, the dioptric length can be calculated using:

$$K' = \frac{n_e}{k'} = \frac{4/3}{0.020833} = +64.00\,D$$

The ocular refraction K can be found using:

$$K' = K + F_e$$

which rearranges to become:

$$K = K' - F_e = +64.00 - (+64.00) = 0.00\,D$$

This non-standard reduced eye is therefore **emmetropic**.

Reduced eye (3)

As long as the refractive index of the eye is 4/3, an eye with a dioptric length of +60.00 D has an axial length of 22.22 mm. This can be confirmed using:

$$k' = \frac{n_e}{K'} = \frac{4/3}{+60.00} = 0.02222\,m$$

The ocular refraction can be found using:

$$K = K' - F_e = +60.00 - (+66.00) = -6.00\,D$$

The ametropia in this case is therefore **refractive myopia**.

Reduced eye (4)

In this case, both the power of the reduced surface and the dioptric length have non-standard values. The ametropia is therefore a **mixture of axial and refractive elements**. The axial length can be found using:

$$k' = \frac{n_e}{K'} = \frac{4/3}{+58.00} = 0.02299\,m$$

The axial length is therefore 22.99 mm. The ocular refraction is found using:

$$K = K' - F_e = +58.00 - (+62.00) = -4.00\,D$$

Reduced eye (5)

In this case, as the axial length of the reduced eye is the standard value of 22.22 mm, the ametropia must be refractive in origin. As the error is caused by the radius of curvature of the reduced surface not being the standard value of 5.56 mm, the ametropia in this example could also be described as **curvature ametropia**. As the given radius of curvature is longer (or flatter) than 5.556 mm, the ametropia is actually **curvature hypermetropia**. The power of the reduced surface F_e can be found using:

$$F_e = \frac{n_e - 1}{r_e} = \frac{4/3 - 1}{+0.006410} = +52.00\,D$$

The dioptric length can be calculated using:

$$K' = \frac{n_e}{k'} = \frac{4/3}{0.02222} = +60.00\,D$$

The ocular refraction can be found using:

$$K = K' - F_e = +60.00 - (+52.00) = +8.00\,D$$

As predicted, the ametropia is **curvature ametropia**.

Reduced eye (6)

In this case, the radius of curvature of the reduced surface has the standard value of 5.556 mm. If the refractive index of the eye is 4/3, the power of the reduced surface is +60.00 D.

$$F_e = \frac{n_e - 1}{r_e} = \frac{4/3 - 1}{+0.005556} = +60.00\,D$$

The ametropia is therefore **axial** in origin. The ocular refraction can be calculated using the far point distance (always entered as metres).

$$K = \frac{1}{k} = \frac{1}{-0.125} = -8.00\,D$$

The ametropia in this case is therefore **axial myopia**.

Example 4.2

A reduced eye with axial myopia is corrected using a −6.00 D thin lens at a vertex distance of 14 mm. Find the ocular refraction and axial length.

It is very important to use all the information given in the question. The most important word here is *axial* because this word tells us the exact type of ametropia in relation to the standard reduced emmetropic eye with which we are dealing. The use of the word axial means that there is something wrong with the length of the eye, but the power and refractive index are both normal. Before we start calculating any values,

therefore, we know that $F_e = +60.00\,D$, $n_e = 4/3$ and $k' \neq 22.22$ mm. As the eye is myopic, $k' > 22.22$ mm. We should therefore expect our answer to be >22.22 mm!

As the power of the spectacle lens (spectacle refraction F_{sp}) is given, we can first of all find the ocular refraction using the expression:

$$K = \frac{F_{sp}}{1 - (dF_{sp})}$$

$$K = \frac{-6.00}{1 - (0.014 \times -6.00)} = -5.53\,D$$

To find the axial length we need to find the dioptric length:

$$K' = K + F_e \qquad K' = -5.53 + (+60.00) = +54.46\,D$$

$$k' = \frac{n_e}{K'} \qquad k' = \frac{4/3}{+54.46} = 0.02448\,m$$

The axial length of the eye is therefore 24.48 mm; based on the assumptions made at the beginning of this example, it is almost certainly correct! Always ask yourself the question: 'Does my answer seem sensible?' It is usual to state an axial length in millimetres.

Example 4.3

A reduced eye with refractive hypermetropia has an ocular refraction of +8.00 D. Find the spectacle refraction at 10 mm and the focal length of the reduced surface.

Once again look carefully at the question. We are told that the eye has refractive hypermetropia, so the power of the reduced surface is not equal to +60.00 D. Hypermetropic eyes are weaker than the standard reduced emmetropic eye, so we should expect F_e to be *less* than +60.00 D. The word refractive tells us that the axial length and refractive index are standard at 22.22 mm and 4/3, respectively.

As we are given the ocular refraction, the first thing that we can do is calculate the spectacle refraction at the given vertex distance:

$$F_{sp} = \frac{K}{1 + (dK)}$$

$$F_{sp} = \frac{+8.00}{1 + (0.010 \times +8.00)} = +7.41\,D$$

39

To find the focal length of the reduced surface, we need to know the power of the reduced surface. We know that:

$$K' = K + F_e$$

so, by simple rearrangement:

$$F_e = K' - K.$$

The use of the word refractive in the question tells us the value for k' (22.22 mm), so we can use the expression:

$$K' = \frac{n_e}{k'}$$

to find the dioptric length and therefore the reduced surface power:

$$K' = \frac{4/3}{0.02222} = +60.00\,D$$

Note that the above equation involves a distance *and* the dioptre so k' has been entered in metres. The power of the reduced surface can now be found using the expression:

$$F_e = K' - K \qquad F_e = +60.00 - (+7.41) + 52.59\,D$$

Once again, the answer corresponds with the prediction at the start of this problem. To find the focal length of the reduced surface, use the expression:

$$f'_e = \frac{n_e}{F_e} \qquad f'_e = \frac{4/3}{+52.59} = +0.02535\,m$$

The focal length of the reduced surface is therefore +25.35 mm. As the axial length of this eye is 22.22 mm (assumed because the error is refractive), the focus occurs behind the eye. The eye is of course hypermetropic! Once again, we have made sure that our answer makes sense.

Example 4.4

A prescription reads +5.00 D at 10 mm. The final lens is to be fitted at a vertex distance of 16 mm. What is the power of the final lens?

In this example the final lens is to be fitted 6 mm further away from the eye. As the vertex distance is being *increased* we need to use the equation:

$$F_{new} = \frac{F_{old}}{1 + (dF_{old})}$$

where $F_{old} = +8.00$ and $d = 0.006\,m$. Substituting into the above equation gives:

$$F_{new} = \frac{+5.00\,D}{1 + (0.006 \times +5.00)} = +4.85\,D$$

The final lens to be ordered therefore has a power of +4.85 D. As predicted in the preceding text, the 'new' lens is **weaker**, to reflect the **longer** focal length produced by **increasing** the vertex distance of a plus lens. Remember that a +4.85 D lens at 16 mm does exactly the same thing as a +5.00 D lens at 10 mm. This is what effectivity calculations are all about – the effect must remain the same!

Example 4.5

A prescription reads −10.00 D at 16 mm. The final lens is to be fitted at a vertex distance of 12 mm. What is the power of the final lens?

In this example the final lens is to be fitted 4 mm closer to the eye. As the vertex distance is being *decreased* we need to use the equation:

$$F_{new} = \frac{F_{old}}{1 - (dF_{old})}$$

where $F_{old} = -10.00$ and $d = 0.004\,m$. Substituting into the above equation gives:

$$F_{new} = \frac{-10.00}{1 - (0.004 \times -10.00)} = -9.61\,D$$

As predicted, the 'new' lens is **weaker** to reflect the **longer** focal length produced by **reducing** the vertex distance of a minus lens.

Example 4.6

Calculate the power of the thin correcting lens at 10 mm for a reduced eye with an axial length of 24.500 mm and refractive index 1.3475, the power of the reduced surface being +59.762 D.

This example is the first so far to use a 'non-standard' refractive index. As the axial length and refractive index are given, the dioptric length can be found. As the power of the reduced surface is also given, K and therefore F_{sp} can be determined.

To find the dioptric length use:

$$K' = \frac{n_e}{k'}$$

$$K' = \frac{1.3475}{0.0245} = +55.00\,D$$

Now:

$$K' = K + F_e$$

so, by simple rearrangement:

$$K = K' - F_e$$

$$K = +55.00 - (+59.762) = -4.762\,D$$

And to convert from K to F_{sp} use:

$$F_{sp} = \frac{K}{1 + (dK)}$$

The vertex distance is given as 10 mm, which must be entered in metres.

$$F_{sp} = \frac{-4.762}{1 + (0.01 \times -4.762)} = -5.00\,D$$

The power of the thin correcting lens at 10 mm is therefore −5.00 D.

Example 4.7

The far point of an axially myopic reduced eye is 10 cm from the reduced surface. Calculate:
- **The ocular refraction**
- **The axial length**
- **The power of the thin spectacle lens fitted at 11.5 mm.**

If the far point distance is known, the ocular refraction is given by:

$$K = \frac{1}{k}$$

In this example, the eye is myopic so $k = -10$ cm. When using the above expression, k is entered in metres.

$$K = \frac{1}{-0.1} = -10.00\,D$$

The ocular refraction is therefore −10.00 D. The axial length is given by:

$$k' = \frac{n_e}{K'}$$

We therefore need to find K':

$$K' = K + F_e$$

As the eye is axially myopic, we can assume that $F_e = +60.00$ D. We can also assume that $n_e = 4/3$:

$$K' = -10.00 + (+60.00) = +50.00\,D$$

The axial length is, therefore:

$$k' = \frac{4/3}{+50.00} = 0.026667\,m$$

The axial length of this myopic eye is therefore 26.67 mm. This answer makes sense because eyes with axial myopic should have an axial length >22.22 mm.

The power of the spectacle lens fitted at 11.5 mm can now be found using the expression:

$$F_{sp} = \frac{K}{1 + (dK)}$$

The vertex distance is given as 11.5 mm, which must be entered in metres:

$$F_{sp} = \frac{-10.00}{1 + (0.0115 \times -10.00)} = -11.30\,D$$

The power of the thin correcting lens at 11.5 mm is therefore −11.30 D. The spectacle refraction is greater than the ocular refraction, which is the expected result for a myopic eye.

Example 4.8

A reduced eye with refractive ametropia is corrected by a thin converging lens of focal length 12.5 cm, at a vertex distance of 12 mm. Calculate:
- **The ocular refraction**
- **The power of the reduced surface.**

The first task is to find the power of the thin spectacle lens. As the lens is a converging lens, the focal length is +12.5 cm. If we know the focal length we can find the lens power using:

$$F_{sp} = \frac{1}{f'}$$

In this example, $f' = +12.5$ cm and is entered in metres:

$$F_{sp} = \frac{1}{+0.125} = +8.00\,D$$

The spectacle refraction is therefore +8.00 D at 12 mm.

The ocular refraction can now be found using:

$$K = \frac{F_{sp}}{1-(dF_{sp})}$$

$$K = \frac{+8.00}{1-(0.012 \times +8.00)} = +8.85\,D$$

The question informs us that the eye has refractive hypermetropia. We can therefore assume that the axial length is 22.22 mm and the refractive index 4/3. The dioptric length is therefore:

$$K' = \frac{n_e}{k'}$$

$$K' = \frac{4/3}{0.02222} = +60.00\,D$$

$$K' = K + F_e$$

which rearranges to give:

$$F_e = K' - K$$

$$F_e = +60.00 - (+8.85) = +51.15\,D$$

The power of the reduced surface is +51.15 D. This result makes sense because the eye is refractively hypermetropic.

Concluding points for Chapter 4

This chapter has:

- explained the ocular and spectacle refractions
- illustrated the correction of spherical ametropia
- introduced examples of elementary reduced eye calculations
- discussed and illustrated with examples of the vertex distance.

Further recommended reading

Tunnacliffe A H (1993) *Introduction to Visual Optics.* Association of the British Dispensing Opticians, London

Rabbetts R B B (1998) *Bennett & Rabbetts' Clinical Visual Optics.* Butterworth-Heinemann, Oxford

The Basic Retinal Image

Andrew Keirl

Introduction

As with any optical system, the eye forms images of objects. These images are formed on the **retina** and are referred to as **retinal images**. The image formed by an eye *without* the aid of a spectacle lens, contact lens or other optical device is known as the unaided or **basic** retinal image, whereas the image formed by an eye with the aid of a spectacle lens or contact lens is known as the **corrected** or **aided** retinal image size. This chapter discusses the formation and size of basic (unaided) retinal images.

Chapter content

Retinal images in the reduced eye

The unaided retinal image size

The size of the unaided (or basic) retinal image is the distance on the retina between the points where the chief rays from the top and bottom of an object intersect the retina (**Figure 5.1**). The unaided retinal image may be clear, as in the emmetropic eye, or blurred, as in the uncorrected ametropic eye. The size of the retinal image formed in the uncorrected (unaided) eye is given the symbol h'_u, where the subscript 'u' stands for uncorrected or unaided. With reference to Figure 5.1:

- The distant object is assumed to be at *infinity* so *no* object distance is given.
- w is the angle between the chief rays from the top and bottom of the object; it is the angle subtended by the distant object at the eye and is essentially the 'size' of the object.
- The chief ray from the bottom of the object passes into the eye without undergoing a change in direction, whereas the ray from the top of the object is refracted according to Snell's law. The chief ray from the top of the object is refracted towards the normal.
- w' is the angle between the two refracted rays.
- k' is the axial length of the reduced eye.
- n_e is the refractive index of the reduced eye.
- As the eye is assumed to be in air, $n = 1$.
- h'_u is the size of the basic, uncorrected or unaided retinal image.

Inspection of Figure 5.1 shows that the size of h'_u can be determined by the application of simple trigonometry:

$$h'_u = -k' \times \tan w'$$

The use of the minus sign in the above expression indicates that the image formed is inverted. This should be evident from Figure 5.1. Remember that h'_u can refer to the *clear* image formed in the emmetropic eye or the *blurred* image formed in the uncorrected ametropic eye. This is why some authors prefer to use the term 'basic' retinal image.

Figure 5.1 Formation of the basic retinal image of a distant object.

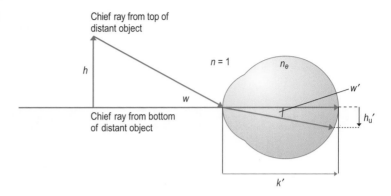

Several equations can be derived to calculate the size of the basic retinal image. Some are suitable only if the reduced eye in question has the standard refractive index of 4/3. It should be noted, however, that some questions may involve the use of a reduced eye with a non-standard refractive index. For this reason, this text uses the expression:

$$h_u' = -\frac{k'}{n_e} \tan w$$

whenever an uncorrected (basic) retinal image size is required.

Example 5.1
Find the size of the retinal image formed in an emmetropic eye if a distant object subtends an angle of 3°. Assume the constants of the standard reduced eye (refer to Figure 5.1).

The constants of the standard reduced emmetropic eye that are of interest here are:

$F_e = +60.00\,D$, $k' = 22.22\,mm$ and $n_e = 4/3$

To find the size of the basic (uncorrected) image, use the expression:

$$h_u' = -\frac{k'}{n_e} \tan w$$

$$h_u' = -\frac{22.22}{4/3} \tan 3° = -0.873\,mm$$

The size of the basic image formed in this standard reduced emmetropic eye is therefore

–0.873 mm. The use of the minus sign indicates that the image is inverted.

Example 5.2
Find the size of the uncorrected retinal image formed in an eye with axial myopia if the ocular refraction is –4.00 D and a distant object subtends an angle of 3°. Assume the constants of the standard reduced eye.

This question is slightly more involved because we have to calculate the axial length using the ocular refraction. As the above eye is axially myopic, we know straight away that the axial length is *longer* than 22.22 mm. However, the refractive index of the eye is 4/3 and the power of the reduced surface, F_e, +60.00 D. To find the axial length we need to find the dioptric length. To find the dioptric length, use:

$$K' = K + F_e$$

$$K' = -4.00 + (+60.00) = +56.00\,D$$

Now we have K', we can use:

$$k' = \frac{n_e}{K'}$$

to find k'

$$k' = \frac{4/3}{+56.00} = 0.02381\,m$$

The axial length of the eye is therefore 23.81 mm. As the eye is axially myopic, this makes sense because the axial length should be greater than the standard 22.22 mm. The size of the uncorrected retinal image can now be found using:

$$h'_u = -\frac{k'}{n_e}\tan w$$

$$h'_u = -\frac{23.81}{4/3}\tan 3° = -0.936\,\text{mm}$$

Again, the use of the minus sign indicates that the image is inverted.

Example 5.3

Find the size of the uncorrected retinal image formed in an eye with axial hypermetropia if the spectacle refraction (F_{sp}) is +6.50 at 13 mm and a distant object subtends an angle of 3°. Assume the constants of the standard reduced eye.

This question is even more involved, because we have to calculate the ocular refraction from the spectacle refraction in order to find the dioptric length and therefore the axial length. As the above eye is **axially** hypermetropic, we know that the axial length is **shorter** than 22.22 mm. However, we can assume that the refractive index of the eye is 4/3 and the power of the reduced surface +60.00 D. To solve this problem we need to find K, K' and hence k'. To begin this problem we need to find K from F_{sp}. The appropriate expression is

$$K = \frac{F_{sp}}{1-(dF_{sp})}$$

(d must be entered in metres)

$$K = \frac{+6.50}{1-(0.013\times+6.50)} = +7.10\,\text{D}$$

Now we have K, we can find K' and k'.

The dioptric length can be found using the expression:

$$K' = K + F_e$$

Substitution into the above gives:

$$K' = +7.10 + (60.00) = +67.10\,\text{D}$$

Finally the axial length can be found using:

$$k' = \frac{n_e}{K'}$$

Substitution gives:

$$k' = \frac{4/3}{+67.10} = 0.01987\,\text{m}$$

The axial length of this axially hypermetropic eye is, therefore, 19.87 mm and this of course makes sense, because it should be shorter than 22.22 mm as the refractive error is axial! The size of the uncorrected retinal image is given by:

$$h'_u = -\frac{k'}{n_e}\tan w$$

$$h'_u = -\frac{19.87}{4/3}\tan 3° = -0.781\,\text{mm}$$

For Examples 5.1, 5.2 and 5.3, the basic retinal image sizes should be examined closely. These examples show that, for the same object size (3°), the **longer** the axial length the **larger** the uncorrected (basic) retinal image.

The formation of the basic (uncorrected, unaided, blurred) retinal image of a distant object

To understand exactly how the basic retinal image is formed, we must discuss the concept of **blur discs**. A basic retinal image is made up of a series of blur discs and an individual blur disc on a retinal image represents a particular point on an object. **Figure 5.2** shows an inverted retinal image, at the top and bottom of which, blur discs are formed. The chief ray from the top of the object is at the centre of the blur disc at the 'top' of the inverted image on the retina, and the chief ray from the bottom of the object is at the centre of the blur disc at the 'bottom' of the inverted retinal image. It should be noted that, for simplicity, only two blur discs are shown in Figure 5.2.

Remember that there is a blur disc formed at *every* point on the retinal image, which represents *every* point on the object. The symbol for blur disc diameter is y and the size of one blur disc can be calculated by the application of similar triangles. The total extent of the blurred retinal image formed is, therefore, the basic (uncorrected) image size plus one blur disc diameter: $|h'_u| + y$. The minus sign is ignored when calculating the total extent of the blurred **45**

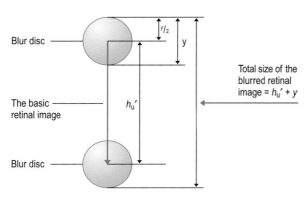

Figure 5.2 The formation of the blurred retinal image.

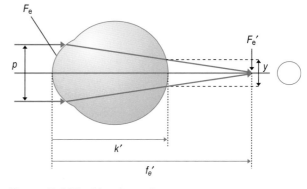

Figure 5.4 The blur disc in hypermetropia (unaccommodated eye).

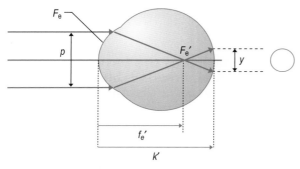

Figure 5.3 The blur disc in myopia (unaccommodated eye).

retinal image. **Figure 5.3** shows the formation of one blur disc in an unaccommodated eye myopic eye. With reference to Figure 5.3 and by using similar triangles we can state that:

$$\frac{y}{p} = \frac{k' - f_e'}{f_e'}$$

which becomes:

$$y = p \times \frac{k' - f_e'}{f_e'}$$

where y is the size of one blur disc.

The blur disc in the unaccommodated hypermetropic eye is illustrated in **Figure 5.4**. Again, the application of similar triangles provides a corresponding expression to calculate the diameter of one blur disc:

$$\frac{y}{p} = \frac{f_e' - k'}{f_e'}$$

which becomes:

$$y = p \times \frac{f_e' - k'}{f_e'}$$

Note that the equation to calculate the size of one blur disc in a hypermetropic eye is different to that stated for a myopic eye.

In emmetropia, $f_e' = k'$ and the above equations go to zero, implying that the blur disc diameter is zero when the emmetropic eye looks at a distant object point, i.e. the image is sharply focused on the retina. In myopia, as the degree of myopia increases, the blur disc diameter also increases. So unaided individuals with higher degrees of myopia have more blur when viewing a distant object than those with lower amounts of myopia, assuming that the pupil diameters are the same. In hypermetropia, if the individual increases the eye's power by accommodating, a point is reached when $F_e = K'$ and therefore $f_e' = k'$. In this situation, the blur disc diameter is 0 and a clear retinal image is formed. On the other hand, if someone with myopia accommodated while viewing a distant object, the power of the eye F_e increases along with the blur disc diameter. It should be noted that larger blur disc diameters have a detrimental effect on vision. If two point objects are to be seen as separate, the blur discs formed within the image must not overlap.

The pinhole disc

The equations for the calculation of the blur disc diameter show that if the pupil diameter, p, is reduced, the blur disc diameter, y, is also reduced. This is why people with unaided myopia often comment that they see better if they 'screw up their eyes'. They often incorrectly refer to this as 'squinting'. What they are actually doing is reducing the diameter of their pupils. An artificial pupil, known as a **pinhole disc**, is often used during a routine eye examination. A pinhole disc is simply an opaque disc with a 1 mm diameter hole at its centre. If placed in front of an uncorrected or partly corrected ametropic eye, it reduces the size of the blur discs formed on the retina and the individual should report an apparent improvement in vision. The pinhole test is used by optometrists to demonstrate that an improvement in vision is possible with spectacle lenses. A negative result with the pinhole test may indicate an underlying pathology. The clinical use of the pinhole disc is discussed in Chapter 10. It is interesting to note that individuals with a smaller than average pupil diameter, say 2 mm, are able to tolerate more blur than those with larger pupils. Someone with a low degree of myopia, but with large pupils, may experience more blur of a distant object than another person with a larger ocular refraction but smaller pupils.

The pinhole disc is illustrated in **Figure 5.5** and **Figure 5.6** shows how a blurred image, in the form of a letter E on the retina, may be easier to recognise with smaller blur discs. In Figure 5.6a, the blur discs are large enough to spread across either side of a limb and overlap. The resulting reduction in image contrast means that the letter may not be recognised. If the blur discs are relatively small, as in Figure 5.6b, the image contrast remains high and the letter is recognised. In summary, blurred images of letters are less easily recognised when the blur disc diameter is large relative to the letter size. This can be described in terms of the blur ratio:

Blur ratio

$$= \frac{\text{Blur disc diameter}}{\text{Retinal image size in the uncorrected eye}}$$

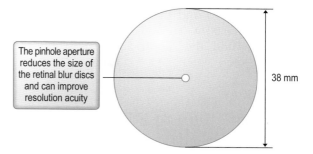

The pinhole aperture reduces the size of the retinal blur discs and can improve resolution acuity

38 mm

Figure 5.5 The pinhole disc.

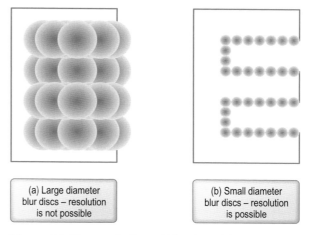

(a) Large diameter blur discs – resolution is not possible

(b) Small diameter blur discs – resolution is possible

Figure 5.6 The visual effect of blur discs. (a) Large diameter blur discs means that resolution is not possible. (b) Small diameter blur discs means that resolution is possible.

Example 5.4
Find the total size of the blurred retinal image formed in an eye with axial myopia if the ocular refraction is −6.00 D and a distant object subtends an angle of 3°. The pupil diameter is 4 mm. Assume the constants of the standard reduced eye.

The question asks for the total size of the blurred retinal image. This is the sum of the basic retinal image size and one blur disc and is given by: $|h_u'| + y$. We therefore have to find the size of the basic retinal image *and* the diameter of one blur disc. We can calculate the axial length using the ocular refraction and information given in the question. As the above eye is axially myopic, we know straight away that the axial length is longer than 22.22 mm. The refractive index of

47

the eye is 4/3 and the power of the reduced surface +60.00 D. To find the axial length we need to find the dioptric length, using the expression:

$$K' = K + F_e$$

$$K' = -6.00 + (+60.00) = +54.00\,D$$

$$k' = \frac{n_e}{K'} \qquad k' = \frac{4/3}{+54.00} = 0.02469\,m$$

The axial length of the eye is therefore 24.69 mm. The size of the uncorrected retinal image is given by:

$$h_u' = -\frac{k'}{n_e}\tan w$$

$$h_u' = -\frac{24.69}{4/3}\tan 3° = -0.970\,mm$$

The minus sign indicates that the image is inverted. The diameter of one blur disc, y, formed in a **myopic** eye is given by:

$$y = p \times \frac{k' - f_e'}{f_e'}$$

where p is the pupil diameter and f_e' the focal length of the reduced surface. The focal length of the reduced surface is given by (do not forget the refractive index):

$$f_e' = \frac{n_e}{F_e} \qquad f_e' = \frac{4/3}{+60.00} = 0.02222\,m$$

In this example, the focal length of the reduced surface is therefore 22.22 mm. The diameter of one blur disc can now be calculated. Note that all values are given in millimetres:

$$y = 4 \times \frac{24.69 - 22.22}{22.22} = 0.445\,mm$$

The total size or extent of the blurred retinal image is therefore:

$$|h_u'| + y = 0.970 + 0.445 = 1.415\,mm$$

Note that, during this final stage of the calculation, the minus sign was ignored.

Concluding points for Chapter 5

This relatively lengthy and important chapter has covered:

- the formation of the basic and corrected retinal image sizes
- the calculation of the basic retinal image size
- the formation and calculation of the blur disc
- the pinhole disc

Further recommended reading

Tunnacliffe A H (1993) *Introduction to Visual Optics*. Association of the British Dispensing Opticians, London
Rabbetts R B B (1998) *Bennett & Rabbetts' Clinical Visual Optics*. Butterworth-Heinemann, Oxford

Spectacle Magnification and the Corrected Retinal Image

Andrew Keirl

Introduction

When a spectacle lens is worn, the retinal image formed is referred to as the **corrected retinal image** and is given the symbol h'_c. As far as corrected retinal image sizes are concerned, positive spectacle lenses produce an **increase** when compared with the basic retinal image size, whereas negative lenses produce a **decrease**. It follows that positive lenses make things appear larger to the individual, and negative lenses have the opposite effect. These effects are exaggerated by increasing the vertex distance.

Chapter content

- Spectacle magnification
- Relative spectacle magnification
- The retinal image of a near object

Spectacle magnification

Spectacle magnification is a term used to describe the ratio of the size of the retinal images formed in the **same** eye when corrected and uncorrected. Spectacle magnification (SM) is the magnification of the retinal image produced by placing a spectacle lens in front of an eye and is defined as:

$$SM = \frac{h'_c}{h'_u}$$

where h'_u is the size of the retinal image formed in the uncorrected ametropic eye and h'_c is the size of the retinal image formed in the same ametropic eye when corrected.

As spectacle magnification is a ratio, it has no units. The simplest way to calculate the size of the retinal image formed when the eye is corrected is to calculate the spectacle magnification and multiply this value by the uncorrected (basic) retinal image size. The size of the corrected retinal image is therefore given by:

$$h'_c = h'_u \times SM$$

Spectacle magnification with thin lenses

Any spectacle magnification produced by a 'thin' lens is the result of the power and position of the correcting lens only, because a thin lens has *no* form or thickness. The magnification produced by a thin spectacle lens is known as the **power factor** (PF). For thin spectacle lenses, spectacle magnification (or power factor) can be given by:

$$\frac{\text{Ocular refraction}}{\text{Spectacle refraction}} \quad \text{or} \quad SM\,(PF) = \frac{K}{F_{sp}}$$

This definition and equation applies to hypermetropia and myopia, both axial and refractive. An alternative expression to find the spectacle magnification or power factor of a thin lens is:

$$SM\,(PF) = \frac{1}{1-(dF_{sp})}$$

where d is the vertex distance in metres and F_{sp} the power of the thin lens in dioptres. If a patient

49

is corrected using a contact lens (assuming a thin contact lens) the contact lens magnification can be given by:

$$\frac{\text{Ocular refraction}}{\text{Contact lens correction}} \quad \text{or} \quad SM_{cl} = \frac{K}{F_{cl}}$$

With the positions of the correcting contact lens and principal point of a reduced eye almost coinciding, we can state that the contact lens magnification approximates to unity (SM = 1). Thus contact lenses do not alter the retinal image size significantly from that of the uncorrected eye. It is often said, although not strictly true, that the power of a contact lens is the same as a patient's ocular refraction.

Spectacle magnification with thick spectacle lenses

Any spectacle magnification produced by a thick or 'real' spectacle lens is the result of the power and position of the lens and *also* the **form** and **thickness** of the lens. The magnification resulting from the form and thickness of a lens is known as the **shape factor** (SF), which is given by:

$$SF = \frac{1}{1 - \left(\frac{t}{n}\right)F_1}$$

where t is the centre thickness of the spectacle lens in metres, n the refractive index of the lens material and F_1 the power of the front surface of the lens in dioptres. An alternative expression to find the shape factor of a thick lens is:

$$SF = \frac{F_v'}{F_E}$$

where F_v' is the back vertex power of the thick lens and F_E the power of the equivalent thin lens. This alternative expression is mainly used in contact lens problems. So, unlike a thin lens, the magnification produced by a *thick* spectacle lens has **two** components: the **power factor** and the **shape factor**. The total spectacle magnification for a thick lens is therefore:

SM = Power factor × Shape factor

Shape factor is important for plus lenses and cannot be ignored. However, it *can* be ignored for minus lenses, because, for such a lens, the centre

thickness is in the region of 1.0 to 1.5 mm. If we take the refractive index as 1.5, t/n will be approximately 1.0 mm. This means that the shape factor value is *always* in the region of 1.0. Put some numbers into the expression and try it! If the spectacle magnification for a thick lens is PF × SF and SF = 1.0, then PF × 1.0 = PF. The shape factor can therefore be ignored for a minus lens. For plus lenses, the shape factor *always* has a significant effect on the SM produced and *cannot* be ignored.

How much spectacle magnification can be noticed? Researchers have found that a difference of as little as 0.25% in spectacle magnifications between the right and left eyes of a patient affected an individual's binocular vision function. On a more practical note, a patient notices the magnification produced by a pair of +1.00 D reading lenses. As a rule of thumb, each 1.00 D of lens power gives approximately 1% magnification. The patient is therefore noticing 1% spectacle magnification.

Some practical effects of spectacle magnification

When dispensing a first-time prescription, e.g. –2.50 DS R & L, the patient should be forewarned that the image is smaller than without spectacles. Sometimes, when not warned, such first-time wearers do not like the smaller image, even though it is clearer.

A difference in the spectacle magnification between the right and left lenses can cause problems with binocular vision. However, such problems occur only if the difference in the right and left spectacle magnifications is greater than about 5%. Binocular vision with spectacles, in the case of monocular aphakia with the other eye being normal, is the most dramatic example of this. In this situation, the difference in spectacle magnifications can be as great as 30%. Problems are caused because the induced **aniseikonia** (a difference in cortical image sizes) is such that images formed in the two eyes do not fall on corresponding areas and cortical cells are not driven binocularly.

In astigmatism, correction with spectacles produces different magnifications in the two principal meridians. This can give rise to **per-**

ceived distortion of space, which can be made worse if the front surface of the lens is toroidal in a plus prescription, because the shape factor compounds the problem.

When changing from spectacles to contact lenses, the retinal image size is larger in myopia but smaller in hypermetropia. The opposite is true when going from contact lenses to spectacles. This is particularly noticeable with higher myopic prescriptions, e.g. a person with myopia of −10.00 D may have a visual acuity of 6/9 with spectacles and 6/6 with contact lenses.

Changing the form of the lens changes the spectacle magnification and therefore the retinal image size, especially with plus lenses because they are thicker than minus lenses. In **anisometropia** (a difference in the right and left prescriptions), where a patient is adapted to particular right and left lens forms, it is important to maintain those forms so that binocular vision remains comfortable in the new spectacles. This could be important when changing from spherical forms to aspheric forms.

Patients prescribed with reading spectacles for the first time think that they are just for 'magnification'. The magnification with a pair of +1.00 DS spectacle lenses is about 1%, which, although perceptible, is of no practical value when compared with magnifications of two times (i.e. 100%) or more, which can be obtained with magnifiers.

One of the advantages of aspheric lenses (especially plus aspheric lenses) is that they exhibit less spectacle magnification. This is because the lens is flatter than the spherical equivalent and therefore produces less shape factor magnification. It should, however, be noted that some patients may dislike the smaller retinal image size produced as a consequence of a reduction in shape factor magnification.

Relative spectacle magnification

Relative spectacle magnification (RSM) is defined as:

$$\frac{\text{The retinal image size in the corrected ametropic eye } (h'_c)}{\text{The retinal image size in the standard emmetropic eye } (h'_{em})}$$

$$RSM = \frac{K}{F_{sp}} \times \frac{K'_{em}}{K'}$$

In the standard, emmetropic eye, K'_{em} is always equal to +60.00 D, so:

$$RSM = \frac{K}{F_{sp}} \times \frac{+60.00}{K'}$$

where K' is the dioptric length of the ametropic eye ($K' = K + F_e$), K is the ocular refraction and F_{sp} is the spectacle refraction. The significance of relative spectacle magnification is seen by comparing corrected retinal image sizes in myopic and hypermetropic eyes with the basic retinal image size in a standard emmetropic eye. The corrected retinal image in an axially hypermetropic eye is increased when wearing the spectacle correction, but it is still smaller than the image in the emmetropic eye. In axially myopic eyes, the minifying correction still leaves an image that is larger than that of the standard emmetropic eye. RSM problems are mainly encountered in contact lens visual optics.

Spectacle magnification: a summary

- For a 'thin' lens, any spectacle magnification present is attributed to the power of the lens only because a thin lens has no centre thickness.
- Spectacle magnification for a thin lens is called the power factor.
- Spectacle magnification with a thick lens can be attributed to the power of the lens and also its form and thickness.
- Spectacle magnification resulting from the form and thickness is known as the shape factor.
- Total spectacle magnification for a thick lens = power factor × shape factor.
- Plus lenses magnify for a plus lens; SM is always >1.0.
- Minus lenses make smaller for a minus lens; SM is always <1.0.
- The shape factor is ignored for a minus lens, because t/n is about 1 mm.
- The shape factor can never be ignored for a plus lens.

Spectacle magnification equations

Basic equations:

$$SM = \frac{h'_c}{h'_u}$$

$$h'_c = h'_u \times SM$$

Power factor:

$$SM \text{ (thin lens)} = PF$$

$$PF = \frac{K}{F_{sp}}$$

or:

$$PF = \frac{1}{1 - (dF_{sp})}$$

Shape factor:

$$SF = \frac{1}{1 - \left(\frac{t}{n} F_1\right)}$$

or:

$$SF = \frac{F'_v}{F_E}$$

The total magnification of a thick lens:

$$SM \text{ (thick lens)} = PF \times SF$$

The retinal image of a near object

When calculating the retinal image formed by an object closer than infinity, trigonometry cannot be used because the angles involved could be large. The size of the retinal image for a near object is therefore found by calculating the magnification of the lens/eye system. With reference to Figure 6.1, the size of the image (h') formed by the near object at a given distance to the left of the spectacle lens is given by:

$$h' = h \times \frac{L_1}{L'_1} \times \frac{L_2}{K'}$$

The symbols in the above equation have their usual meaning and are illustrated in Figure 6.1.

Example 6.1

Find the size of the uncorrected and corrected retinal images formed in an eye with axial myopia if the ocular refraction is −8.00 D and a distant object subtends an angle of 3°. The thin spectacle lens is to be positioned at a vertex distance of 14 mm. Assume the constants of the standard reduced eye.

This question now asks us to find the size of the basic or uncorrected retinal image size *and* the size of the retinal image formed when the correcting lens is worn. As the lens is thin, we have to find only the power factor. However, to do this we must find the spectacle refraction F_{sp}.

The axial length of the eye and the size of the uncorrected retinal image must be found first. As the eye is axially myopic, $F_e = +60.00$ D.

$$K' = K + F_e \qquad K' = -8.00 + (60.00) = +52.00 \text{ D}$$

$$k' = \frac{n_e}{K'} \qquad k' = \frac{4/3}{+52.00} = 0.02564 \text{ m}$$

The axial length of the eye is therefore 25.64 mm. The size of the uncorrected retinal image is given by:

$$h'_u = -\frac{k'}{n_e} \tan w$$

$$h'_u = -\frac{25.64}{4/3} \tan 3° = -1.008 \text{ mm}$$

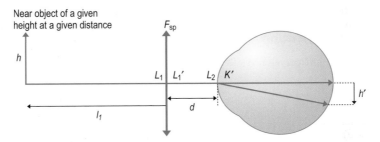

Figure 6.1 Magnification for a near object.

As the lens in question is thin, the spectacle magnification is equal to the power factor only, which is given by:

$$PF = \frac{K}{F_{sp}}$$

As the ocular refraction is given in the question, we need to find F_{sp}:

$$F_{sp} = \frac{K}{1+(dK)}$$

$$F_{sp} = \frac{-8.00}{1+(0.014 \times -8.00)} = -9.01\,D$$

(d must be entered in metres)

$$PF = \frac{K}{F_{sp}} \qquad PF = \frac{-8.00}{-9.01} = 0.888\times$$

$$SM\,(\text{thin lens}) = PF = 0.888\times$$

A spectacle magnification of 0.888 is equivalent to an 11.2% ($[1.000 - 0.888] \times 100$) *decrease* in image size. To find the size of the corrected retinal image, use:

$$h_c' = h_u' \times SM$$

$$h_c' = -1.008 \times 0.888 = -0.895\,mm$$

Example 6.2

Find the size of the uncorrected and corrected retinal images formed in an eye with axial hypermetropia if the spectacle refraction is +8.00 at 14 mm and a distant object subtends an angle of 3°. The correcting lens is thick with F_1 = +12.00 D, n = 1.50 and t = 8.00 mm. Assume the constants of the standard reduced eye.

This question also asks for the size of the retinal image both with and without the correcting spectacle lens. In this case, as the correcting lens is *thick*, we need to calculate both the power factor and the shape factor. The total spectacle magnification can then be calculated and the corrected retinal image size found. As the spectacle refraction is given in the question, we must find the ocular refraction:

$$K = \frac{F_{sp}}{1-(dF_{sp})}$$

$$K = \frac{+8.00}{1-(0.014 \times +8.00)} = +9.01\,D$$

(d must be entered in metres)

Now we have K, we can find K' and k'. As the eye is axially hypermetropic, $F_e = +60.00\,D$:

$$K' = K + F_e \qquad K' = +9.01 + (+60.00) = +69.01\,D$$

$$k' = \frac{n_e}{K'} \qquad k' = \frac{4/3}{+69.01} = 0.01932\,m$$

$$h_u' = -\frac{k'}{n_e}\tan w$$

$$h_u' = -\frac{19.32}{4/3}\tan 3° = -0.759\,mm$$

As the lens is thick, to find the spectacle magnification, we need to find both the power factor and the shape factor:

$$PF = \frac{K}{F_{sp}} \qquad PF = \frac{+9.01}{+8.00} = 1.126\times$$

$$SF = \frac{1}{1-\left(\dfrac{t}{n}F_1\right)}$$

$$SF = \frac{1}{1-\left(\dfrac{0.008}{1.5} \times +12.00\right)} = 1.068\times$$

For a thick lens:

$$SM = PF \times SF \qquad SM = 1.126 \times 1.068 = 1.202\times$$

A spectacle magnification of 1.202 is equivalent to a 20.2% ($[1.202 - 1.000] \times 100$) increase in image size.

$$h_c' = h_u' \times SM$$

$$h_c' = -0.759 \times 1.202 = -0.912\,mm$$

Example 6.3

An axially myopia reduced eye can be corrected with a thin −5.00 D lens placed at a vertex distance of 12 mm. When wearing the spectacle correction, this eye views a 15 mm tall object placed at one-third of a metre from the spectacle lens. Calculate the size of the retinal image formed.

This is an example of a retinal image formed by a near object. Note that the size of the object is *not* given by stating the angle subtended at the eye (refer to **Figures** 6.1 and 6.2).

To find the retinal image size for a near object we need to use the equation:

$$h' = h \times \frac{L_1}{L_1'} \times \frac{L_2}{K'}$$

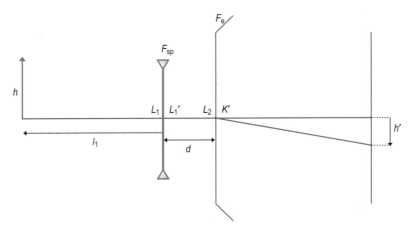

Figure 6.2 Diagram for Example 6.3.

As the question gives us F_{sp}, we can find K and therefore K'. The object distance, l_1, can be used to find L_1 and therefore L_1'. Finally, step-along is needed to find L_2. The appropriate values can then be substituted into the above equation:

$$K = \frac{F_{sp}}{1-(dF_{sp})}$$

$$K = \frac{-5.00}{1-(0.012 \times -5.00)} = -4.72\,\text{D}$$

(*d* must be entered in metres)

Now we have K, we can find K' and k'. As the eye is axially myopic, $F_e = +60.00\,\text{D}$.

$$K' = K + F_e \qquad K' = -4.72 + (+60.00) = +55.28\,\text{D}$$

Step-along is now used to find L_2. The computation takes place in two columns: one headed 'vergences' and the other 'distances'.

Vergences (D)   Distances (m)

$$l_1 = -1/3\,\text{m}$$

$$L_1 = \frac{n}{l_1}$$

$$L_1 = -3.00\,\text{D} \qquad \leftarrow \qquad L_1 = \frac{1}{-1/3}$$

$$F_1 = -5.00\,\text{D}$$

$$L_1' = -3.00 + (-5.00) = -8.00\,\text{D} \qquad \rightarrow \qquad l_1' = \frac{n}{L_1'}$$

$$l_1' = \frac{1}{-8.00} = -0.125\,\text{m}$$

$$l_2 = l_1' - d$$

$$L_2 = \frac{n}{l_2} \qquad \leftarrow \qquad l_2 = -0.125 - 0.012 = -0.137\,\text{m}$$

$$L_2 = \frac{1}{-0.137} = -7.299\,\text{D}$$

We can now use:

$$h' = h \times \frac{L_1}{L_1'} \times \frac{L_2}{K'}$$

to find the size of the retinal image when $h = 15\,\text{mm}$.

$$h' = 15 \times \frac{-3.00}{-8.00} \times \frac{-7.299}{+55.280} = -0.7427\,\text{mm}$$

The size of the image formed by the near object is therefore $-0.7427\,\text{mm}$. The minus sign indicates that the image is inverted.

Example 6.4

A reduced eye with axial ametropia has an axial length of 25.641 mm, the other parameters being standard. Calculate the ocular refraction and the thin correcting lens power at 13.9 mm. Find the uncorrected and corrected retinal image sizes if a distant object subtends 7^Δ.

As the ametropia is axial, we can assume that the power of the reduced surface is the standard +60.00 D. As the axial length is given, we can easily find K', K and therefore the power of the correcting lens at 13.9 mm. So, starting with k' we have:

$$K' = \frac{n_e}{k'} \qquad K' = \frac{4/3}{+0.025641} = +52.00\,\text{D}$$

Now:

$$K' = K + F_e$$

This becomes:

$$K = K' - F_e \qquad K = +52.00 - (+60.00) = -8.00\,\text{D}$$

Now we have K, we can find F_{sp}. We need to use:

$$F_{sp} = \frac{K}{1 + (dK)}$$

$$F_{sp} = \frac{-8.00}{1 + (0.0139 \times -8.00)} = -9.00\,\text{D}$$

In this example, the angle subtended at the eye is given in prism dioptres as opposed to degrees. The prism dioptre (P^Δ) is simply another way of stating an angle and is given by:

$$P^\Delta = 100 \tan \text{angle}$$

So, in the case of the angle subtended at eye:

$$P^\Delta = 100 \tan w$$

As $w = 7^\Delta$

$$7^\Delta = 100 \tan w \quad \text{and} \quad \tan w = \frac{7}{100} = 0.07$$

We can now find the uncorrected retinal image size h'_u using:

$$h'_u = -\frac{k'}{n_e} \tan w$$

$$h'_u = -\frac{25.641}{4/3} 0.07 = -1.346\,\text{mm}$$

As the lens is thin, only the power factor is required:

$$PF = \frac{K}{F_{sp}} \qquad PF = \frac{-8.00}{-9.00} = 0.888\times$$

$$\text{SM (thin lens)} = PF = 0.888\times$$

A spectacle magnification of 0.888 is equivalent to an 11.11% ([1.000 − 0.888] × 100) *decrease* in image size. To find the size of the corrected retinal image use:

$$h'_c = h'_u \times \text{SM}$$

$$h'_c = -1.346 \times 0.888 = -1.196\,\text{mm}$$

Concluding points for Chapter 6

This relatively lengthy chapter has covered:

- spectacle magnification
- the calculation of the corrected retinal image size
- the retinal image size for a near object.

The material in this chapter must not be considered to be 'dry' and 'academic' because an understanding of the clinical significance of spectacle magnification is often necessary in optometric and dispensing practice, particularly when dealing with a non-tolerance to a pair of spectacles.

Further recommended reading

Rabbetts R B B (1998) *Bennett & Rabbetts' Clinical Visual Optics*. Butterworth-Heinemann, Oxford
Tunnacliffe A H (1993) *Introduction to Visual Optics*. Association of the British Dispensing Opticians, London

Astigmatism

Andrew Keirl

Introduction

Astigmatism is a condition in which the refracting surfaces of an eye consist of two principal powers. In practice, most individuals, even those who do not wear spectacles or contact lenses exhibit a small amount of astigmatism. Myopic and hypermetropic corrections are often prescribed with an astigmatic element. In other individuals, the refractive error can be entirely astigmatic and often substantial.

Chapter content

- Astigmatism: types and characteristics
- Regular and irregular astigmatism
- Classification of regular astigmatism
- The ocular and spectacle refractions in astigmatism
- The size of the retinal image in the uncorrected astigmatic eye
- The detection and correction of astigmatism

Astigmatism: types and characteristics

Eyes that suffer from **astigmatism** need a correcting lens with a power that differs along the principal meridians of the lens, whereas spherical ametropia is corrected using lenses that have the same power along all meridians. A patient with astigmatism may complain that vertical and horizontal lines look different, e.g. the bars in a window frame may be clear horizontally but blurred vertically. You can demonstrate this yourself by focusing the light from a window through a plus spherical trial lens, say +3.00 DS, on to a piece of white paper. The vertical and horizontal bars of the window are in focus at the same time. If you attempt the same procedure with a +3.00 DC trial lens held with its axis either 90° or 180°, only the vertical or horizontal bars are in focus, not both. **Figure 7.1a** shows what an uncorrected eye, with an ocular refraction of plano/−2.00 DC × 180, viewing a fan chart would see. Notice that the vertical line is in focus and the horizontal line is very blurred. The amount of blur varies from a minimum along the vertical to a maximum along the horizontal. What is happening inside the eye is illustrated in **Figure 7.1b**.

Astigmatic eyes need to be corrected with an **astigmatic lens**, which has a *minimum* power along one meridian and a *maximum* power at 90° to the minimum meridian. These maximum and minimum meridians are known as the **principal meridians** of the lens. Astigmatism in the eye is usually caused by one or more of the refracting surfaces of the eye being toroidal in shape. The word **astigmatic** literally means 'not point forming' whereas the word **stigmatic** means 'point forming'. An astigmatic lens (or eye) is simply a lens (or eye) that has two different powers at right angles to each other.

An astigmatic surface can be described as a refracting surface of different curvatures in different meridians, progressing from a minimum

(a)

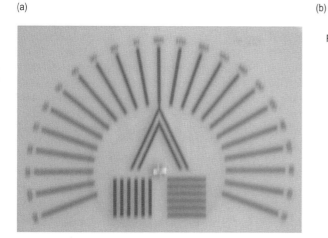

(b)

Rx = −2.00 DC × 180 Uncorrected eye

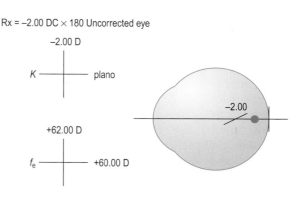

Figure 7.1 (a) The appearance of a fan-and-block chart, as seen by an uncorrected individual with the prescription plano/−2.00 DC × 180. Notice that both the vertical line of the fan chart and the vertical line of the blocks are the clearest. The horizontal lines of the fan chart and the blocks are the most blurred and the lines that make up each side of the Maddox V are equally blurred. (b) This diagram shows what is happening inside the eye. Note that the line images are formed at 90 degrees to the orientation of the power meridians.

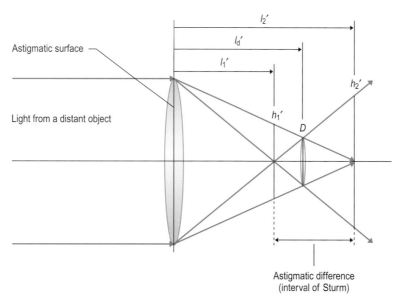

Figure 7.2 Astigmatic lens – position and lengths of the line foci and the disc of least confusion.

curvature to a maximum curvature in a mutually perpendicular meridian. As mentioned, an astigmatic surface does not produce a point image. The refracted pencil of light formed by an astigmatic surface is called an **astigmatic pencil** (Figure 7.2), which is composed of several image 'shapes', including lines, a disc called the **disc of least confusion** and blur ellipses. In exactly the same way as an astigmatic lens forms an astigmatic pencil, the more *powerful* meridian of the eye always forms its image *nearer* to the reduced surface. The shape of the image formed is a line that is always formed at *right angles* to the power meridian, e.g. a *vertical* power meridian produces a line image that is oriented *horizontally* whereas a *horizontal* power meridian produces a *vertically* oriented line image. From the reduced surface (or lens) the various images within the astigmatic pencil are formed in the following sequence: ellipse, line, ellipse, disc, ellipse, line and ellipse. The disc of least confusion is always located *dioptrically halfway* between the two line images. The astigmatic pencil is simply illustrated in **Figure 7.2**, in which l_1' and l_2' represent the positions of the two line images, h_1' and h_2' represent the sizes of the two line images, l_d' represents the position of the disc of least confusion and D represents the size of the disc of least confusion. The two line images are always formed at right angles to each other.

Astigmatic ametropia

Astigmatism in the eye can originate at the cornea (**corneal astigmatism**), the crystalline lens (**lenticular astigmatism**) or be a combination. Corneal astigmatism occurs when one or both of the corneal surfaces are toroidal in nature and can arise from both the anterior and posterior corneal surfaces. Lenticular astigmatism can be the result of the refracting surfaces of the crystalline lens being toroidal in form or the crystalline lens being tilted (oblique astigmatism).

Astigmatism can therefore be divided into two groups: corneal astigmatism and lenticular astigmatism. Corneal astigmatism is the major source of astigmatism in the eye and arises from the cornea being toroidal in shape. The astigmatic error that is measured during an eye examination is known as the **total** astigmatism, which does not differentiate between corneal and lenticular astigmatism. However, an instrument called a keratometer (see Chapter 18) can be used to measure corneal astigmatism and any difference between total and corneal astigmatism must be the result of lenticular astigmatism.

Most corneal power is a result of its front surface. With corneal astigmatism, the majority of any cylinder present is also a result of its front surface being toroidal in form. This has important practical consequences for the correction of astigmatism using contact lenses (see Chapter 22). Astigmatism caused by the back surface of the cornea is unlikely to be more than 1.00 D, and even this would be uncommon.

Lenticular astigmatism is astigmatism caused by the crystalline lens which could be due to one or both surfaces of the crystalline lens being toroidal in shape, the lens being tilted or lens opacities. If present, it is generally about 0.50–0.75 D in magnitude among youth although in patients aged over 50 years much higher values can often be found because of the formation of age-related opacities in the crystalline lens. As far as spectacles are concerned, it simply becomes part of the spectacle refraction. With contact lenses, any corneal astigmatism is very nearly neutralised if a rigid gas-permeable (RGP) contact lens is employed and any *residual* astigmatism is usually attributed to the crystalline lens. Lenticular astigmatism is almost invariably *against the rule* and usually does not exceed 1.50 D.

With-the-rule and against-the-rule astigmatism

These terms are occasionally encountered in practice, particularly in relation to contact lens fitting. Very simply, **with-the-rule astigmatism** occurs when the flattest corneal meridian is nearer the horizontal and **against-the-rule astigmatism** occurs when the flattest corneal meridian is closer to the vertical. In minus cylinder form, with-the-rule astigmatism has the axis of the correcting cylinder closest to 180° whereas the axis of the minus cylinder is closest to 90° in against-the-rule astigmatism.

Signs and symptoms of astigmatism

Symptoms (reported by an individual):

- As an astigmatic error is usually (but not always) the same for both distance and near, blurred vision for all distances can be a symptom of uncorrected astigmatism.

Signs (detected by a practitioner):

• Meridionial blur on a fan chart
• Keratometer readings that reveal an astigmatic (toroidal) cornea.

It is interesting to note that meridionial blur on an astigmatic fan chart and a spherical corneal radius, when using the keratometer, gives an indication of the presence of lenticular astigmatism.

The correction of astigmatic ametropia

If a correction is required in one meridian only, a cylindrical lens may be used (**Figure 7.3**). A cylindrical surface is plane (flat) along a meridian parallel with the axis of revolution of the cylinder but circular at 90° to the axis meridian. The flat or plane meridian of a cylinder is called the **axis meridian** and the meridian of maximum curvature (at 90° to the axis) is called the **power meridian**. The power of a cylindrical lens is therefore at 90° to its axis. A

cylindrical lens is seldom used in the form of a spectacle lens to correct a patient's refractive error. However, cylindrical lenses are extensively used in optometry because they are used together with spherical lenses to determine the refractive error of a patient during an eye examination.

When using a cylindrical lens for the correction of a refractive error or during an eye examination, the axis direction of the cylinder must always be specified. In UK optometric practice, the axis direction of astigmatic (or cylindrical) lenses must be specified in Standard Notation (**Figure 7.4**). The main points to consider are:

• Cylinder axis direction is specified in degrees, starting on the right-hand side of each eye and moving in an anti-clockwise direction.
• The axis direction of the cylinder must be between 0 and 180 as specified in BS EN ISO 8429 but without the inclusion of the degree symbol (°). The horizontal meridian is always referred to as the 180 meridian, never the 0 meridian.

In contrast to a cylindrical lens, an astigmatic lens has two powers: the **principal powers**. There are two basic types of astigmatic lens: a **sphere–cylindrical** lens and a **toric** lens. Both types have the same function (to correct astigmatism) but differ in form. In practice, prescriptions are usually dispensed in a curved form (one convex surface and one concave surface) to provide better quality images when

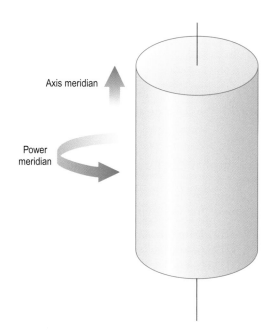

Figure 7.3 The cylindrical surface.

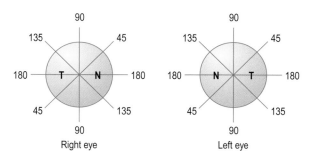

Figure 7.4 Standard axis notation.

the eye views through off-axis portions of the lens. Astigmatic lenses made in curved form are known as **toric lenses** where one (or very occasionally both) of the curved surfaces is **toroidal** in form.

Illustrating astigmatism in the reduced eye

To understand the correction of astigmatic ametropia, it is necessary to be able to visualise and illustrate the dioptric positions of the line images formed in an uncorrected and partially corrected astigmatic eye. This is *always* carried out with the aid of diagrams.

Example 7.1

A reduced eye with astigmatic ametropia has reduced surface powers, F_e = +62.00 D along 90 and F_e = +58.00 D along 180. The refractive index of the eye, n_e, is 4/3 and the axial length, k', is 22.22 mm. Find the dioptric positions of the line images for a distant point object.

The important points to grasp here is that, in the uncorrected eye:

- the **vertical** power meridian forms the **horizontal** line focus
- the **horizontal** power median forms the **vertical** line focus.

Remembering that the power of the standard reduced emmetropic eye is +60.00 D, the horizontal line image is formed 2.00 D in front of the retina (too strong) and the vertical line image is formed 2.00 D behind the retina (too weak). The horizontal line image is effectively 'myopic' and requires a correction of −2.00 D to place the image on the retina. Hence, in **Figure 7.5**, this image has been labelled '−2.00 D'. The vertical line image is effectively 'hypermetropic' and requires a correction of +2.00 D to place the image on the retina. In **Figure 7.6**, this image has been labelled +2.00 D. The labels inform us of the two powers required to place the line images on the retina and thus produce a single point image. Such labelling is sometimes referred to as 'modelling', and this technique is used throughout this chapter. A

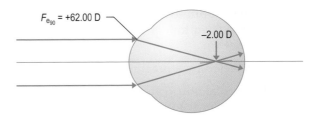

Figure 7.5 Horizontal line focus formed by the vertical power meridian.

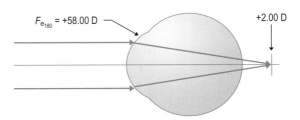

Figure 7.6 Vertical line focus formed by the horizontal power meridian.

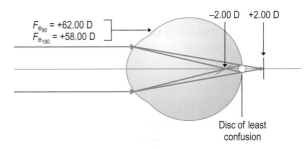

Figure 7.7 The complete astigmatic pencil.

composite diagram for this example is shown in Figure 7.7.

In this example, the ocular refraction is, therefore, one of the following:

- −2.00 D along 90/+2.00 D along 180
- −2.00 DS/+4.00 DC × 90
- +2.00 DS/−4.00 DC × 180

Example 7.2

Find the dioptric positions of the line images for a distant point object.

F_e = +66.00 D along 150 and +64.00 D along 60
n_e = 4/3
k' = 22.22 mm

Even though the power meridians given in this example are oblique, we approach the problem in exactly the same way. As the power of the standard reduced emmetropic eye is +60.00 D, both line images in the uncorrected eye in this example are formed in front of the retina, because both meridians are effectively myopic. The 150 power meridian produces an image oriented at 60° and the 60 power meridian produces an image oriented at 150°. The image formed by the 150 meridian is formed 6.00 D in front of the retina and requires a −6.00 D lens to place it on the retina. The image formed by the 60 meridian is formed 4.00 D in front of the retina and requires a −4.00 D lens to place it on the retina. The ocular refraction is, therefore, one of the following:

- −6.00 D along 150/−4.00 D along 60
- −6.00 DS/+2.00 DC × 150
- −4.00 DS/−2.00 DC × 60

Example 7.3
Consider the following three uncorrected astigmatic reduced eyes:

1. F_e = +59.00 D along 180 and +61.00 D along 90
2. F_e = +60.00 D along 180 and +62.00 D along 90
3. F_e = +58.00 D along 180 and +56.00 D along 90

Describe the nature of the retinal image formed in each case. Assume refractive ametropia and the parameters of the standard reduced eye.

In (1), as the horizontal power meridian is 1.00 D weaker than +60.00 D, it produces a focus +1.00 D behind the retina and forms a vertical line image. As the vertical power meridian is 1.00 D stronger than +60.00 D, it produces a focus −1.00 D in front of the retina and forms a horizontal line image. The disc of least confusion is located dioptrically halfway between the two line images, and its dioptric position can be located by taking the mean of the two principal powers:

$$\frac{F_{e_1} + F_{e_2}}{2}$$

which, in this case is:

$$\frac{F_{e_{180}} + F_{e_{90}}}{2} = \frac{+59.00 + (+61.00)}{2} = +60.00\,D$$

If the above eye has refractive ametropia, k' = 22.22 mm and n_e = 4/3, so:

$$K' = \frac{n_e}{k'} \qquad K' = \frac{4/3}{0.0222} = +60.00\,D$$

So for an image to be formed on the retina of this eye, the vergence leaving the reduced surface must be +60.00 D. As the dioptric position of the disc of least confusion is also +60.00 D, the disc of least confusion is formed on the retina. The retinal image formed in this eye is therefore a blurred disc or circle of light. In summary:

- The vertical line image is formed 1.00 D behind the retina
- The disc of least confusion is formed on the retina
- The horizontal line image is formed 1.00 D in front of the retina.

The disc of last confusion is always formed dioptrically halfway between the two line images.

In (2), the power of the horizontal meridian is +60.00 D and, as previously determined, a vergence of +60.00 D is required to place an image on the retina of this eye. The horizontal power meridian therefore forms a vertical line image on the retina. As the vertical power meridian is 2.00 D stronger than +60.00 D, it produces a focus −2.00 D in front of the retina and forms a horizontal line image. The retinal image formed in this eye is therefore a vertical line image. The disc of least confusion is located dioptrically halfway between the two line images and its dioptric position is given by:

$$\frac{F_{e_{180}} + F_{e_{90}}}{2} = \frac{+60.00 + (+62.00)}{2} = +61.00\,D$$

For the eye in (2) the disc of least confusion is therefore located −1.00 D in front of the retina. In summary:

- The vertical line image is formed on the retina

- The disc of least confusion is formed 1.00 D in front of the retina
- The horizontal line image is formed 2.00 D in front of the retina.

In (3), the horizontal power meridian (+58.00 D) is 2.00 D weaker than +60.00 D. It forms a vertical line image +2.00 D behind the retina. The vertical power meridian (+56.00 D) is 4.00 weaker than +60.00 D, and forms a horizontal line image +4.00 D behind the retina. As always, the disc of least confusion is located dioptrically halfway between the two line images and its dioptric position is, again, given by:

$$\frac{F_{e_{180}} + F_{e_{90}}}{2} = \frac{+58.00 + (56.00)}{2} = +57.00\,D$$

For the eye in (3) the disc of least confusion is therefore located +3.00 D behind the retina and dioptrically halfway between the two line images. If both line images and the disc of least confusion are formed off the retina, the retinal image must take the form of an out-of-focus ellipse. The orientation of such an ellipse must always be stated. If an ellipse is formed on the retina its orientation always matches the orientation of the closest line image. For the eye in (3), the vertical line image is formed closest to the retina, so the orientation of the ellipse is therefore vertical. In summary:

- The vertical line image is formed 2.00 D behind the retina
- The disc of least confusion is formed 3.00 D behind the retina
- The horizontal line image is formed 4.00 D behind the retina
- The retinal image is a blurred vertical ellipse.

So, for the three eyes in this example:

1. The retinal image is the disc of least confusion.
2. The retinal image is a vertical line.
3. The retinal image is an ellipse oriented vertically.

From a practical point of view, it is interesting to note that, in the vast majority of cases of uncorrected astigmatism, the retinal image formed is a blurred ellipse. Occasionally, the retinal image is a line focus or the disc of least confusion.

Regular and irregular astigmatism

The terms **regular** and **irregular** astigmatism are occasionally encountered in practice and are often differentiated as follows: 'In regular astigmatism, the line foci are 90° apart whereas in irregular astigmatism they are not 90° apart.' It is a common mistake to think that **irregular refraction**, sometimes called **irregular astigmatism**, means that the line foci are not orthogonal. This is not true; the line foci are *always* orthogonal.

During a routine refraction there are occasions when an individual's best corrected visual acuity falls below that expected. It may also be extremely difficult for the person to give clear-cut responses. In such cases, a pinhole disc (see Chapter 5) is placed in front of the eye under test. The result is often a marked improvement in visual acuity, which on occasions can give rise to the suspicion of an ocular pathology. Even in normal eyes, the refracting surfaces are not perfectly symmetrical about an optical axis. In addition, the cornea (and to some extent the crystalline lens) flattens towards the periphery. There may also be irregularities in the refracting surfaces and in the refractive indices of the eye as a result of ageing, injury or infection. The cornea and the crystalline lens may possess pathological conditions, producing irregular surface topography known as **keratoconus** and **lenticonus**.

In all of the above examples the refraction result obtained with a normal pupil diameter of around 4–5 mm may vary from that obtained with a small limiting pupil of, say, 1 mm in diameter. This is one reason that the determination of the refractive error should be undertaken in normal room lighting where the pupil is assumed to be an average diameter. Where two or more clear or regular regions occupy the pupil area, these usually result in two or more refractive errors, a different refraction being found for each clear area, as a result of local variations in the surface topography. If a pinhole disc is used during the refraction of an eye with irregular refracting surfaces, the individual is able to use a small, clear, regular region of the optical system with which to view the target. If there are two or more clear regions, each with its own refractive error, then two or more results may be possible and the one that gives the best acuity when the pinhole is removed is used for the spectacle prescription.

It is important to appreciate that a refractive result obtained using a pinhole disc may produce good acuity, which is not maintained when the disc is removed. The use of the pinhole reduces the refracting or scattering of light from areas of irregular refraction. If light from these areas reaches the retina, it degrades the contrast in the image and therefore reduces visual acuity. Where two or more refractive errors are found with a refraction using a pinhole, the astigmatism is likely to be different in amount and axes. This leads to the description of **irregular astigmatism**. However, it must again be em-phasised that the principal meridians in each astigmatic correction are orthogonal. Where the irregularity is in the cornea, e.g. in keratoconus, a rigid contact lens is often prescribed because the tear layer that is trapped between the back surface of the contact lens and the cornea 'fills in' any surface defect. Significant improvements in visual acuity can be obtained. In summary:

- Regular astigmatism: refractive error is constant over the refracting surface. The principal meridians of the refracting surface are 90° apart.
- Irregular astigmatism (irregular refraction): refractive error varies over the refracting surface. The principal meridians of the refracting surface are 90° apart. It is detected with a pinhole disc and sometimes corrected with a rigid contact lens.

Possible causes of irregular refraction include:

- corneal scars
- lenticonus
- pterygium
- dislocation of the crystalline lens
- keratoconus
- eyelid lesions that press on the cornea
- cataract
- refractive index changes.

Classification of regular astigmatism

The classification of regular astigmatism is based on the position of the line images formed in the uncorrected astigmatic eye, and also on the powers of the principal powers of the correcting lens. There are five types of regular astigmatism.

Compound myopic astigmatism (CMA) (Figure 7.8)

- Both line images are formed in front of the retina.
- Both principal powers of the correcting lens are negative.
- An example prescription for CMA is −2.00 DS/−2.00 DC × 90.

Simple myopic astigmatism (SMA) (Figure 7.9)

- One line image is formed in front of the retina and one on the retina.
- One principal power of the correcting lens is negative, the other plano.
- An example prescription for SMA is plano/−2.00 DC × 90.

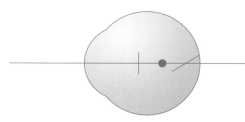

Figure 7.8 CMA −2.00 DS/−2.00 DC × 90

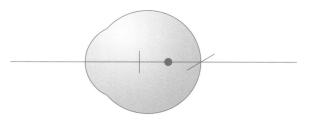

Figure 7.9 SMA plano/−2.00 DC × 90

Mixed astigmatism (MA) (Figure 7.10)

- One line image is formed in front of the retina and one behind the retina. The disc of least confusion may be formed on the retina, but it does not have to be!
- One principal power of the correcting lens is negative, the other positive.
- An example prescription for MA is −2.00 DS/+4.00 DC × 180.

Simple hypermetropic astigmatism (SHA) (Figure 7.11)

- One line image is formed behind the retina and one on the retina.
- One principal power of the correcting lens is positive, the other plano.
- An example prescription for SHA is plano/+2.00 DC × 180.

Compound hypermetropic astigmatism (CHA) (Figure 7.12)

- Both line images are formed behind the retina.

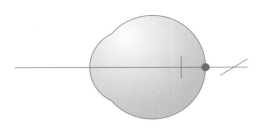

Figure 7.10 MA −2.00 DS/+4.00 DC × 180

- Both principal powers of the correcting lens are positive.
- An example prescription for CHA is +2.00 DS/+2.00 DC × 180.

The ocular and spectacle refractions in astigmatism

An astigmatic eye displays two ocular refractions (K), one for each principal meridian. It also has two spectacle refractions (F_{sp}), again one for each principal meridian. The equations given in Chapter 4 for K and F_{sp} should be applied to each meridian in turn and *not* to the sphere and the cylinder separately.

Example 7.4
F_{sp} = +4.50 DS/+2.50 DC × 90 at 10 mm. Find the ocular refraction.

$$K = \frac{F_{sp}}{1-(dF_{sp})}$$

must be applied to *each* meridian in turn. The principal powers are $F_{sp90} = +5.00\,D$ and $F_{sp180} = +7.00\,D$.

$$K_{90} = \frac{F_{sp90}}{1-(dF_{sp90})}$$

$$K_{90} = \frac{+4.50}{1-(0.01\times+4.50)} = +4.71\,D$$

$$K_{180} = \frac{F_{sp180}}{1-(dF_{sp180})}$$

$$K_{180} = \frac{+7.00}{1-(0.01\times+7.00)} = +7.53\,D$$

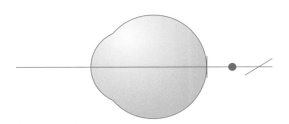

  Figure 7.11 SHA plano/+2.00 DC × 180

Figure 7.12 CHA +2.00 DS/+2.00 DC × 180

The ocular refraction is therefore +4.71 D along 90 and +7.53 D along 180, which, in sph–cyl form, is +4.71 DS/+2.82 DC × 90.

Example 7.5

If an ocular refraction is −6.00 DS/−1.50 DC × 30, find the spectacle refraction at 14 mm.

$$F_{sp} = \frac{K}{1 + (dK)}$$

must be applied to each meridian in turn. The principal powers are $K_{30} = -6.00\,D$ and $K_{120} = -7.50\,D$.

$$F_{sp30} = \frac{K_{30}}{1 + (dK_{30})}$$

$$F_{sp30} = \frac{-6.00}{1 + (0.014 \times -6.00)} = -6.55\,D$$

$$F_{sp120} = \frac{K_{120}}{1 + (dK_{120})}$$

$$F_{sp120} = \frac{-7.50}{1 + (0.014 \times -7.50)} = -8.38\,D$$

The required spectacle refraction is therefore −6.55 D along 30 and −8.38 D along 120, which, in sph–cyl form, is −6.55 DS/−1.83 DC × 30.

In cases of astigmatism, when converting from K to F_{sp} and F_{sp} to K, *always* apply the above expressions to each principal meridian in turn and *never* the sphere and the cylinder separately. This also applies to vertex distance compensation in astigmatism because each principal power must be compensated in turn. The final result can then be re-written in the usual sph–cyl form.

Vertex distance compensation in astigmatism

When dealing with an astigmatic prescription that requires a change, in vertex distance, do *not* compensate for the sphere and the cylinder as each principal meridian must be considered in turn. The following equations were given in Chapter 4 and can be used here.

If the vertex distance is decreased:

$$F_{new} = \frac{F_{old}}{1 - (dF_{old})}$$

If the vertex distance is increased:

$$F_{new} = \frac{F_{old}}{1 + (dF_{old})}$$

Example 7.6

A prescription reads −8.00 DS/−2.00 DC × 180 at 10 mm. The final lens is to be fitted at a vertex distance of 16 mm. What is the power of the final lens?

The principal powers are −8.00 D along 180 and −10.00 D along 90, so two vertex distance compensations are required. The change in vertex distance, d, is an *increase* of 6 mm. First, the −8.00 D meridian using the equation with d entered in metres:

$$F_{new} = \frac{F_{old}}{1 + (dF_{old})}$$

$$F_{new} = \frac{-8.00}{1 + (0.006 \times -8.00)} = -8.40\,D$$

The above is repeated for the −10.00 D meridian:

$$F_{new} = \frac{F_{old}}{1 + (dF_{old})}$$

$$F_{new} = \frac{-10.00}{1 + (0.006 \times -10.00)} = -10.64\,D$$

The power of the final lens, in sph–cyl form, is:

−8.40 DS/−2.24 DC × 180 at 16 mm

As always, the cylinder is the *difference* in the two principal powers. The 'new' lens is *stronger*, to reflect the *shorter* focal length produced by *increasing* the vertex distance of a minus lens.

Size of the retinal image in the uncorrected astigmatic eye

In Chapter 5, the equation to find the size of a basic or unaided retinal image was given as:

$$h'_u = -\frac{k'}{n_e} \tan w$$

Careful inspection of the above equation shows that the value of h'_u depends on the *axial length* of the eye and *not* the power of the reduced surface because F_e does not appear in this equation. As the standard reduced eye has only one axial length, an astigmatic eye has only *one* value for h'_u and not two, as one might expect. The size of the basic retinal image in an astigmatic eye is therefore determined using the methods described in Chapter 4.

Size of the retinal image in the corrected astigmatic eye

This is determined by using the spectacle magnification produced by a correcting lens in exactly the same way as for spherical ametropia, i.e.

$$h'_c = h'_u \times \text{SM}$$

However, as an astigmatic eye has *two* values for K and *two* values for F_{sp}, it has two spectacle magnifications and, therefore, *two* corrected retinal image sizes. The magnifications and retinal images correspond to the principal meridians of the corrected astigmatic eye.

Example 7.7

A reduced eye with refractive ametropia has the spectacle prescription −6.00 DS/−2.00 DC × 180 at 12 mm. Calculate the ocular refraction and the power of the reduced surface. Calculate the size of the retinal image in the uncorrected and corrected eye when the distant object is a square with horizontal and vertical sides subtending an angle of 2°. Assume: (1) the constants of the standard reduced eye, (2) that the visual axis is perpendicular to the plane of the square and (3) that the correcting spectacle lens is thin.

As the question informs us that the eye has refractive ametropia, we can therefore assume that the axial length is the standard 22.22 mm. As the refractive index of the eye is the standard 4/3, the dioptric length is +60.00 D. As the question provides us with the spectacle prescription and vertex distance, we can easily find the two ocular refractions K_{90} and K_{180}. If we know K_{90}, K_{180} and K', we can then find the powers of the reduced surface F_{e90} and F_{e180}. So, the first thing to do is to find K_{90} and K_{180}. The principal powers of the correcting lens F_{sp} are −6.00 D *along* 180 and −8.00 D *along* 90.

$$K_{180} = \frac{F_{sp180}}{1-(dF_{sp180})}$$

$$K_{180} = \frac{-600}{1-(0.012 \times -6.00)} = -5.60\,\text{D}$$

$$K_{90} = \frac{F_{sp90}}{1-(dF_{sp90})}$$

$$K_{90} = \frac{-8.00}{1-(0.012 \times -8.00)} = -7.30\,\text{D}$$

The power of the reduced surface F_e is given by:

$$F_e = K' - K$$

$$F_{e180} = K' - K_{180} = +60.00 - (-5.60) = +65.60\,\text{D}$$

$$F_{e90} = K' - K_{90} = +60.00 - (-7.30) = +67.30\,\text{D}$$

Note that, as the eye is refractively ametropic, K' has the same value in both meridians, so as previously mentioned the size of the uncorrected retinal image is the same in both meridians. To find the size of the basic or uncorrected retinal image use:

$$h'_u = -\frac{k'}{n_e}\tan w$$

$$h'_u = -\frac{22.22}{4/3}\tan 2° = -0.582\,\text{mm}$$

To find the size of the corrected retinal images, the spectacle magnification needs to be found. As the spectacle lens is thin we can use:

$$\text{SM} = \frac{K}{F_{sp}}$$

for each meridian.

$$\text{SM}_{180} = \frac{K_{180}}{F_{sp180}} = \frac{-5.60}{-6.00} = 0.9333\times$$

$$\text{SM}_{90} = \frac{K_{90}}{F_{sp90}} = \frac{-7.30}{-8.00} = 0.9125\times$$

To find the size of the corrected retinal image we can now use:

$$h'_c = h'_u \times \text{SM}$$

twice!

$$h'_{c\,180} = h'_u \times \text{SM}_{180} = -0.582 \times 0.93 = -0.543\,\text{mm}$$

$$h'_{c\,90} = h'_u \times \text{SM}_{90} = -0.582 \times 0.9125 = -0.531\,\text{mm}$$

The corrected retinal image is therefore in the form of a square with vertical sides measuring −0.531 mm and horizontal sides measuring −0.543 mm.

Example 7.8

A refractively ametropic reduced eye has a reduced surface power of +58.00 D along 180 and +56.00 D along 90. Assume the constants of the standard reduced eye. What type of astigmatism does this eye have? Illustrate the dioptric position and orientation of the focal lines, the dioptric position of the disc of least confusion and the shape of the retinal image if this unaccommodated eye views a distant point object. This eye is now corrected using a thin spectacle lens placed at a vertex distance of 16 mm. State the power of the correcting lens.

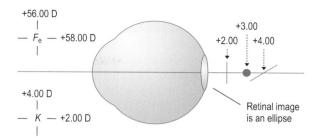

Figure 7.13 Rx = −2.00 DC × 180 (uncorrected eye). See text for explanation.

Once again, as the question informs us that the eye has refractive ametropia, we can assume that the axial length is the standard 22.22 mm and, as the refractive index of the eye is the standard 4/3, the dioptric length is +60.00 D.

The ocular refraction K is given by:

$$K = K' - F_e$$

$$K_{180} = K' - F_{e180} = +60.00 - (+58.00) = +2.00\,D$$

$$K_{90} = K' - F_{e90} = +60.00 - (+56.00) = +4.00\,D$$

The ocular refractions are therefore +2.00 D along 180 and +4.00 D along 90. The ocular astigmatism is 2.00 DC.

The type of astigmatism here is **compound hypermetropic astigmatism** (against the rule) because both ocular refractions are positive. Both line images and the disc of least confusion is formed *behind* the eye. The *horizontal* meridian of the eye forms a *vertical* image 2.00 D behind the eye. The *vertical* meridian of the eye forms a *horizontal* image 4.00 D behind the eye. The disc of least confusion is always formed dioptrically *halfway* between the two line images. In this case it is formed:

$$\frac{K_{180} + K_{90}}{2} = \frac{+2.00 + (+4.00)}{2} = +3.00\,D$$

behind the eye. As both line images and the disc of least confusion are behind the eye, the only remaining shape in the astigmatic pencil left to form the retinal image is the **ellipse**. The orien-

tation of this ellipse matches the orientation of the line image closest to the retina, which in this case is **vertical**. The shape of the retinal image formed is therefore a **vertical ellipse**. This description is illustrated in **Figure 7.13**. The final task is to calculate the power of the thin correcting lens:

$$F_{sp180} = \frac{K_{180}}{1 + (dK_{180})}$$

$$F_{sp180} = \frac{+2.00}{1 + (0.016 \times +2.00)} = +1.94\,D$$

$$F_{sp90} = \frac{K_{90}}{1 + (dK_{90})}$$

$$F_{sp90} = \frac{+4.00}{1 + (0.016 \times +4.00)} = +3.76\,D$$

The correcting spectacle lens therefore has principal powers of **+1.94 D along 180** and **+3.76 D along 90**. The power in sph–cyl form is **+1.94 DS/+1.82 DC × 180**.

Example 7.9

A refractively ametropia reduced eye has an ocular refraction of −6.00 DS/−4.00 DC × 90. What type of astigmatism does this eye have? Find the power of the reduced surface and the axial length of the eye. What is the shape of the blurred retinal image if the uncorrected eye views a distant point object? If the pupil diameter is 4.50 mm, calculate the size of the blurred retinal image formed when this eye views a distant point object. Show the positions of the focal lines in (a) the uncorrected eye, and if (b) a −6.00 D lens is placed in contact with the reduced surface, (c) a −8.00 D lens is placed in contact with the reduced surface and (d) a −10.00 D lens is placed in contact with the reduced surface.

This eye has compound myopic astigmatism (against the rule). In questions involving astigmatism, the ametropia is usually refractive in origin. Once again, we can assume that the axial length is the standard 22.22 mm and, as the refractive index of the eye is the standard 4/3, the dioptric length is +60.00 D. The power of the reduced surface F_e is given by:

$$F_e = K' - K$$

$$F_{e180} = K' - K_{180} = +60.00 - (-10.00) = +70.00\,D$$

$$F_{e90} = K' - K_{90} = +60.00 - (-6.00) = +66.00\,D$$

The vertical meridian forms a horizontal line image that is the closest line image to the retina. The image formed on the retinal is therefore an out-of-focus ellipse with its major axis horizontal.

In this example, the distant object is a point object so, in order to find its size, we need to calculate the diameter of one blur disc. The size of a blur disc, y, formed in a myopic eye is given by:

$$y = p \times \frac{k' - f'_e}{f'_e}$$

where p is the pupil diameter and f'_e the focal length of the reduced surface. As is often the case in astigmatism, we are doing everything twice.

The focal length of the reduced surface is given by:

$$f'_e = \frac{n_e}{F_e} \qquad f'_e = \frac{4/3}{+60.00} = 0.02222\,m$$

Don't forget the refractive index! Taking each meridian in turn:

Horizontal meridian

$$f'_e = \frac{n_e}{F_e} \qquad f'_e = \frac{4/3}{+70.00} = 0.01905\,m$$

$$y = p \times \frac{k' - f'_e}{f'_e}$$

$$y = 4.5 \times \frac{22.22 - 19.045}{19.045} = 0.750\,mm$$

Vertical meridian

$$f'_e = \frac{n_e}{F_e} \qquad f'_e = \frac{4/3}{+66.00} = 0.02020\,m$$

$$y = p \times \frac{k' - f'_e}{f'_e}$$

$$y = 4.5 \times \frac{22.22 - 20.202}{20.202} = 0.450\,mm$$

The blurred retinal image is therefore a horizontal ellipse with the dimensions 0.750 mm × 0.459 mm.

For the last part of this question, the four situations are shown in Figures 7.14a–d.

Example 7.10

An uncorrected and unaccommodated reduced eye with refractive ametropia views a cross with vertical and horizontal limbs. The ocular refraction is −2.00 DS/−2.00 DC × 90. The cross is positioned at distances of ½ m, ⅓ m and ¼ m from the reduced surface. Find the nature of the retinal image in each case.

In the uncorrected eye, a vertical line image is formed −4.00 D in front of the retina and a horizontal line image formed −2.00 D in front of the retina. The disc of least confusion is formed −3.00 D in front of the retina (dioptrically halfway between the two line images). In the uncorrected eye the retinal image is in the form of a blurred ellipse oriented horizontally (the orientation of the ellipse matches the orientation of the line image closest to the retina). For a point image to be formed on the retina, the vergence arriving at the reduced surface must be −2.00 D along 90 and −4.00 D along 180.

Cross at ½ m

With the cross placed at ½ m from the eye, the vergence L arriving at the reduced surface is:

$$L = \frac{1}{l} = \frac{1}{-0.50} = -2.00\,D$$

This has the effect of moving the complete astigmatic pencil 2.00 D to the right. Now the horizontal line image is formed on the retina. The vertical line image is −2.00 D to the left of the retina and the disc of least confusion −1.00 D

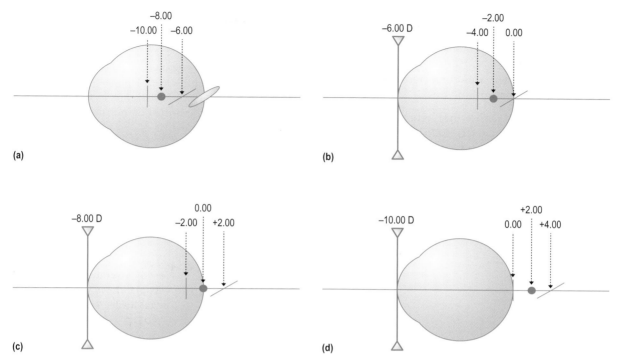

Figure 7.14 (a) The uncorrected eye (the retinal image is a blur ellipse); (b) The uncorrected eye with a −6.00 D lens in contact with the reduced surface (the retinal image is a horizontal line). (c) The uncorrected eye with a −8.00 D lens in contact with the reduced surface (the retinal image is a blurred disc). (d) The uncorrected eye with a −10.00 D lens in contact with the reduced surface (the retinal image is a vertical line).

to the left of the retina. The image of the cross therefore has a clear horizontal limb and a blurred vertical limb.

Cross at ⅓ m

With the cross placed at ⅓ m from the eye, the vergence arriving at the reduced surface is −3.00 D. In this situation, the disc of least confusion is formed on the retina with the horizontal line image 1.00 D behind the retina and the vertical line image 1.00 D in front of the retina. The image of the cross is out of focus but equally blurred in both meridians. This is because the two line images are dioptrically equidistant either side of the retina.

Cross at ¼ m

With the cross placed at ¼ m from the eye, the vergence arriving at the reduced surface is −4.00 D. This has the effect of moving the complete astigmatic pencil 4.00 D to the right, placing the vertical line image on the retina. The

horizontal line image is +2.00 D to the right of the retina and the disc of least confusion +1.00 D to the right of the retina. The image of the cross therefore has a clear vertical limb and a blurred horizontal limb.

The detection and correction of astigmatic ametropia

Astigmatic ametropia can be detected and measured using either the **fan and block** or the **cross-cylinder** methods of refraction. Both techniques are described in Chapter 11.

Concluding points for Chapter 7

This lengthy and perhaps complicated chapter has covered:

- the signs, symptoms, causes and correction of astigmatism
- the classification of astigmatism
- regular and irregular astigmatism

- the formation of the retinal image in the uncorrected astigmatic eye
- the ocular and spectacle refractions in astigmatism
- the uncorrected and corrected retinal image sizes in the astigmatic eye

It is important to realise that, when dealing with problems involving astigmatism, we are basically doing everything twice!

Further recommended reading

Freeman M H, Hull C C (2003) *Optics*. Butterworth-Heinemann, Oxford

Rabbetts R B B (1998) *Bennett & Rabbetts' Clinical Visual Optics*. Butterworth-Heinemann, Oxford

Tunnacliffe A H (1993) *Introduction to Visual Optics*. Association of the British Dispensing Opticians, London

Retinoscopy

Andrew Franklin

Introduction

Retinoscopy is an objective method of determining a patient's refractive error and it is probably the most common objective refractive technique employed in optometric practice. As it is a purely objective technique, the patient has no say in the outcome. An objective measurement of a refractive error is the only assessment available in patients who are unable to cooperate in a subjective refraction, such as young children and those with developmental delay. It is also heavily relied upon when subjective responses are limited (non-English-speaking patients) and in patients whose subjective responses are poor (including some low vision patients) or unreliable (malingerers). In more routine patients, it provides an objective first measure of refractive error, which is usually refined by subjective refraction.

Although autorefractors are gaining popularity for measuring refractive errors, because ancillary staff can operate them, a competent retinoscopist is still more likely to outperform an autorefractor during the course of a working day in terms of both speed and accuracy. The results obtained with an autorefractor can be unreliable or unobtainable in patients with high refractive errors, small pupils, media opacities or pseudophakia.

Retinoscopy is normally employed together with subjective methods, but, as mentioned above, there are occasions when it is required to provide an endpoint without the benefit of subjective refinement. These include:

- Patients whose visual limitations make subjective methods unreliable, e.g. people with severe amblyopia.
- Those whose limited cognitive or communicative abilities render subjective findings unreliable. Very young children, those with Alzheimer's disease and those with learning disabilities may also fall into this category.

During retinoscopy, a cone or cylinder of light is shone into the patient's eye and the practitioner observes the movement of light reflected from the patient's retina. The technique of retinoscopy can be usefully compared to the neutralisation (by hand) of a spectacle lens (Table 8.1). In retinoscopy, trial lenses placed in a trial frame are used to neutralise a patient's refractive error.

There are two basic methods of retinoscopy. The most common is static retinoscopy in which the patient's accommodation is controlled or suppressed. The second method is dynamic retinoscopy in which accommodation is allowed to occur. This can be useful in the examination of young children.

Even when used with subjective methods, retinoscopy may reveal information that is beyond the reach of subjective refraction. People with hypermetropia often do not accept a full positive correction subjectively, and retinoscopy may reveal the latent element. In addition, retinoscopy may be performed before ophthalmoscopy, so it is often the first chance to view the internal structures of the eye. A number of conditions may alter the appearance of the light reflex, as follows:

- The light reflected from the retina retro-illuminates the lens, iris and cornea. Opacities in these structures can be seen as dark areas against the red background. The same effect may be observed with an ophthalmoscope held about 30–40 cm from the patient's eye. Opacities in the crystalline lens and vitreous may be easier to see by retro-illumination than by direct observation with the ophthalmoscope.

- Where extensive transillumination defects are present in uveitis or pigment dispersion syndrome, it may be possible to see them as bright radial streaks on the iris. Similar effects are also observed in patients with iridotomies. In such cases the iris is being 'back-lighted' using light reflected from the retina (retro-illumination). However, the slitlamp is a better instrument with which observe this.
- Keratoconus distorts the reflex and produces a swirling motion.

- Retinal detachment involving the central area distorts the reflecting surface and a grey reflex may be seen.
- A tight, soft, contact lens has apical clearance in the central area that causes distortion of the reflex.

The use of an autorefractor does not allow the above points to be observed. It is interesting to note that it is in fact possible to perform indirect ophthalmoscopy with a retinoscope and a high plus lens, provided that the instrument is bright enough.

Chapter content

General principles and clinical techniques used in retinoscopy

The earliest retinoscopes consisted of little more than a mirror on a stick, although modern designs are a little more sophisticated. With modern instruments, there is an internal light source and an angled mirror, which either has a central sight hole or may itself be semi-silvered to allow observation of the reflex in the patient's pupil. In addition, there is a collar containing a positive lens, which may be moved up and down to change the distance of the lens from the light source.

During retinoscopy, a cone of light is shone into the eye. Some of the light passes through the pupil and is reflected off the patient's retina. The remainder of the light falls on the ocular

Table 8.1 Hand neutralisation versus retinoscopy

Hand neutralisation	Retinoscopy
Stationary test object	Test object moves across the retina (retinal light patch or reflex)
Lens under test is moved	The 'lens' under test is the eye, which is stationary
Lens power is indicated by the direction and speed of the image movement	The prescription is indicated by the direction and speed of the reflex movement
Trial lenses of known power are placed in contact with the unknown lens until neutralisation occurs	Trial lenses of known power are placed in front of the eye by means of a trial frame until reversal is obtained
End-point is no image movement	End-point is an extremely rapid movement and disappearance of the reflex – the pupil fills with light

After Rabbetts (1998).

adnexa, forming a facial light patch. The practitioner oscillates (tilts) the retinoscope about any axis (vertical, horizontal or any oblique axis) and observes the movement of the patch of light reflected from the patient's retina. This reflected light is known as the **reflex**. When performing retinoscopy, the movement of the retinal reflex is compared with the movement of the facial light patch. There are three movement/comparative options:

1. Whatever the oscillation of the retinoscope, no movement of the retinal reflex is seen. The whole pupil glows with light. This is seen in emmetropia.
2. An 'against' movement of the retinal reflex is seen when compared with the movement of the patch of light on the patient's face. This is seen in myopia.
3. A 'with' movement the retinal reflex is seen when compared with the movement of the patch of light on the patient's face. This is seen in hypermetropia and, in certain circumstances, low degrees of myopia.

The accuracy of retinoscopy as a clinical technique relies on the ability of the observer to interpret the movement of the light reflex seen within the patient's pupil.

In summary:

- The retinoscope places a patch of light on the patient's retina.
- This patch of light acts as a 'secondary source' and light from this source is refracted by the patient's eye.

Light can emerge from the patient's eye as one of the following:

- A divergent beam in hypermetropia – seen as a with movement.
- A parallel beam in emmetropia – the pupil is 'filled' with light.
- A convergent beam in myopia – seen as an against movement.

The endpoint of retinoscopy is the 'neutralisation' of the eye's refractive error. This is called **reversal** and occurs when the pupil appears to 'fill' or 'glow' with light. To obtain reversal, the appropriate trial case lenses are placed in front

of the eye to neutralise the 'with' or 'against' movements. If an **against** movement is seen, the practitioner adds **minus** lenses to the trial frame. If a **with** movement is seen, **plus** lenses are added to the trial frame. In both cases, trial case lenses are added until reversal is obtained.

The phenomenon of 'reversal' occurs when the far point of the patient's eye coincides with the pupil plane of the practitioner. As this point is approached the reflex movement speeds up, but at the point of reversal itself no movement is seen because the whole of the practitioner's field is simultaneously illuminated. The observed speed of the reflex has great practical value because the experienced practitioner can estimate the refractive error simply by observing the direction of the reflex and its speed. The lower the refractive error, the faster reflex appears to move. Higher degrees of ametropia therefore produce slower reflex movements.

The instrument is normally used with the collar fully down. This means that the light leaving the instrument is divergent. By moving the collar up the beam of light leaving the instrument becomes less divergent (and eventually convergent) and the reflex is brighter. However, the direction of reflex movement is reversed, i.e. 'with' for myopia and 'against' for hypermetropia.

The formation of the reflex movements observed in retinoscopy

The formation of the reflex movements observed in retinoscopy is explained in **Figures 8.1–8.4**. To improve clarity and understanding of what can be a difficult concept, the figures have been simplified somewhat.

In Figure 8.1, the retinoscope is positioned at some distance in front of a myopic eye. This distance is called the working distance and is usually in the region of 50–66 cm. The retinoscope is held in a vertical plane (no tilt) and a beam of light shone into the patient's eye. The retinoscope places a patch of light (S) on the patient's retina. This patch of light acts as a secondary source of light and light leaving S is refracted by the patient's eye. As the eye is positive in power, a convergent beam of light leaves

73

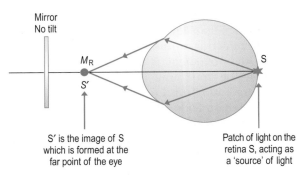

Figure 8.1 Myopia: see text for explanation.

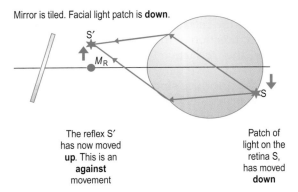

Figure 8.2 Myopia: see text for explanation.

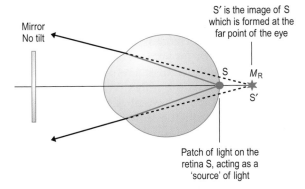

Figure 8.3 Hypermetropia: see text for explanation.

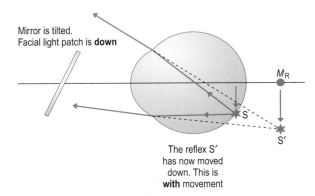

Figure 8.4 Hypermetropia: see text for explanation.

the eye and, because the power of the eye is greater than +60.00 D (the eye is myopic), the light leaving the eye is converged to a focus at S', in front of the eye. The light comes to a focus at the **far point** of the eye and S' therefore coincides with the far point M_R.

Figure 8.2 shows the same myopic eye. This time the retinoscope is tilted producing a downward movement of the facial light patch. The retinal light patch S also moves downwards. Light, is refracted by the eye in exactly the same way as described in Figure 8.1 and the focus produced (S') again coincides with the plane of the eye's far point, although this time *above* it. So, the facial light patch has moved *down* and the focus of light in the far point plane has moved *up*. This is the apparent **against** movement of the reflex seen in myopia.

In Figure 8.3 a retinoscope is positioned in front of a hypermetropic eye. The retinoscope is held in a vertical plane (no tilt) and a beam of light shone into the patient's eye. The retinoscope places a patch of light (S) on the patient's retina. This patch of light acts as a secondary source of light and light leaving S is refracted by the patient's eye. As the eye is positive in power, convergent light once again leaves the eye. This time, because the power of the eye is *less* than +60.00 D (hypermetropia), the light leaving the eye has not come to a point focus. We have to trace the rays of light leaving the eye backwards, to find the virtual focal point S'. This virtual focal point is formed at the **far point** of the eye and once again S' coincides with M_R.

Figure 8.4 shows the same hypermetropic eye. This time the retinoscope is tilted, producing a downward movement of the facial light patch. The retinal light patch S also moves downwards. Light, is refracted by the eye in exactly the same way as described in Figure 8.3 and the focus produced (S') again coincides with the plane of the eye's far point but this time *below* it. So, the facial light patch has moved *down* and the focus of light in the far point plane has also moved *down*. This is the apparent **with** movement of the reflex seen in myopia.

The working distance

When performing retinoscopy, it is impossible and impractical for the practitioner to work at infinity. The usual *working distance* is at an 'arm's length' distance in the region of 50–66 cm. As the retinoscope is effectively acting as a near object, it generates negative vergence in the plane of the eye, e.g. if the retinoscope is held at a distance of 50 cm, a vergence of -2.00 D is incident at the eye. This working distance vergence needs to be neutralised *before* retinoscopy begins *or* be allowed for in the final prescription.

As an example, let us take a patient with a prescription of -3.00 D and assume that we are going to perform retinoscopy at a distance of -50 cm. At this distance a vergence of -2.00 D arrives at the eye from the retinoscope. To neutralise this negative vergence; a $+2.00$ D trial lens must be placed in the trial frame before retinoscopy is performed. This is known as the **working distance lens**. Retinoscopy can now be performed and neutralisation (reversal) occurs when a -3.00 D trial lens is placed before the eye. If a working distance lens ($+2.00$ D) were not used, neutralisation (reversal) would occur when a -1.00 D trial lens were placed before the eye. Remember that the instrument itself is generating a negative vergence of -2.00 DS, which when added to the -1.00 D trial lens gives a combined power of -3.00 D. The combination of vergences is therefore producing reversal. When writing down the final result the practitioner, working without a working distance lens, must remember to add the appropriate negative power, in this case -2.00 D, to the power of the lens in the trial frame or the patient's vision will be quite blurred!

The most common working distance is two-thirds of a metre, corresponding to a working distance lens of $+1.50$ D. This is about arm's length for many practitioners. Whatever working distance is chosen, it is essential to maintain that distance throughout the procedure or errors may be introduced. Occasionally, it is necessary to modify the working distance for certain situations, e.g. if the patient's medium is cloudy as a result of cataracts, it may be difficult to see the reflex without moving closer, say to 30 cm (-3.00 D). A closer working distance produces a brighter and therefore easier to see reflex, courtesy of the inverse square law.

Some practitioners always use a working distance lens to neutralise the negative vergence whereas others do not. Each method has its advantages and disadvantages. If a working distance lens is not used, practitioners have been known to forget to add the appropriate minus power to the trial case lens to obtain the final result. Another disadvantage of not using a working distance lens is that, in low degrees of myopia, a *with* movement may be observed. This occurs if the working distance vergence is greater than the refractive error, which effectively produces an over-correction of the refractive error, rendering the myopic eye artificially hypermetropic, hence the *with* movement observed. If a working distance lens is used, it is yet another element to produce reflections and reduce the amount of light reaching the practitioner's eye. It also adds extra weight to an already uncomfortable trial frame. On balance, it is the author's preference to work without using a working distance lens.

Table 8.2 shows various working distances and corresponding working distance lenses.

Table 8.2 Appropriate lens allowance for working distance

Working distance (cm)	Working distance lens (D)
25	+4.00
33	+3.00
40	+2.50
50	+2.00
66	+1.50
80	+1.25
100	+1.00

Example 8.1

Retinoscopy is performed at a working distance of ⅓ m with no working distance lens in the trial frame. Reversal is achieved when the sph–cyl (sphero–cylindical) combination −4.00 DS/ −2.00 DC × 90 is placed in the trial frame. Calculate the spectacle refraction.

At a working distance of ⅓ m, the negative vergence generated by the retinoscope in the plane of the trial frame is −3.00 D in all meridians. The total vergence in the plane of the trial frame is therefore:

$$-3.00 + (-4.00) = -7.00 \text{ D along } 90$$

$$-3.00 + (-6.00) = -9.00 \text{ D along } 180$$

The full objective spectacle refraction is therefore:

$$-7.00 \text{ DS}/-2.00 \text{ DC} \times 90$$

Notice that the cylinder value has not changed. This is because the retinoscope is adding −3.00 D to both meridians, which results in a 2.00 D difference in the two principal powers. It follows that there is no change in the cylinder value. Remember that the cylinder value is simply the difference in the two principal powers.

Example 8.2

A patient has a refractive error of +3.00 DS/ −5.00 DC × 180. Retinoscopy is performed at a working distance of ⅔ m with no working distance lens in the trial frame. Calculate the trial lens power required to produce reversal.

At a working distance of ⅔ m, the negative vergence generated by the retinoscope in the plane of the trial frame is −1.50 D in all meridians. In the horizontal meridian, the refractive error is hypermetropia (+3.00 D). The power required to produce reversal in the horizontal meridian is therefore +1.50 D to neutralise the working distance vergence and +3.00 D to correct the refractive error:

$$+1.50 + (+3.00) = +4.50 \text{ D along } 180$$

In the vertical meridian, the refractive error is myopia (−2.00 D). As the retinoscope is generating a vergence of −1.50 D in the plane of the trial frame, the power required to produce reversal in the vertical meridian is −0.50 DS.

The power of the lenses placed in the trial frame to produce reversal without a working distance lens is therefore:

$$+4.50 \text{ D along } 180 \text{ and } -0.50 \text{ D along } 90$$

In sph–cyl form this is:

$$+4.50 \text{ DS}/-5.00 \text{ DC} \times 180$$

Notice once again that the cylinder value has not changed. This is because the retinoscope is adding −1.50 D to both meridians, which results in no change in the cylinder value. Remember that the cylinder value is simply the difference in the two principal powers.

From the above examples we can conclude that, if performing retinoscopy *without* a working distance lens, the trial lens power required to produced reversal is *greater* than the actual objective result in hypermetropia, and *less* in myopia. In both cases, the difference is equivalent to the working distance vergence.

Age-related errors in retinoscopy

It is known that retinoscopy tends to overestimate hypermetropia in young patients and underestimate it in older ones. Millodot and O'Leary (1978) found that the discrepancy between retinoscopy and subjective findings had a linear relationship with the age of the patient. Retinoscopy is subject to spherical aberration and, when the patient has a large pupil, this can cause a difference in movement of the peripheral portion of the reflex (which should be ignored) compared with the central portion. Chromatic aberration is also present, but neither type of aberration is of sufficient magnitude to account for this age-related discrepancy. Millodot and O'Leary suggested that in young people, most of the light forming the reflex has been reflected from the internal limiting membrane, which is in front of the photoreceptors. With increasing age, this layer displays a lower reflectance whereas a greater percentage of the light is reflected from the deeper Bruch's membrane.

Streak and spot retinoscopy

Spot retinoscopes project a round spot of light on to the patient's retina. Streak retinoscopes produce an elongated image of their line filament bulb, which can be rotated through 360°. Either type of retinoscope is perfectly adequate. Streak retinoscopes are currently fashionable and, where there is a high cylinder, axis determination is slightly easier with a streak (although spot users may disagree with this). However, spot retinoscopes are probably better for lower levels of astigmatism. Some instruments now come with interchangeable bulbs, so practitioners can have the best of both worlds.

Preparing for retinoscopy

Plus or minus cylinders?

The use of plus cylinders does tend to give a clearer, more easily neutralised reflex if a streak retinoscope is used. For this reason, some practitioners who need to rely on retinoscopy as a stand-alone technique (e.g. working with patients with special needs) favour plus cylinders. However, if minus cylinders are employed, accommodation is better controlled. Minus cylinders are more commonly used in routine refraction and most refractor heads have only minus cylinders.

Measuring pupillary distance

It is important that the trial frame be set up accurately, because failure to do so may introduce significant artefacts into the result obtained. The distance pupillary distance (DPD) should be measured using an appropriate technique. The near centration distance (NCD) may be measured by instructing the patient to look at the bridge of the practitioner's nose with the ruler placed against the trial frame. The zero is lined up using the patient's right eye and the NCD is recorded using the patient's left eye. However, the actual NCD varies with both the DPD and the working distance of the patient, so unless you know these and place your nose in precisely the correct position the measurement is not totally accurate. If the DPD and working distance are known, the required NCD can be obtained from Table 8.3. The term 'fitting distance' refers to the distance from the back vertex of the lens to the centre of rotation of the eye.

Adjusting the trial frame

Before retinoscopy starts, the centres of the trial frame aperture should be set to match the DPD. If there is marked facial asymmetry, it may be necessary to measure the monocular pupillary distances and adjust the trial frame accordingly.

Table 8.3 Near centration distance (NCD) as a function of distance pupillary distance (DPD) and working distance

DPD (mm)	Working distance (cm)						
	20	25	30	35	40	45	50
54	48	49	50	50	51	51	51
56	49	51	51	52	52	53	53
58	51	52	53	54	54	55	55
60	53	54	55	56	56	57	57
62	55	56	57	58	58	58	59
64	56	58	59	59	60	60	61
66	58	60	61	61	62	62	63
68	60	61	62	63	64	64	65
70	62	62	64	65	66	66	66
72	63	65	66	67	67	68	68
74	65	67	68	69	69	70	70
76	67	69	70	71	71	72	72

The fitting distance is 27 mm.

If near fixation retinoscopy is to be performed the centres of the trial frame apertures can be adjusted to equal the NCD, and dropped slightly by adjusting the nosepiece of the trial frame.

The trial frame should be level and allowance should be made for any facial asymmetries that may be present. If the frame is not level, the cylinder axis found may be wrong, and vertical prismatic effects may cause artefacts on subsequent subjective binocular tests. The **pantoscopic angle** and **vertex distance** of the frame should both be at sensible values. If the frame is inappropriately tilted, high-powered prescriptions may throw up significant errors in both sphere and cylinder. If the power of either principal meridian is over ±5.00 D, the vertex distance should be measured and noted on the record card.

Adding lenses to the trial frame

Spheres should be placed in the back cells of the trial frame. When more than one sphere is in the trial frame the most powerful should be at the back to minimise vertex distance effects. However, if the practitioner employs an Oculus or similar trial frame that incorporates a built-in vertex distance scale, the most powerful sphere should be placed in the rear cell nearest to the back plane of the front. This is the cell to which the scale is referenced.

When changing spherical lenses, it should be ensured that the patient is never grossly underplussed, because this can stimulate accommodation. It is best to add the next plus lens before the current one is removed. It can be tricky with modern trial frames, but becomes easier with practice. It is essential to make sure that all lenses remain clean throughout the procedure. Experience suggests that this is often not the case, so it is useful to cultivate the habit of cleaning each lens with a lens cloth before adding it to the trial frame.

The target for use in retinoscopy

The ideal target promotes accurate and steady fixation but provides no stimulus to accommodation. Various targets are used and they probably make little difference to the end-result, but there is a little evidence that the rings on the green filter of the duochrome test might be the one that produces least accommodation (McBrien and Taylor 1986). In the absence of any contradictory evidence, the rings on the green filter would be the recommended fixation target for retinoscopy.

Light conditions

A darkened room causes pupil dilation and makes the reflex more visible, although complete darkness can stimulate accommodation. It might also be difficult to find the trial lenses or indeed the patient!

Working on the visual axis

It is best to work within 5° of the visual axis, both horizontally and vertically. The chair height should be adjusted for vertical alignment, allowing for the fact that the test chart may be above the patient, resulting in the patient possibly looking slightly upwards. Errors of the order of −0.50 DC × 90 can occur if the practitioner is working 10° off-axis horizontally. Unless the practitioner has reduced vision in one eye, it is customary to use the right eye to test the patient's right eye and the left eye for the patient's left. If this is impossible, the Barrett method should be employed (see below). For horizontal alignment, the patient should be instructed to look at the green target on the duochrome test. The practitioner then interposes the head so that the target is obscured, and slowly moves it back out of the way until the target is just visible. The patient should be asked to tell the practitioner if the head gets in the way, although this should not be unduly emphasised unless the practitioner is exceptionally tolerant of interruption.

Practitioners should work at a distance that allows them to change the lenses in the trial frame without changing body position, and for most people this means that the working distance is less than the ⅔ m that seems to be the expected norm. Only the very tall or rather simian can reach if they work at ⅔ m and for many, ½ m is more realistic. It does not matter what the distance is, provided that the appropriate working distance lens is used and the same distance is maintained throughout the test. The customary working distance should be measured so that the appropriate working distance lens is used. Each time that the lenses are changed, practitioners can ensure that they return to the same distance by measuring with their arm. Usually the base of the fingers or the wrist is used as a reference point, because this allows lenses to be changed without a change of body position. If the working distance allowance is wrong, errors in the power of the sphere result, e.g. if the working distance is longer or shorter than the distance predicted by the working distance lens by 10 cm then at ⅔ m the sphere power is about 0.25 D in error. It is important to note that, the shorter the working distance, the greater the error that is introduced by a 10 cm variation.

Fogging

During retinoscopy, it is the fixating eye that controls accommodation, so it must be fogged to ensure that accommodation is relaxed. However, if this is overdone it can induce accommodation, so the fogging should be less than 2.00 D. Initially both eyes should be corrected with what is likely to be the full correction, based on the patient's existing correction (if available), distance visual acuity, symptoms and history, and the working distance allowance. Before starting the procedure, check with the retinoscope that there is an 'against' movement along all meridians in the fixating eye. From time to time during retinoscopy, the retinoscope beam should be passed across the fixating eye to make sure that it is still fogged and shows an 'against' movement in all meridians. This is particularly important with hyperopic patients.

A basic method for retinoscopy

Essentially retinoscopy consists of three phases:

1. Initial lens selection to enable the formation of a reflex that is easy to interpret.
2. Identification of the principal meridians, and therefore the astigmatic axis.

3. Neutralisation of the refractive error along the two principal meridians to establish the powers of the spherical and cylindrical components of the patient's refractive error.

Initial lens

If the patient's last spectacle prescription is available, this is a good starting point. If patients have lost their spectacles or no previous prescription is available, consider the unaided vision and far point. A little thought at this stage can save a lot of time and effort. There is nothing to stop the practitioner checking the visual acuity (VA) when the more positive meridian is neutralised, to get an idea of the minus cylinder power required. Unaided distance vision is related to the refractive error in people with myopia and those with manifest hyperopia. This is illustrated in Table 8.4.

In purely astigmatic refractive errors, or with the best vision sphere in place, the required cylinder can be estimated from Table 8.5.

Table 8.4 Expected vision for any uncorrected mean sphere (myopia or manifest hypermetropia)

Vision	Equivalent sphere (myopia or manifest hypermetropia)
6/5	Plano
6/6	0.25–0.50 DS
6/9	0.50–0.75 DS
6/12	0.75–1.00 DS
6/18	1.00–1.25 DS
6/24	1.25–1.75 DS
6/36	1.75–2.25 DS

Table 8.5 Expected vision for any uncorrected cylinder

Vision	Astigmatic error with the best vision sphere in place
6/5	0.25 DC
6/6	0.50–0.75 DC
6/9	1.00–1.25 DC
6/12	1.50–1.75 DC
6/18	2.00–2.25 DC
6/24	2.50–3.00 DC
6/36	3.25–4.00 DC

Table 8.6 The true far point in uncorrected myopia

Uncorrected spherical refractive error (D)	Position of the true far point (cm)
−2.00	50
−4.00	25
−6.00	16.7
−8.00	12.5
−10.00	10
−12.00	8.3

It is important to remember that the values given in Tables 8.4 and 8.5 are only average values. Patients with small pupils (usually those with presbyopia) experience less blur per dioptre compared with those with large pupils. The axis of a cylinder also affects vision because an axis of 90/180 gives less blur than an oblique axis of, say, 45. For those with uncorrected myopia, the true far point at which small print can be seen clearly varies inversely with refractive error (Table 8.6). This information can also provide a useful starting point for retinoscopy.

The working lens should be incorporated into the correcting sphere. The use of a separate working lens introduces an extra set of reflections and it uses up one of the trial frame spaces that may be needed for the patient's prescription. For the ideal starting point, we would want the patient slightly fogged (over-plussed) for distance to discourage accommodation, but rather less than 2.00 D. The easiest type of reflex to interpret is a quick 'with' movement, which should occur if the patient is slightly under-plussed for your working distance. Patients are still somewhat fogged for the distance that they are fixating because the retinoscope and the patient are separated by 1.50 D.

Identifying the principal meridians

When performing retinoscopy on any eye with astigmatism, there are two meridians along which the pupillary reflex moves parallel to the direction of the retinoscope beam. These are the principal meridians of the eye and they are the meridians along which the least and greatest dioptric powers are found. Thus, in an eye that has the sphero-cylindrical refractive error of right + 1.00 DS/− 2.00 DC × 180, the principal meridians are found at 180 and 90. The greatest plus power is found along 180 (i.e. the minus cylinder axis) and the least plus/greatest minus power along 90.

To find the principal meridians, the retinoscope beam is moved across the eye in a series of meridians until the central reflex is moving with the beam. The direction of movement now corresponds with one of the principal meridians and, unless the astigmatism is of the rare irregular type, the second principal meridian is found at 90 degrees to the first. If the retinoscope beam is moved along any other meridian, the movement of the reflex deviates from the direction of motion of the beam. If a streak retinoscope is employed and medium-to-high astigmatic ametropia is present, the deviation may even be apparent when the beam is static (**Figure 8.5**). Rotating the streak aligns the reflex and identifies the direction of sweep (**Figure 8.6**).

When the beam is swept along a non-principal meridian, the reflex travels obliquely at right angles to its long axis. The static rotation of the reflex can be corrected by rotating the beam to align it with the reflex. At this point the beam and reflex lie along one of the principal meridians, and the second one is at right angles to the first. When the reflex is not aligned with the beam along one of the principal meridians, it also appears thicker than it does when aligned. This effect has been used by Francis (1973) in a method that works particularly well when small degrees of astigmatism are present. The maximum thickening occurs when the eye is made artificially myopic by about 1.00 D in its most hyperopic meridian. This is achieved by roughly identifying and neutralising the most hypermetropic meridian, with the collar of the retinoscope fully down. An extra +0.50 DS is added to the trial frame. The beam is then rotated through 90 degrees. If no astigmatism is present the beam stays narrow. Even a small amount of astigmatism causes the reflex to fill the pupil as the beam rotates to the minus cylinder axis. Once the axis is identified, the extra 0.50 D is removed and the second meridian is neutralised.

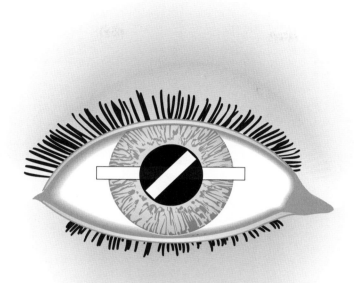

Figure 8.5 With medium-to-high degrees of astigmatism, this deviation is apparent even when the beam is static. (After Harvey and Franklin 2005, with permission of Elsevier Ltd.)

Neutralising the meridians

Once the principal meridians have been identified, the spherical lens in the trial frame is adjusted until 'reversal' is obtained, at least in theory. In practice, reversal is not always as obvious as we would wish, and instead of creeping up to reversal with 0.25 D changes, it is often quicker to use larger intervals (0.50 D or more). If +2.00 gives a 'with' and +2.50 an 'against' movement there is little point in confirming what is already obvious, i.e. that the end-point is +2.25. If we were working with minus cylinders we would neutralise the more positive or least negative meridian first.

- If there is a 'with' movement in both meridians, the meridian showing the slowest movement is the more positive.
- If there is 'with' movement in one meridian, and 'against' in the other, the meridian showing the 'with' movement is the more positive.

- If there is an 'against' movement in both meridians, that showing the faster movement is the more positive or least negative.

When reversal appears to have been obtained, a bracketing technique may be used to check the power. The following are two possible bracketing techniques:

1. Move slightly backwards and forwards. The reflex should change from 'against' to 'with'.
2. Use ±0.25 D twirls. Again the reflex should change from 'against' to 'with'.

If using plus cylinders, the least hyperopic meridian is neutralised first. It is usually recommended in textbooks that one meridian be neutralised to reversal before the cylindrical element is introduced. However, it is often easier to determine the required cylinder if both principal meridians are showing a 'with' movement.

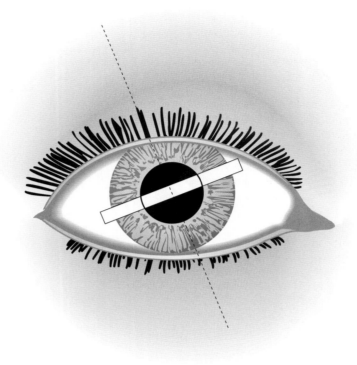

Figure 8.6 Rotating the streak will align the reflex and identify the direction of sweep. (After Harvey and Franklin 2005, with permission of Elsevier Ltd.)

To obtain this state, the less hypermetropic meridian is neutralised first. If the practitioner now leans in a little towards the patient or −0.25 DS is added to the spherical correction in the trial frame, a fast 'with' movement is seen along the less hypermetropic meridian and a slower 'with' movement at right angles to it. Negative cylindrical power is added until the speed of the reflex is equalised in both meridians, while adding enough plus sphere (half the power of the cylinder) to keep the 'with' movement fast. Once the reflex is the same speed in the two meridians, a small addition of plus spherical power achieves reversal.

An alternative and interesting method of neutralisation using a streak retinoscope has been described by Parker (1966). The principal meridians are identified conventionally. The streak is set along the meridians and the collar is adjusted to give a streak of minimum thickness. As the ametropia is corrected, the width of the streak increases to the point of neutralisation when the reflex fills the pupil.

Another bracketing technique (Lindner's method) may be used to verify the cylinder axis. If the beam is swept across at +45 degrees and −45 degrees to the minus cylinder axis, the reflex obtained should be identical. However, if one is 'with' and one is 'against' the minus cylinder axis should be moved towards the meridian that shows the 'with' movement. This technique can be used with both spot and streak retinoscopes.

If using a spot retinoscope, the cylinder axis can be refined by 'driving' the beam along the minus cylinder axis. If the reflex moves anticlockwise to the trial cylinder, rotate the trial cylinder anticlockwise. If the reflex moves clockwise to the trial cylinder, rotate the trial cylinder clockwise. When the reflex moves along the trial cylinder axis, the axis is correct.

Difficult retinoscopy

Split reflex and scissors movement

Sometimes one area of the reflex shows different movement from another, and the dioptric power of the lens required to neutralise the movement differs for different areas (irregalar refraction). This could be caused by keratoconus, corneal scarring or lens changes. Spherical aberration and coma could also be factors that affect the appearance of the reflex. When the pupil is large, positive spherical aberration and coma ensure that the eye's refractive power is greater through the periphery of the pupil. The result may be seen as a 'with' movement in the centre of the pupil, with an against movement seen in the more peripheral zone. This has been termed a 'scissors movement'. It may be particularly confusing when the pupils are dilated. It is worth checking that the trial case lenses are clean and correctly centred, and that you are working on axis. It is best to ignore the peripheral reflex and concentrate on the central part of the reflex (Charman and Walsh 1989). The use of a bracketing technique is also useful.

Crystalline lens opacities

It may be possible to work around opacities by moving off-axis. It might also be necessary to work closer to obtain a brighter reflex courtesy of the inverse square law.

Ocular abnormalities

Localised bulges or asymmetries may mean that the fovea is on a different plane to the slightly off-axis point that forms the reflex. Therefore, the sphere power may be somewhat inaccurate.

Accommodative tonus

The ciliary muscle, similar to all smooth muscles in the body, has both a dependent and an independent tone. As this terminology suggests, the dependent tone is conditional on an intact nerve supply whereas the independent tone in not. The independent tone in the ciliary muscle is very small, does not give rise to symptoms and is not affected by cycloplegics. The dependent tone of the ciliary muscle is totally abolished by complete atropine cycloplegia, but *not* by the full effects of other cycloplegics such as cyclopentolate hydrochloride. An allowance has to considered *only* in the case of complete atropine cycloplegia for the return of the dependent tone to the ciliary muscle on its recovery from the effects of atropine paralysis. This tonus allowance is an adjustment to the spherical element of only the objective findings, to take into account the fact that the eyes, when fully returned to normal, have their overall refractive power slightly increased (more plus) as a result of the constant effect of the dependent tone on the recovered and fully active ciliary muscle. Traditionally, the objective result obtained after full atropine cycloplegia is reduced by 1.00 D as an allowance for tonus. Cyclopentolate hydrochloride does *not* produce complete cycloplegia, but leaves a residual amount of accommodation of about 1.50 D or less. This depth of cycloplegia is adequate in most cases and also means that a tonus allowance *does not* need to be considered when calculating the final prescription.

In young people with hyperopia, the retinoscope result is often considerably more positive than the eventual subjective refraction as a result of high accommodative tonus. Patience is a virtue here because, if it is suspected that there might be more plus to add, it is worth continuing to sweep the beam across, reminding the patient to look at the circles on the green, and eventually a 'with' movement may be seen, albeit a transient one. This is neutralised and repeated until the practitioner is satisfied that all the hyperopia is corrected. If retinoscopy is consistently under-plussing the correction, the working distance should be checked, and slowing down might also be wise.

Near fixation retinoscopy

Barrat's method

With this technique, the patient fixates a bright luminous fixation object binocularly. Alternatively the practitioner's forehead may act as the target. The claimed advantages of this technique include the following:

- Working closer to the visual axis
- Smaller pupil caused by near reflex: fewer aberrations as a result (but also a less bright reflex)
- Only one of the practitioner's eyes is used. This makes the method particularly useful for those optometrists who have reduced acuity in one eye.

The major disadvantage of this technique is that the patient accommodates, which is particularly true for younger patients. The sphere must be checked with distance fixation in one eye either before or after using Barratt's method and the final result adjusted accordingly.

Mohindra's technique

Mohindra's (1975) technique is a development of near-fixation retinoscopy that allows the refraction of infants and young children without the use of cycloplegics. The room lights are slowly extinguished and the child is encouraged to look at the retinoscope light. It is usual to ask the parent to occlude one eye, although opinions vary as to whether this makes a significant difference. Gradually adding positive power tends to relax accommodation. The pupil initially constricts, but after a few seconds dilation occurs. At this point the refractive error may be neutralised. Retinoscopy racks (small spherical lenses mounted in a holder) may be used for speed, each meridian being neutralised separately. Accurate fixation may be encouraged in older children by asking children when they can see 'the black spot in the light' (i.e. the sight hole in the mirror, on a spot retinoscope).

The working distance with this technique is usually 0.5 m, so the expected allowance for the working distance would be 2.00 D. However, near retinoscopy does tend to underestimate hypermetropia, so a correction factor of 1.25 D is used for adult patients. It has been suggested that a correction factor of 1.00 D is appropriate for children older than 2 years and 0.75 D for those younger than 2 years. Opinions vary on

the accuracy of this technique, particularly in infants and those with higher refractive errors.

Cycloplegic refraction

The term 'cycloplegic refraction' usually refers to an objective refraction being performed with the aid of a pharmaceutical agent designed to prevent or reduce accommodation during refraction, thus making latent refractive errors manifest. Cycloplegics are drugs that paralyse the ciliary muscle by blocking the muscarinic receptors normally stimulated by the release of acetylcholine from the nerve endings of the parasympathetic nervous system. As the parasympathetic nervous system also innervates the pupil sphincter muscle, cycloplegia is accompanied by mydriasis (pupil dilation), which indicates paralysis of the pupil sphincter.

Many antimuscarinic agents have been used in the past but only three are used regularly: atropine sulphate, cyclopentolate hydrochloride and tropicamide. Changes in medicines legislation came into force in April 2005 (SI 2005 766), which resulted in atropine sulphate being made available only to optometrists who have successfully completed extended training and achieved level 2 exemption. Both tropicamide and cyclopentolate hydrochloride have level 1 exemption and can be used by all optometrists. Tropicamide in a 1% solution can be used to give satisfactory cycloplegia in older children and adults, but it is generally accepted that it produces inadequate cycloplegia in younger children. We are therefore left with cyclopentolate hydrochloride as the cycloplegic of choice. This preparation is available in single dose minims form in two concentrations: 0.5% and 1.0%.

One drop of the 1.0% solution in each eye is usually all that is necessary to produce adequate cycloplegia. The 1% solution is suitable for most patients. Although one drop is usually enough, for patients with dark irides a second drop may be needed if nothing seems to be happening after 15 minutes. It does not produce absolute cycloplegia, but the residual accommodative tonus is less than 1.50 D. No 'tonus allowance' needs to be made, so the 'full cyclo' can be prescribed. The 0.5% solution is needed for children aged less than 3 months, although few patients of this young age would be encountered in the general ophthalmic services. Tropicamide 1% has been found to be a useful, if short-acting, cycloplegic for patients in their late teens. In adult patients, the short duration is a virtue and this is the ideal agent to use with the adult patient thought to have latent hypermetropia or pseudomyopia. Two drops, about 5 minutes apart, after proxymetacaine, is usually sufficient.

When should cycloplegic refraction be done?

This is included in this chapter rather than the chapter devoted to subjective refraction on the grounds that most 'cyclos' are performed on children whose subjective responses are not entirely reliable. Some practitioners advocate cycloplegic examination of all new young patients. This has the advantage of providing more reliable baseline data on the refractive error at the expense of time and some trauma for the patient. In general optometric practice, most practitioners tend to use cycloplegics when:

- there is undiagnosed manifest esotropia
- an esotropia has been noticed by the parent or guardian
- there is unstable or uncompensated esophoria
- there are significant risk factors for esotropia and amblyopia, e.g. family history, significant refractive error, premature birth, etc.
- a satisfactory level of acuity is not obtained in one or both eyes.
- retinoscopy suggests that accommodation is fluctuating significantly
- the retinoscopy findings differ significantly from the subjective findings
- there is a suspicion of an anomaly of accommodation, e.g. accommodation insufficiency, accommodative fatigue or spasm of accommodation
- stereoscopic acuity is unsatisfactory and/or absent

- latent hypermetropia or pseudomyopia is suspected.

The use of local anaesthetics

Cycloplegics sting and this can make cycloplegic refraction a somewhat unpleasant experience for all concerned, especially those on the receiving end. This can be ameliorated somewhat by the use of proxymetacaine 0.5%, a local anaesthetic, which stings rather less than the other local anaesthetics when used on the eye, and removes the sting of the subsequent cycloplegic entirely. A further advantage is that absorption of the cycloplegic is enhanced. Proxymetacaine is available in minims, but it needs to be stored in a refridgorator, which is not possible at all practices. The only other drawback would be if the patient did not like the first drop and decided to resist the instillation of the second. However, this is rarely a problem, and the phrase 'this drop will feel a bit cold' seems to work well.

Instillation of cycloplegics

Children are rarely tremendously keen to have eye drops. It is therefore important to explain what is going to happen in a calm way, avoiding words such as 'sting' and 'pain'. Although actual lying is to be avoided if possible, children are fairly suggestible. The patient should be told that the drops 'might feel a bit funny (or cold)'. This works surprisingly well. Body language needs to reinforce the verbal message, because children are rather good at reading it. Sitting on mother's lap is a safe place to be for most small children. The child might move fairly suddenly, and it would not be good practice to stick a minim in the child's eye. If children cooperate, ask them to look down, raise the upper lid gently with your thumb and keep the neck of the minim against the thumb while you instil the drop. If the patient moves, so does your thumb and so does the minim. If the child will not open the eyes, a variation of an old contact lens trick can be useful. While trying to raise the upper lid with your thumb, out of the blue say 'now, open your mouth as wide as you can!' It's almost impossible to open your mouth wide and close your eyes tight at the same time. If children close their eyelids and steadfastly refuse to open them, the trick is to put about three drops of the drug on the upper lashes at the lid margin. They have to open their eyes eventually.

As with most drugs there is potential for undesirable side effects to occur when using cycloplegics. The listed side effects of cyclopentolate hydrochloride include local irritation, allergic reactions, increased intraocular pressure, disturbances of the central nervous system and blurred vision. Young children often become sleepy after the use of cyclopentolate hydrochloride. Having mentioned these possible side effects, it must be stressed that the benefits of a cycloplegic refraction vastly outweigh potential problems.

How accurate is retinoscopy?

It is difficult to say exactly how accurate retinoscopy is because the major variable is the skill of the practitioner. Safir et al (1970) and Hyams et al (1971) found a 50% probability of two consecutive measurements of sphere power being within 0.40 DS of each other. Interestingly, repeatability was best for cylinder axis, cylinder power and sphere in that order. Freeman and Hodd (1955) found that an experienced retinoscopist should be able to obtain repeatability of ±0.25 D in each principal meridian. These findings suggest that retinoscopy is not very accurate, but is probably as accurate as it needs to be. It is rarely an end in itself and is best seen as a good method to obtain a quick approximation of the refractive error.

Retinoscopy in contact lens fitting

Retinoscopy has many uses in contact lens practice. Briefly, these include:

- over-refraction of contact lenses during fitting and aftercare
- assessment of soft lens fit
- location of the optic zone of a rigid lens relative to the pupil
- location of the position of the segment of a bifocal contact lens

- assessment of the stability of soft and rigid toric lenses

In contact lens aftercare, retinoscopy can show:

- areas of non-wetting on the lens surface
- epithelial waves and dimples
- stromal haze (oedema)
- deterioration of soft lens surfaces
- imperfections in the preocular tear film
- the presence of keratoconus

Concluding points for Chapter 8

For optometrists, retinoscopy is, without doubt, an essential skill. In fact most optometrists rely on their 'ret' more than any other item of equipment. Retinoscopy also has a role in contact lens practice, not just in over-refraction but also in other aspects of contact lens fitting and aftercare. The technique of retinoscopy takes time to master but, once proficient, the results can be surprisingly accurate. However, in general optometric practice, objective methods are not usually required to provide the final prescription. The objective result usually gets us to the point from which subjective methods can take us to the end-point accurately and quickly.

References

Charman W N, Walsh G (1989) Variations in the local refractive correction of the eye across it's entrance pupil. *Optometry and Vision Science* **66**(1):34–40

Freeman H L, Hodd F A B (1955) Comparative analysis of retinoscopic and subjective refraction. *British Journal of Physiological Optics* **12**:8–19

Francis J L (1973) The axis of astigmatism with special reference to streak retinoscopy. *British Journal of Physiological Optics* **28**:11–22

Harvey W, Franklin A (2005) *Eye Essentrials: Rontine Eye Examination.* Butterworth-Heinmann, Oxford

Hyams L, Safir A, Philpott J (1971) Studies in refraction. II. Bias and accuracy of refraction. *Archives of Ophthalmology* **85**:33–41

McBrien N, Taylor S P (1986) Effect of fixation target on objective refraction. *American Journal of Optometry and Physiological Optics* **63**:346–50

Millodot M, O'Leary D (1978) The discrepancy between retinoscopic and subjective measurements: Effect of age. *American Journal of Optometry and Physiological Optics* **55**:309–16

Mohindra I (1975) A technique for infant visual examination. *American Journal of Optometry* **52**:867–70

Parker J A (1966) Stationary streak retinoscopy. *Canadian Journal of Ophthalmology* **1**:228–39

Safir A, Hyams L, Philpott J (1970) Studies in refraction. I. The precision of retinoscopy. *Archives of Ophthalmology* **84**:49–61.

Further recommended reading

Elliott D B (2003) *Clinical Procedures in Primary Eyecare*. Butterworth-Heinemann, Oxford

Eperjesi F, Jones K (2005) Cycloplegic refraction in optometric practice. *Optometry in Practice* **6**:107–20

Hopkins G, Pearson R (1998) *O'Connor Davies's Ophthalmic Drugs: Diagnostic and therapeutic uses.* Butterworth-Heinemann, Oxford

Rabbetts R B (1998) *Bennett & Rabbetts' Clinical Visual Optics*. Butterworth-Heinemann, Oxford

Tunnacliffe A H (1993) *Introduction to Visual Optics*. Association of the British Dispensing Opticians, London

Visual Acuity and the Measurement of Visual Function

Andrew Keirl

Introduction

This relatively lengthy and clinically important chapter discusses and defines terminology associated with the measurement of visual function, and also reviews a selection of test charts and testing methods used in optometric practice. The chapter also examines the relevant British Standards associated with optometric test types and charts.

Chapter content

- The Snellen letter
- Resolution, Snellen and decimal acuity
- The Bailey–Lovie logMAR chart
- Standards for the design of test charts

Visual acuity

Visual acuity is a term used to describe the eye's ability to discriminate *detail* in an object, e.g. the limbs and spaces in a letter E or the gap in a letter C. The smallest angle (w) subtended at the nodal point of the eye by two points, which can just be seen or **resolved** as separate, is a measure of the eye's ability to discriminate detail (**Figure 9.1**). The angle w is known as the **minimum angle of resolution** (MAR). Figure 9.2 is a simple illustration of two point objects that are easily resolved (seen) as being separate. In such a situation w must be *greater* than the minimum angle of resolution for this particular eye and S$_1$ and S$_2$ are *easily* resolved. In **Figure 9.3** w is less than the MAR and so the two objects S$_1$ and S$_2$ are *not* resolved. In **Figure 9.4**, the angle is equal to the MAR for this particular eye and S$_1$ and S$_2$ are *just* resolved.

It is in this situation that visual acuity, or to be precise **resolution acuity**, is measured. It must be noted that this description of resolution is somewhat simplistic because, ultimately, limitation in resolution is a result of the overlapping of the diffraction patterns in the neighbouring images of their conjugate object points. When the emmetropic eye is in perfect focus, the image of a point object formed at the macula, is not a perfect point but a blur circle, surrounded by a series of concentric rings. This phenomenon is known as **diffraction** and was first explained by Sir George Airy in 1835. If an eye were perfectly focused and free of aberrations, the only optical factor limiting the resolution of the eye would be diffraction. The extent of overlap in these diffraction patterns dictates whether or not the images of two point objects are resolved. In addition, there is also a neural limit on resolution imposed by the anatomical structure of the retinal photoreceptors. For images of two point images to be resolved, they should fall on individual cones, separated by one cone that is not stimulated.

As w is an angle, it needs to be given an appropriate unit. It is usually expressed in minutes of arc or arcminutes and resolution or visual acuity (VA) is defined as:

$$VA_{res} = \frac{1}{w}$$

where 'res' = resolution acuity.

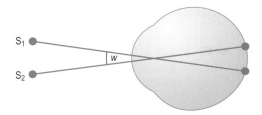

Figure 9.1 The minimum angle of resolution (*w*).

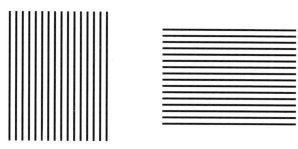

Figure 9.5 The bar grating: bars are resolved.

Figure 9.2 *w* greater than MAR. S_1 and S_2 are easily resolved.

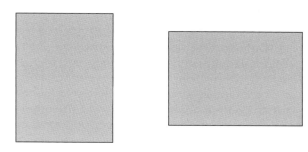

Figure 9.6 The bar grating: bars are not resolved.

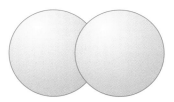

Figure 9.3 *w* less than MAR. S_1 and S_2 are not resolved.

Figure 9.4 *w* = MAR. S_1 and S_2 are just resolved.

terms. Vision (V) refers to the visual standard obtained *without* a correction, whereas visual acuity (VA) refers to the visual standard obtained *with* a correction. In clinical practice, visual or resolution acuity is a measure of the eye's ability to discriminate some particular detail in the image of a test object e.g. a letter, picture or a bar grating. A simple bar grating can be used to measure resolution acuity and **Figure 9.5** shows such a grating when the individual bars and spaces are resolved (seen as separate). **Figure 9.6**, however, shows the same grating when the individual detail (bars and spaces) have not been resolved. A bar grating could, in theory, be used to measure the visual (resolution) acuity of patients in practice. An appropriate method is as follows:

1. View the grating at a distance where it appears uniformly grey (cannot be resolved).
2. The individual approaches the grating, stopping when it can be just resolved.

In practice, **visual acuity** is *synonymous* with **resolution acuity**. In addition the terms 'vision' and 'visual acuity' are constantly used and recorded on patients' record cards. There is, however, an important difference between these

89

3. Repeat this three times and calculate the average distance.
4. Use trigonometry to calculate w using the width of the bar and the average distance.

It should be obvious why this method has not caught on! However, letter charts measure resolution acuity in exactly the same way as a bar grating. Letters are probably easier to recognise and do not have to be moved.

The Snellen letter

F C Donders (1818–1889), a professor of physiology and ophthalmology, suggested to Herman Snellen (1834–1908) that an angular subtense of 1 arcminute should form the basis of the construction of a letter chart. This 1 arcminute angular separation was, and still is, taken as the 'norm' for a measure of visual acuity using letters, although most patients aged under 60 achieve better (smaller) than this. In the design of the original Snellen letter, the limbs and spaces subtend 1 arcminute at specified distances, and the overall letter height subtended 5 arcminutes. The design of the Snellen letter is illustrated in **Figure 9.7**. The original Snellen letter was designed on a 5×5 grid and subtended 5 arcminutes at distances of 6, 9, 12, 18, 24, 36 and 60 metres (**Figure 9.8**). The limb widths were a fifth of the letter height. A typical Snellen letter chart is shown in **Figure 9.9**.

Snellen acuity

The purpose of measuring and recording visual acuity is to obtain a figure that indicates the state of the imaging system of the eye. Snellen defined visual acuity as:

$$\frac{\text{Testing distance in metres}}{\begin{array}{c}\text{Distance in metres at which letters}\\\text{on the best line subtended } 5'\end{array}}$$

The usual testing distance is 6 m and an individual with 'normal' visual acuity could resolve a Snellen letter placed at 6 m from the eye if the letter subtended an angle of 5 arcminutes at the eye. In this case the VA would be recorded as 6/6. Most patients aged under 60 achieve a VA better than 6/6.

Most consulting rooms are not 6 m long and a test chart placed 6 m from a patient is actually not very convenient. The most common clinical arrangement is that a chart designed for use at

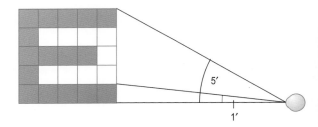

Figure 9.7 The Snellen letter (5×5).

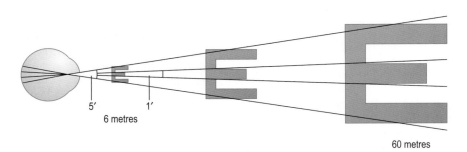

Figure 9.8 The Snellen letter.

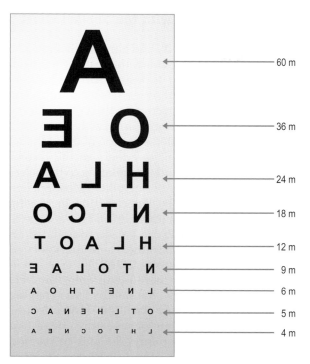

Figure 9.9 A typical distance Snellen letter chart constructed using sanserif letters for use at 6 m.

a distance of 6 m is used together with a plane mirror. The image produced by a plane mirror is of course formed 'behind' the mirror at a distance equal to the object distance. So if the object (test chart) is 3 m in front of the mirror, its image is formed 3 m behind and, if an individual views a test chart in this way, the testing distance is effectively 6 m. This arrangement is known as an **indirect** method, whereas the term **direct** is used if the chart is used without the aid of a mirror.

An interesting question to ask is why a testing distance of 6 m? Well, a testing distance of 6 m produces an incident vergence of –1/6 D at the eye. The eye's depth of field is greater than this and therefore does not need to accommodate (increase in power) to counteract this small negative vergence. A shorter testing distance of, say, 1 m would cause problems because the vergence incident at the eye would be approximately –1.00 D, which would stimulate the eye to accommodate. When measuring distance

vision or distance VA the eye must be assumed to be in its weakest dioptric state (unaccommodated). A testing distance of 6 m is therefore effectively infinity. Equipment manufacturers, such as Nidek, have introduced so-called 'short-form' testing equipment where the test 'chart' is placed 1 m from the patient and, according to the manufacturer, the test target viewed by the patient is effectively at infinity. The advantage of such a system is that rooms smaller than the traditional 3 m can be used. However, users of such charts have commented that patients are often over-plussed and allowances in the final prescription have to be made. Current British Standards (BSI 4274-1 2003) state that the testing distance should be no less than 4 m.

When recording Snellen acuity, the very smallest row of letters that the individual can read should always be recorded because only then can we tell whether there has been deterioration between consultations. If the patient were able to read the 4 m row but we stop recording once the 5 m row has been reached, we would not detect a drop in VA if the individual could manage only the 5 m row on the next visit. This is important, because there are certain pathological ocular conditions that can cause a reduction in VA. When examining an individual, vision and VA, both monocular and binocular, must be recorded. If an individual reads only part of a row, this fact must also be recorded and if an individual's vision is less that 6/60 this must also be recorded.

Resolution, Snellen and decimal acuity

There is a simple link between resolution, Snellen and decimal acuity. Remember that all three terms are stating the same thing, the ability of the eye to discriminate or resolve detail in an object! Resolution acuity is defined as:

$$VA_{res} = \frac{1}{w} \ (w \text{ is in arcminutes})$$

To convert from resolution (VA_{res}) to Snellen acuity, simply multiply $1/w$ by 6/6. So

$$\text{Snellen acuity} = \frac{1}{w} \times \frac{6}{6}$$

For example, if $w = 2'$

$$VA_{res} = \frac{1}{2}$$

In Snellen form, the acuity is:

$$Snellen\ acuity = \frac{1}{2} \times \frac{6}{6} = \frac{6}{12}$$

The decimal form of the acuity is, of course, 0.5.

Example 9.1

A subject can read a 23.27 mm Snellen letter at a distance of 4 m. What is the visual acuity in Snellen and decimal notations?

The diagram for this calculation is shown in Figure 9.10.

Remember that, in the design of the Snellen letter, the limb size is a fifth of the letter height. The limb size is therefore:

$$\frac{23.27}{5} = 4.654\,mm$$

$$Tan\ w = \frac{Limb\ size}{Dis\tan ce} = \frac{4.654}{4000} = 1.1635^{-1}$$

(4 m is equivalent to 4000 mm)

w is therefore 0.06666°.

Of course, w has to be stated in arcminutes, and to convert from degrees to arcminutes multiply by 60.

$$0.06666 \times 60 = 4.00'$$

$$VA_{res} = \frac{1}{4}$$

$$Snellen\ acuity = \frac{1}{4} \times \frac{6}{6} = \frac{6}{12}$$

Figure 9.10 Diagram for Example 9.1.

Design problems with the Snellen chart

Despite the popularity and almost universal use of the Snellen letter chart, the test suffers from a number of flaws in its design. Although letters provide a convenient series of test stimuli, some letters are easier to recognise than others, some combinations of letters are easily confused, most use only a limited selection of the alphabet and many charts fail to adhere to the recommendations and standards relating to the selection of letters based on legibility. Most Snellen charts have one 6/60 letter and an increasing number of letters on lower lines. This is done to ensure that the chart can be accommodated on one small rectangular frame. There are, however, a number of problems with this.

With most charts, there are different numbers of letters on each row, the smaller letters having more letters per row than the larger letters, whereas the larger letters have only one or two letters per row. Therefore, the task of reading the letters is not equally difficult for all letter sizes. Patients with poor acuity are required to read fewer letters than those with good acuity. This does not give a fair assessment of the acuity of a low-vision patient because they have only one chance of getting it right. The Snellen chart is therefore of little use in the assessment of low-vision patients.

The letters on the lower lines are more crowded than the letters on the upper lines. It is well known that the crowding of letters increases the difficulty of the task, particularly for those with amblyopia. To make matters worse, the spacing between each letter and each row of letters bears no systematic relationship to the width or height of the letters. The task required of the patient therefore changes as they read down the chart. For this reason, VA measured at a distance of less than 6 m cannot be easily equated to a 6 m equivalent because the patient will most probably read further down the chart, effectively having carried out a different task.

Most Snellen charts contain a 60 m line, a 6 m line and a varied selection of other lines. The progression of letter sizes is approximate (non-linear and non-geometric), with the letter size doubling every other line. However, the progression is not uniform from chart to chart because extra lines are sometimes added at the bottom whereas others are omitted from the top. There are usually no letters between 36 m and 60 m so an individual whose VA is actually 6/48 will have a recorded VA of 6/60. It is of course important that accurate VAs are recorded because deterioration in VA can sometimes indicate active pathology.

Noting the lowest line of letters that can be read provides the 'score' for the Snellen chart. However, in practice patients can often manage only part of a line and the end-point may even be spread over several lines. As there are no agreed standards for the exact recording of acuity in these situations, there is scope for confusion. A recorded acuity of 6/6^{-1} on one chart cannot possibly mean the same as 6/6^{-1} on another chart if the two charts have a different number of letters on the 6 m line!

The Bailey–Lovie logMAR chart

There have been various attempts to improve the design of the Snellen chart. These include a chart design proposed by Ian Bailey and Jan Lovie in 1976 (**Figure 9.11**), which has become the test of choice in vision research and is beginning to appear in the high street, particularly with the introduction of electronic and computer-based test charts. The underlying principle of the Bailey–Lovie chart is the use of a logarithmic scale. A logarithmic scale is one where a change in a certain number of steps is a change in constant ratio. On the scale chosen for this chart, a change of three steps indicates a doubling of size and each step indicates an increase of 25% in letter size. The progression of letter sizes is uniform, increasing in a constant ratio of 1.26 (0.1 log unit steps) from the bottom to the top of the chart. So, working from the bottom of the chart, each line is 1.26 times larger than the preceding line. Other examples of the use of a logarithmic scale include the Richter scale for the strength of earthquakes and the pH scale of acidity.

The Bailey–Lovie chart incorporates five letters on every row. The letter spacing on each row is equal to one letter width and the row spacing is equal to the height of the letters on the lower row, i.e. each row is simply a scaled-down version of the row above. This means that the task remains the same as the patient reads down the chart and results obtained at different distances can be easily equated. The letters chosen for the Bailey–Lovie chart have almost equal legibility.

Remember that Snellen letters are designed to measure resolution acuity and that a certain MAR is required to resolve the elements of a letter, e.g. a 6/6 letter equates to an MAR of 1', 6/12 equates to 2', etc. LogMAR is simply $\log_{10}$ of the MAR. Table 9.1 shows the relationship between different acuity scales.

When the Bailey–Lovie chart is used to assess VA, the result is usually recorded in terms of a logMAR score. With this notation, 6/6 (MAR = 1') is equivalent to a logMAR of **zero** ($\log_{10}1 = 0$), whereas smaller letters have a **negative** logMAR score ($\log_{10}$ of any number <1 is negative) and larger numbers a **positive** score. As each letter size changes by 0.1 logMAR units per row, and there are five letters on each row, each individual letter can be assigned a score of 0.02. If all five letters on the 6/6 line are correctly read, the logMAR score is 0. If one letter is missed on the 0 line, the logMAR score is +0.02, two letters missed +0.04, etc., i.e. 0.02 is added for each letter **incorrectly** read. Thus, the final logMAR score takes account of every letter that has been correctly read, which avoids the confusion that can occur with Snellen notation. The Bailey–Lovie chart was designed to achieve valid, reliable and repeatable scores of VA when compared with the traditional Snellen chart. The critical feature of the chart is that the task is essentially the same at each size level.

One of the major advantages of the Bailey–Lovie chart is that VA results at different distances can easily be equated. This advantage is a result of the use of a logarithmic scale. The logMAR scale for Snellen notation is shown in Tables 9.2 and 9.3. The key to understanding this advantage is in the number of steps within the logarithmic scale. As an example, let us assume that at 2.4 m a patient can read all five of the letters on the 24 m line of a Bailey–Lovie chart. On the logMAR scale, changing from 6 m to 2.4 m is a change of four steps (Table 9.4). So,

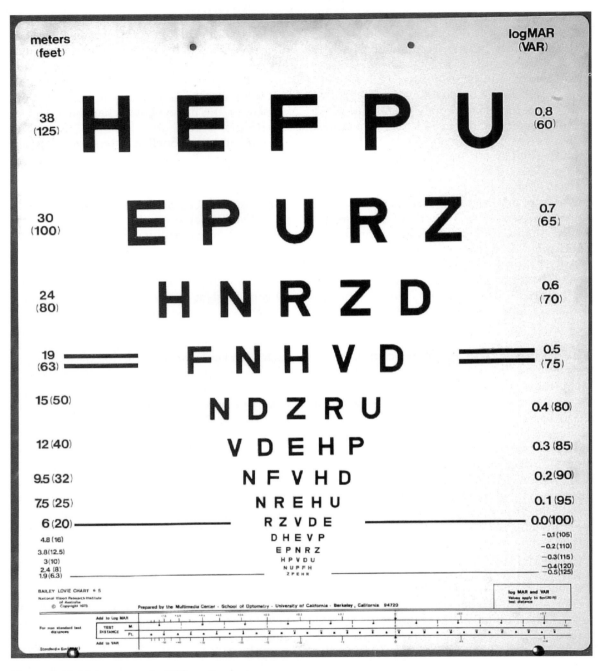

Figure 9.11 High-contrast logMAR acuity chart.

moving four steps from 24 m gives 60 m (Table 9.2). Therefore, at a testing distance of 2.4 m, 2.4/24 is equivalent to 6/60. A patient with a VA of 6/60 on a traditional Snellen chart at 6 m could be either 3/36 or 3/24 at 3 m. Determin-

ing visual acuities at shorter testing distances is even easier if logMAR notation is used. Each time the testing distance is halved, the individual should be able to read three more lines, so simply add 0.3 to the individual's logMAR score

Table 9.1 The relationship between different acuity scales

Snellen	Decimal	MAR	LogMAR
6/60	0.10	10	1.00
6/24	0.25	4	0.602
6/12	0.50	2	0.301
6/6	1.00	1	0.000
6/4	1.50	0.667	−0.176

Table 9.2 The logMAR scale for Snellen notation

Snellen	LogMAR
60	1.0
48	0.9
38	0.8
30	0.7
24	0.6
19	0.5
15	0.4
12	0.3
9.5	0.2
7.6	0.1
6.0	0.0

Table 9.3 The logMAR scale for acuities less than 6/60

Snellen	logMAR
60	1.0
75	1.1
95	1.2
120	1.3
150	1.4
190	1.5
240	1.6

Table 9.4 The logMAR scale for acuities better than 6/6

Snellen	logMAR
6.0	0.0
4.8	−0.1
3.8	−0.2
3.0	−0.3
2.4	−0.4

determined at the shorter distance. This gives the individual's true logMAR score for the chart's design distance.

The Bailey–Lovie chart therefore has several important advantages over the conventional Snellen letter chart. It should, however, be appreciated that both charts measure the same aspect of visual capability, which is the patient's ability to resolve high-contrast letters. However, as the ability to resolve small high-contrast letters does not always relate to the 'real world' low contrast (10%) Bailey–Lovie charts are available. The author routinely uses low contrast charts in the assessment of patients presenting with cataract. As an example, a patient recently presented vehemently complaining of poor vision in her left eye when compared with the right. She was rather surprised when she achieved VAs of right 6/6 and left 6/7.6 when assessed using a high-contrast letter chart. However, when assessed with a 10% low-contrast chart she achieved acuities of right 6/7.5 and left 6/24. A subsequent examination at the slitlamp biomicroscope revealed the presence of cataract and the patient was referred for left phacoemulsification and intraocular lens implantation.

When referring patients with cataract, the inclusion of low-contrast acuities provides useful additional information about the patient's visual function and can occasionally be the justification for referral. High- and low-contrast Bailey–Lovie logMAR charts are shown in Figures 9.11 and 9.12. Note the triangular arrangement of the letters that is characteristic of a logarithmic design. Other low contrast and contrast sensitivity charts are widely available, but the use of electronic computer-based test charts such as Test Chart 2000 and Test Chart 2000 Pro (from Thompson Software Solutions) make the assessment of low-contrast acuity particularly straightforward.

Visual acuity measurements using a logMAR chart have been shown to be twice as repeatable as those obtained using a Snellen chart; in addition, measurements have been shown to be over three times more sensitive to interocular differences in VA. A logarithmic progression and proportional spacing of letters allows VA levels to be monitored accurate and consistently.

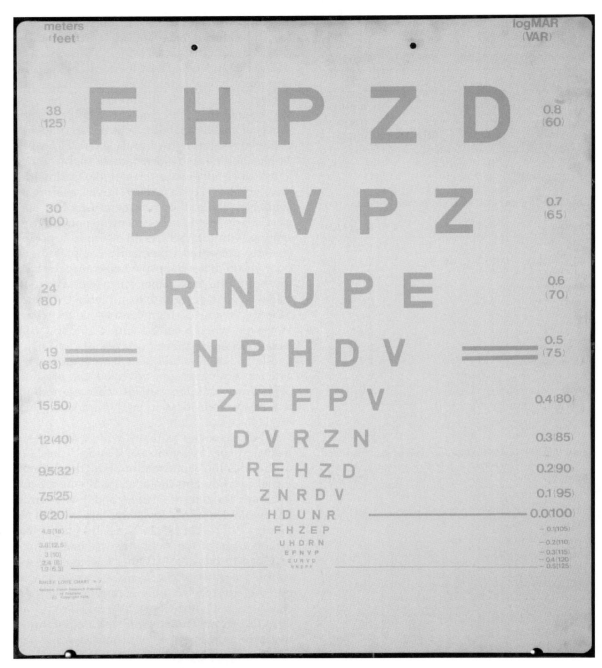

Figure 9.12 Low-contrast logMAR acuity chart.

A logarithmic chart also provides a better interpretation of the significance of recorded visual changes. This is because a three-line change in the higher acuity levels would represent the same degree of change as a three-line difference at lower acuity levels. This cannot occur with a traditional Snellen chart because the difference in acuity from 6/60 to 6/36 (two lines) is not the same as 6/6 to 6/5 (again two lines). As distance vision is often the only measurement of visual function taken in clinical practice, the author feels that visual acuity should always be measured as accurately as possible. For clinicians to monitor changes in ocular health and visual status, VA measurements should be made using appropriate test charts viewed under controlled conditions and credit should be given for every letter that is correctly read. It is interesting to note that a paper published in the *British Medical Journal* warned that doctors must be extremely cautious when assigning clinical importance to changes in acuity of two lines or less because of the inherent variability of the Snellen chart (McGraw et al 1995)!

Standards for the design of test charts

The following factors should be considered in the design of the modern Snellen letter chart:

- The letter design and relative legibility
- The progression of letter sizes
- Letter and row spacing
- Contrast between the letters and the background
- The chart luminance

Two documents published by the British Standards Institution and titled *Visual Acuity Test Types* give guidelines and recommendations about the design of test charts for the clinical determination of distance visual acuity. The original standard, BS 4274-1 1968, has now been withdrawn and replaced with BS 4274-1 2003.

Letter design

With regard to the design and selection of letters for a test chart, BS 4274-1 1968 recommends the following 5 × 4 sanserif letters:

D E F H N P R U V Z

Not all letters are of equal legibility and legibility varies according to font style. The term **relative legibility** is used when comparing and standardising letter styles and is defined as:

$$\frac{Distance}{Mean\ distance}$$

The relative legibilities of the 5 × 4 sanserif letters recommended by BS 4274-1 1968 is:

D = 0.95	E = 1.09	F = 1.04
H = 1.02	N = 0.97	P = 1.01
R = 0.91	U = 1.08	V = 0.94
Z = 1.05		

The most easily recognised letters have higher relative legibility values. The legibility of all the above falls between the BS recommended values of 0.9 and 1.1, and are chosen because the letters used should have the same level of difficulty with regard to recognition. It should be remembered that these letters are recommendations and manufacturers of test charts can (and often do) use any letters that they wish. This last comment is particularly relevant to older charts. It should be appreciated that charts can last a very long time because the only things that can go wrong are the lamp or the fuse!

BS 4274-1 2003 recommends that the following 5 × 5 sanserif letters be used:

C D E F H K N P R U V Z

and states that no letter should be repeated on any one line.

The progression of letter sizes and the spacing of lines and letters

Traditional Snellen charts (see Figure 9.9) usually display letters with the following conventional progression: 60 m, 36 m, 24 m, 18 m, 12 m, 9 m, 6 m and 5 m. BS 4274-1 1968 stated that a test chart should comprise 10 lines of letters as detailed in Table 9.5. Some charts have other letter sizes such as 30 m, 10 m and 7.5 m.

According to BS 4274-1 1968, the rows of letters should be separated by at least 20 mm and the letters on the same row should be evenly spaced, although no exact recommendations are given.

It is reassuring to note that BS 4274-1 2003 recognises the need for a test chart with a logarithmic progression. The standard states that the minimum range of letter sizes shall be from logMAR −0.1 to logMAR 1.0. The minimum graduation of letter sizes shall be in not more than 0.1 logMAR steps between logMAR −0.1 and logMAR 0.4, and in not more than

Table 9.5 Range of letter sizes and number of letters per line (BS 4274-1 1968)

Letter size	Number of letters per line	Linear height of letter
60	1	87.3
36	2	52.4
24	3	34.9
18	4	26.2
12	5	17.5
9	6	13.1
6	8	8.73
5	8	7.27
4	8	5.82
3	8	4.37

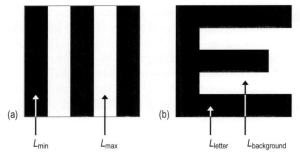

Figure 9.13 High-contrast targets and the definitions of contrast: (a) Michelson contrast $(L_{max} - L_{min})/(L_{max} + L_{min})$ and (b) Weber contrast $(L_{background} - L_{letter})/L_{background}$

Table 9.6 Range of letter sizes and number of letters per line (BS 4274-1 2003) for a 6 m testing distance

Letter size		Minimum number of letters per line	Linear height of letter (mm)
LogMAR	Snellen		
1.0	60	1	87.3
0.8	38	2	55.08
0.6	24	3	34.75
0.4	15	4	21.93
0.3	12	5	17.42
0.2	9.5	5	13.84
0.1	7.5	5	10.99
0	6	5	8.73
−0.1	4.8	5	7.03

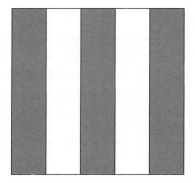

Figure 9.14 A low-contrast target.

- If additional lines are included, their size must be in accordance with the logMAR notation and the number of letters per line must correspond to the number of letters in the immediately adjacent larger line

Letter contrast

Contrast is given either by the Michelson or the Weber definitions of contrast (**Figure 9.13**). Black letters on a white background are an example of a high-contrast target whereas grey letters on a slightly lighter background describe a low-contrast target (**Figure 9.14**). The British Standards recommendation (both BS 4274-1 1968 and BS 4271-1 2003) is a minimum contrast of 0.9 (90%). This is close to the highest attainable. Snellen charts therefore measure **high contrast resolution acuity**.

0.2 logMAR steps thereafter to logMAR 1.0. These recommendations are shown in Table 9.6.

The letter size must be indicated in both the familiar Snellen denominator and the logMAR notation. According to the 2003 standard:

- The letters on each line should be evenly spaced and centrally disposed
- The gap between the letters must be equal to the width of the letters
- The space exposed at the ends of each line, as presented, must not be less than the limb width of the letters

Chart luminance

Both BS 4274-1 1968 and BS 4271-1 2003 state that internally illuminated charts should have a luminance of 120 cd m⁻². Visual acuity increases considerably with luminance up to about 120–140 cd m⁻², above which level the rate of improvement becomes insignificant. Externally illuminated charts should have a minimum luminance of 600 lux. BS 4274-1 1968 also gives recommendations about the area surrounding the test chart and recommendations about room illumination. However, these recommendations do not appear in BS 4274-1 2003.

The forewords to both BS 4274-1 1968 and BS 4274-1 2003 make interesting reading. The following is taken from the 1968 standard:

> Since their introduction by Snellen in 1862, distance test charts have become so diverse that they have ceased to provide a uniform test of visual acuity.

> The object of this British Standard is to encourage the production and use of test charts to yield comparable results in the recording of visual acuity.

> Much thought was given to another vexing question, the progression of letter sizes. Although a true geometric progression of sizes would be desirable on scientific grounds, the practical disadvantages of departing from the traditional range of sizes were considered too heavy a price to pay.

The following is taken from the 2003 standard:

> This British Standard utilises the logMAR notation (logarithm of the minimum angle of resolution) as opposed to the traditional Snellen notation. Use of the logMAR notation facilitates analysis of visual acuity scores, and changes in such scores, more effectively than other notations because equal linear steps on the logMAR scale represent equal ratios in the standard size sequence. This minimum standard, which in itself does not represent a full logMAR chart, is intended as a stepped approach to the goal of the introduction of a full logMAR chart. This British Standard takes into account the need for these charts to fit into existing hardware/test cabinets. The physical size of a full logMAR chart would exclude this possibility. Nevertheless, although a full chart exceeds the minimum requirements of this British Standard, this British Standard does not preclude the use of a full chart.

So the British Standards Institution has recognised that flaws exist in the design of the traditional Snellen chart and that logMAR charts are superior in design. The recommendations in BS 4274-1 2003 are both necessary and welcome, but the gold standard chart for the scientific and clinical measurement of visual acuity is the Bailey–Lovie chart.

The Landolt ring

This target measures resolution acuity in the exactly the same way as a Snellen letter because the outer diameter of the ring subtends 5′ at a specified distance and the width of the ring and the gap both subtend 1′ (Figure 9.15). The Landolt ring is sometimes described as a 'four-way forced choice test' because the patient has to identify the position of the gap as up, down, left or right. If the gap is not resolved the blur circles overlap and the patient sees a ring. This target is sometimes useful if communication is a problem. Landolt rings for various testing distances are available (Figure 9.16).

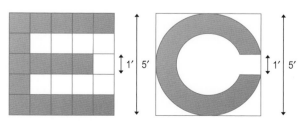

Figure 9.15 The Snellen E and the Landolt ring.

Figure 9.16 The Landolt ring in six Snellen equivalent sizes.

Other methods of assessing vision

Although beyond the scope of this text, readers need to be aware of other methods of assessing vision, particularly in young children. These include:

- Preferential looking (Keeler Acuity Cards, Cardiff Cards)
- Pictures and symbols (Kay pictures, Lea symbols, Tumbling Es)
- Tests that employ letters (Sonksen-Silver, Cambridge Acuity Cards)
- Contrast sensitivity
- Electrodiagnostic tests

Concluding points for Chapter 9

This relatively lengthy and perhaps controversial chapter has covered:

- resolution acuity
- visual acuity and vision
- Snellen acuity and the Snellen letter
- a selection of test charts used in optometric practice

References

McGraw P, Winn B, Whitaker D (1995) Reliability of the Snellen chart. *BMJ* **310**:1481-1

Further recommended reading

Bailey I L, Lovie J E (1976) New design principles for visual acuity test charts. *American Journal of Optometry* **53**:745–53

Doshi S, Harvey W (2003) *Investigative Techniques and Ocular Examination.* Butterworth Heinemann, Oxford

Elliot D (2003) *Clinical Procedures in Primary Eye Care.* Butterworth Heinemann, Oxford

Harvey W, Gilmartin B (2004) *Paediatric Optometry.* Butterworth Heinemann, Oxford

Jackson A J, Bailey I L (2004) Visual acuity. *Optometry in Practice* **5**:53–68

Rabbetts R B B (1998) *Bennett & Rabbetts' Clinical Visual Optics.* Butterworth Heinemann, Oxford

Tunnacliffe A H (1993) *Introduction to Visual Optics.* Association of the British Dispensing Opticians, London

Zadnik K (1997) *The Ocular Examination – Measurements and findings.* Saunders, Philadelphia

Subjective Refraction: Principles and Techniques for the Correction of Spherical Ametropia

Andrew Franklin

Introduction

There are two methods of evaluating the refractive error of an eye:

1. **A subjective refraction** where the result depends on the patient's ability to discern changes in clarity. This process relies on the cooperation of the patient.
2. **An objective refraction** (usually retinoscopy) where the result depends purely on the examiner's judgement to determine the optimum optical correction. Retinoscopy has been covered in detail in Chapter 8. An autorefractor (see Chapter 16) can also be used to obtain an objective refraction.

Subjective refraction

Subjective refraction consists of three distinct phases. The first is designed to correct the spherical element of the refractive error in such a way as to facilitate the accurate determination of any astigmatic element present. It should be remembered that, although astigmatism is often present, a refractive error may be entirely spherical. The second phase is the determination of the astigmatic error (see Chapter 11) and the third phase involves the balancing and/or modification of the refractive correction to ensure optimal visual performance and patient comfort (see Chapter 12).

As always, the patient's history and symptoms are important and can be used to help predict a refractive error. Remember that a symptom is a patient's complaint whereas a sign is a practitioner's observation. The symptoms of uncorrected myopia may include:

- blurred distance vision
- headaches from screwing up the eyes to try to obtain clearer vision by the pinhole effect
- clear near vision.

The signs of uncorrected myopia may include:

- poor distance vision on a letter chart
- good near vision on a near test chart.

The symptoms of uncorrected hypermetropia may include:

- eyestrain, especially for close work, caused by the accommodative effort to form a clear image
- blurred vision with medium-to-high amounts of hypermetropia and in advancing age (blurred vision is not usually a problem with low amounts of hypermetropia).

The signs of uncorrected hypermetropia may include:

- usually no signs in low hypermetropia – screwing up the eyes and wrinkling of the brow may be a sign of high amounts of uncorrected hypermetropia
- a nasalward deviation (esotropia) of one eye in high amounts of uncorrected hypermetropia.

Objective refraction

Objective refraction (retinoscopy) is often used to determine the initial spherical element of refraction. However, when a patient has a recent correction that is providing good acuity, say 6/9 or better, this can be taken as the starting point.

The purpose of the first phase of a subjective refraction is to determine the best vision sphere (BVS). This can be defined as the most positive (or least negative) spherical lens that provides best visual acuity. During a subjective refraction accommodation must be not be allowed to fluctuate randomly. The eye should be as relaxed as possible so that changes in the accommodative state do not influence the end-result. As the accuracy of any subjective test or routine relies on the individual patient's ability to discriminate and communicate accurately, the potential for error must be kept to a minimum. The ability to discriminate and communicate will of course vary widely from person to person but, in general, the simpler the task, the more likely it will be performed well. To determine the BVS a suitable letter chart (logMAR or Snellen), trial frame and trial case of lenses (or a refractor head) are required.

Chapter content

- Determination of the best vision sphere
- The use of the pinhole disc
- The duochrome test
- Patients with poor visual acuity (VA)
- The Scheiner disc

Determination of the best vision sphere

The following discussion takes place in two parts: the first assumes that retinoscopy has not been performed whereas, in the second, the determination of the BVS follows retinoscopy. If a good retinoscopy has been performed, the technique of finding the BVS in isolation should, in theory, be redundant. However, it can prove useful when retinoscopy is difficult (small pupils or media opacities). It is usual to start with the right eye, the left being occluded (**Figure 10.1**). This is called a *monocular refraction*. The procedure is repeated on the left eye with the right occluded. However, it is possible and often preferable to refract under binocular conditions. Binocular refraction is discussed in Chapter 12.

In both binocular and monocular refraction, it is important to control accommodation in the person with pre-presbyopia, so a 'fogging' technique is employed whereby the spherical element is deliberately over-plussed and then reduced to find the final spherical power. Traditionally, the right eye is usually refracted first, because it is the nearest one to the practitioner in most consulting rooms. However, when the left eye has significantly worse acuity as a result of amblyopia or pathology, or if the right eye is markedly dominant, the left eye should be refracted first.

Figure 10.1 A trial frame shown with the left eye occluded.

When using the best sphere technique without the aid of retinoscopy, the practitioner must find the maximum amount of positive power or the minimum amount of negative power that can be tolerated by the eye, without causing blurring of the retinal image. After occlusion, the first task is to measure the unaided vision. This is useful because unaided vision can give a reasonable estimate of the magnitude of any uncorrected myopia or manifest hypermetropia (Table 10.1). It is important to note that the estimates in Table 10.1 are of no use if the patient is accommodating to 'correct' any hypermetropia.

The questioning technique is very important throughout subjective refraction because the use of appropriately phrased questions can make the difference between a quick, precise refraction and a long-winded, potentially inaccurate refraction.

The patient's attention is directed towards the letter chart. Whenever a positive lens is held before the eye, the question to the patient should take the following format:

Table 10.1 Expected vision for any uncorrected mean sphere (myopia or manifest hypermetropia)

Vision	Equivalent sphere (myopia or manifest hypermetropia)
6/5	Plano
6/6	0.25–0.50 DS
6/9	0.50–0.75 DS
6/12	0.75–1.00 DS
6/18	1.00–1.25 DS
6/24	1.25–1.75 DS
6/36	1.75–2.25 DS

Is the target better with or without this lens, or is there no real difference?

Positive lenses either blur the retinal image, indicating that the maximum amount of positive power is already in place, or relax accommodation (where it is in use). Therefore an answer of 'no difference' indicates the need to add positive power to the trial frame combination of lenses until no more can be tolerated. The initial plus lens may be in the region of +1.00 DS. Later in the procedure, a +0.50 DS may be used. When the total power of the trial lenses in the trial frame is close to the endpoint; the practitioner should add spherical lens power in ±0.25 DS steps. The lens must be held in the plane of the trial frame and along the visual axis to avoid inducing off-axis aberrations. It should also be moved quickly and precisely, allowing enough time in each position for the patient to make a decision. Some patients require longer than others. Practitioners usually need to repeat this process a number of times to confirm the result and continue to add positive power until the addition of an extra +0.25 DS results in blurring. This final +0.25 DS is discarded and the remaining lens is the **best vision sphere**.

When negative lenses are required care must be taken not to over-minus the patient, which results in a stimulation of the patient's accommodation. The question required when adding negative power should be altered to:

Is the target clearer or just darker with this lens?

Should the target appear darker but not clearer, extra minus power should not be added because this just stimulates accommodation. Also, extra minus power should not be added if the target appears smaller but not clearer. A negative lens should be added to the trial frame only if the patient can resolve a greater number of letters on the letter chart. The results are often rechecked and confirmed throughout the test using the same or a different technique, e.g. best sphere and duochrome because patient's answers are frequently inconsistent! When the BVS has been reached the point focus (in the case of spherical ametropia) or disc of least confusion (in the case of astigmatic ametropia) should be on or very close to the retina. The distance vision with this correction should be measured and recorded because it is useful for estimating the magnitude of any uncorrected astigmatic error.

The use of ±0.25 DS 'twirls' (see below) helps in fine-tuning the BVS. Remember that the endpoint is the maximum plus or minimum minus that the patient will tolerate without causing blurring of the retinal image.

Summary of the procedure to find the BVS without the use of retinoscopy

1. Occlude the left eye.
2. Measure the unaided vision (V).
3. If possible, estimate the ametropia. This is particularly helpful in the case of uncorrected myopia. Also in myopia, the position of the true far point can be used to estimate the refractive error (see Table 8.6, Chapter 8), e.g. a person with −8.00 D myopia sees clearly if a target is placed at approximately 12.5 cm from the eye.
4. Add a +1.00 D sphere.
5. Is the vision worse?
6. No: add more plus spherical power until the vision blurs. From the blur point, reduce by +0.25 DS. The BVS should be the maximum plus that the eye can tolerate without causing blur on a letter chart.
7. Yes: add minus spherical power until the best line can be resolved. Make sure that each addition actually increases VA and does not just make the letters smaller.

8. If possible, adjust the final sphere on the letter chart and/or the duochrome (see later) using a ±0.25 DS twirl.
9. Record the VA.
10. Occlude the right eye and repeat the procedure for the left eye.

After retinoscopy, the procedure is as follows:

1. If the working distance lenses are +1.50 DS, the over-correction provided should blur the acuity back to 6/24 in a young patient. However, in a patient with small pupils the acuity may be rather better. An occluder is placed before the left eye.
2. The acuity of the right eye may now be checked with the left eye occluded and the right working distance lens still in place. In a young patient this should be around 6/24 and, if it is better than this, the retinoscopy result may be under-plussed. In older patients the effect of the over-correction may be less, but it is also less likely that retinoscopy under-plussed the correction if accommodation is inactive.
3. If the VA with the working distance lens in place is around 6/24, remove the working distance lens and refine the BVS. If the acuity with the working lens in place is better than 6/24, the retinoscopy result may be under-plussed and a smaller amount of positive power should be removed.

Refinement of the BVS after retinoscopy

After retinoscopy the spherical correction may be refined by use of the duochrome (see later) or by using plus and minus spheres of equal but opposite power. The two methods give statistically similar results (Jennings and Charman 1973), although this does not mean that they necessarily agree on every patient. It may also be determined by a 'coincidence' method employing the Scheiner disk (see later). This method has been commonly employed in optometers and some autorefractors. It is also employed as the focusing mechanism in the one-position keratometer (see Chapter 18).

Simultan technique (using plus and minus Freeman twirls)

This technique relies on the sequential presentation of plus and minus spheres, which are usually mounted together in a 'twirl' with a handle, although individual trial lenses may be used. The lenses presented are normally ±0.25 D (Figure 10.2). However, if the VA of the eye after retinoscopy is less than 6/9, it is unlikely that the patient is able to differentiate reliably between these low-powered lenses, so ±0.50, 0.75 or 1.00 D twirls may be required. Using this technique, the plus lens must be presented first for at least 1 second to relax accommodation. The minus lens should not be held for more than 1 second, which is the reaction time plus response time for accommodation. If this time is exceeded, it is likely that the patient will accommodate.

The patient should be asked: 'Are the letters clearer with the first lens or the second lens or are they both the same?' It is useful to split the two halves of this question to avoid asking a multiple question. The initial comparison should be between more plus and more minus. The third option should be offered only if the patient could not differentiate between the first two. If the first lens is clearer or they are both the same, +0.25 DS is added to the trial frame. Additional +0.25 DS lenses are added until the VA first blurs. The end-point is the most plus or least minus that does not blur the VA. If the second lens is clearer (as opposed to

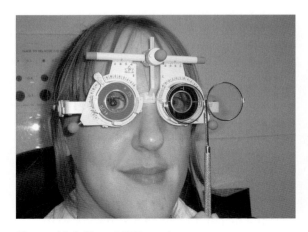

Figure 10.2 The ±0.25 D twirl.

just smaller and darker), −0.25 DS is added. If acuity improves, add further minus lenses in 0.25 D steps only for as long as the acuity continues to improve. Ask: 'Do the letters actually appear clearer or just smaller and blacker?' If the letters look smaller and blacker but not clearer, do not add the −0.25 DS. Also, if the patient reports no change or a drop in acuity, do not add the −0.25 DS lens. The end-point is the most plus or least minus that does not blur the VA.

The rapidity with which the minus lens must be withdrawn can cause problems when a patient is slow to react. For this reason, many practitioners have modified the Simultan technique to eliminate the minus lens completely.

Adding plus only

After initially determining that the sphere is a little (not more than 0.50 DS) under-plussed by the duochrome or Simultan techniques, +0.25 DS is introduced and one of the following questions asked:

> Are the letters clearer with the lens, without it or just the same?
> Are the letters the same with the lens or worse?

The first variant has the disadvantage of being a compound question. The second may confuse because the +0.25 DS is often clearer. It is necessary to pick the question to suit the patient, and it is sometimes necessary to change the question once the practitioner has got used to the patient. If in doubt, try both variations in succession. If the patient finds the vision clearer or identical with the plus lens, +0.25 DS is added to the sphere in the trial frame and the sequence is repeated by presenting the +0.25 DS lens once more. If the patient rejects the plus lens, −0.25 DS is added to the sphere in the trial frame and again the sequence is repeated. With this method, accommodation may be induced when minus power is added to the sphere in the trial frame, but we are always adding plus, and therefore relaxing accommodation, immediately before the comparison is made. Both this technique and the unmodified Simultan method are repeated until the patient accepts *no more plus without losing clarity*. At this point the inves-

tigation of the astigmatic element of the refractive error can proceed.

The use of the pinhole disc

Where there is uncorrected ametropia, a distance point source of light produces a blurred image on the retina composed of a series of blurred discs. The dimensions of a single blurred disc depend on the degree of ametropia present, the diameter of the individual's pupil and the distance of the point source from the eye. A pinhole may be employed to reduce the diameter of these burred discs and thus improve the VA. The pinhole disc (discussed in Chapter 5) is an opaque disc with a central circular aperture of about 1 mm in diameter. A pinhole with an aperture smaller than 1 mm would cause diffraction effects and also a reduction in retinal illumination. This would result in a dim, unfocused image. An aperture larger than 2 mm approaches the size of some human pupils and so might not significantly reduce the blur circle produced by an uncorrected refractive error. A diameter of 1.32 mm was recommended by Lebesohn (1950) as optimally balancing the opposing demands of reduction of the blur circle and diffraction, although the aperture most commonly seen is rather smaller (1 mm).

If the pinhole is placed before an uncorrected ametropic eye, the VA should increase. Normally correction of the refractive error should improve the VA by at least as much as that produced by the pinhole. The pinhole disc can therefore be used to estimate the maximum VA that the eye would obtain if the refractive error were to be corrected. If acuity does not improve through the pinhole, it is unlikely that reduced acuity is caused by an uncorrected refractive error and pathology is suspected, e.g. the VA in amblyopia, macular disease and central media opacities is not improved by using the pinhole disc; in fact the pinhole disc may actually reduce acuity in such cases. However, if the patient has an irregular cornea or peripheral media opacities, the pinhole may give a better result than can be achieved by refraction. If the pinhole fails to improve VA, the reason for the reduced acuity is unlikely to be purely refractive. In practice, the pinhole disc test can prove very useful, especially if subjective techniques are unsuccessful and VA does not improve with the addition of lenses.

The duochrome test

The eye, in common with most optical systems, displays a certain amount of axial chromatic aberration (ACA). The refractive indices of the various optical components of the eye vary with the wavelength of the incident light, with light of longer wavelength (i.e. towards the red end of the spectrum) resulting in longer focal lengths than shorter wavelength light. The total amount of ocular chromatic aberration present has been estimated as approximately 2.50 D (Bedford and Wyrszecki 1957). With white light, this should cause some defocus, although placing an achromatic doublet before the eye does not appear to improve VA significantly. Chromatic aberration appears to be slightly reduced (about 0.30 D) with smaller pupils and rather more (about 1.00 D) with accommodation (Jenkins 1962), although the reasons for this are not entirely clear. It was for many years assumed that the eye would focus light from the middle of the spectrum on the retina in order to attain maximum VA. Rabbetts (1998) estimated that yellow light with a wavelength of 570 nm was preferred by the eye. If this wavelength is used as a reference point, as it often is for tungsten light, green light with a wavelength of 535 nm focuses 0.25 D in front of the retina and red light with a wavelength 620 nm focuses 0.25 D behind it. So, by using appropriate filters a test may be constructed that, by comparing the clarity of targets presented on red and green backgrounds, allows the practitioner to focus the yellow reference wavelength accurately on the retina and achieve maximum acuity. Such a test is known as the **duochrome test** (Figure 10.3). The exact filters used with the duochrome test are specified in BS 3668:1963.

Unfortunately, things are not quite that simple. Ivanof (1949) reported that the wavelength focused on the retina appeared to change with target distance although the amount of ACA remained fairly constant. Millodot and

Figure 10.3 The duochrome test (arrowed).

Sivak (1973) confirmed these findings with and without cycloplegia, and suggested a conditioned response designed to lessen accommodative demand. Red light is focused preferentially when the target is remote, but the preferred wavelength shifts progressively towards the blue end of the spectrum as the target distance becomes shorter. This could have implications for the practitioner in the determination of the spherical refractive error. If, as is customary, the power of the correcting lens is adjusted to give equal clarity of red and green targets (equalisation), it may result in a slight under-plussing or over-minusing for a distant target. However, most distance test charts are situated at 6 or 3 m rather than true infinity, so a little under-plussing probably works out quite well. For modern collimated charts, however, which do present the target at true infinity, there may be a slight tendency to under-plus if the duochrome is equalised.

When using the duochrome test, the two colours should appear equally bright. For the colours used, the eye's dioptric power differs by about 0.50 D. It is important to note that the coloured filters are simply used to present the targets (black circles, dots or letters) to the patient and the judgement is based on defocus blur *not* brightness. Patients can get confused over this, so it is important that the practitioner carefully phrases the instructions to the patient. If the black targets appear equally clear to the patient, the red and green foci are dioptrically equal either side of the retina. This is a result of the 0.50 D interval between the eye's powers for the two colours (red and green) and the black targets are equally clear (or blurred) to the patient. From the position of equality (**Figures 10.4 and 10.5**), if a +0.25 D sphere were placed before the eye, both the

red and green foci would move to the left by 0.25 D. As a result of the 0.50 D interval, the red focus is on the retina and the patient reports that the targets on the red background appear clearer and those on the green appear blurred. Adding a +0.25 DS therefore moves the red focus on to the retina and the black targets will appear clearer on the red (**Figures 10.6 and 10.7**).

Again from the position of equality (**Figures 10.4 and 10.5**), if a −0.25 D sphere is placed before the eye, both the red and green foci move to the right by 0.25 D. As a result of the 0.50 D interval, the green focus is now on the retina and the patient now reports that the targets on the green background appear clearer and those on the red appear blurred. Addition

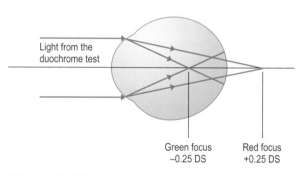

Figure 10.4 The black targets will be equally clear (or blurred) to the individual.

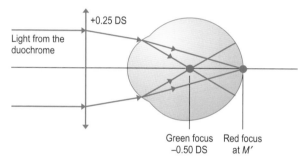

Figure 10.6 Adding a +0.25 DS to bring the red focus onto the retina. The black targets will be clearer on the red.

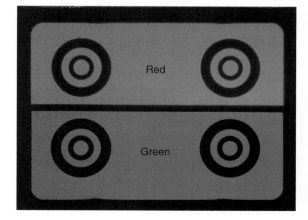

Figure 10.5 The circles on the red and the green backgrounds appear equally clear.

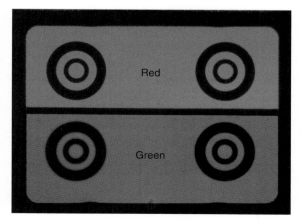

Figure 10.7 The circles on the red appear clearer than the circles on the green.

of a −0.25 DS therefore moves the green focus onto the retina and black targets are clearer on the green (Figures 10.8 and 10.9). As with all subjective techniques, the questioning is critical. The patient should be asked whether the circles look 'darker and clearer on the red background or the green background, or both about the same'.

When using the duochrome, there may not be an exact endpoint because the chances of the spectacle refraction falling into a 0.25 DS endpoint are fairly remote. If it is not possible to balance the red and the green equally, the duochrome is usually left 'on the red' to avoid over-minusing, as long as this does not have a detrimental effect on the VA. When refining the near addition, it is common to leave the patient 'on the green'. Another situation where a bias

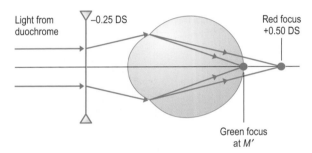

Figure 10.8 Adding a −0.25 DS to bring the green focus onto the retina. The black targets will be clearer on the green.

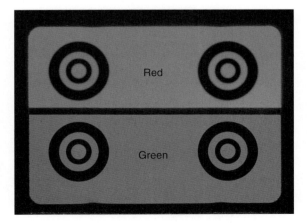

Figure 10.9 The circles on the green appear clearer than the circles on the red.

towards the green may be useful is a patient who needs spectacles for night driving because there is a tendency for the eye to become more myopic under conditions of low illumination. If the duochrome test is used before checking for astigmatism, the end-point of the test depends on the chosen method of astigmatic refinement. For the cross-cylinder technique, the targets on the duochrome should be balanced or marginally 'on the green', so that the patient accommodates to place the disc of least confusion on the retina. For the fan and block method, it is usual to leave the patient 'on the red'.

The duochrome test was first described by Brown in 1927. It was reintroduced by Freeman (1955) after an extended lapse into disuse. The modern duochrome test used in the consulting room uses circles, letters or numbers on red and green backgrounds. Polasky (1991) noted that darkening the room dilates the pupil and therefore slightly increases the chromatic aberration of the eye. It also reduces veiling glare somewhat, although few practitioners routinely dim the lights when employing the duochrome test.

There are a number of other factors that may influence the end-point obtained. The filters are chosen such that their focal points are 0.50 D equidistant from the focal point of the 'reference wavelength', which for tungsten light is 570 nm. However, not all duochrome tests use the same peak wavelengths, so the dioptric interval may vary a little from test to test (Mandell and Allen 1960), and the relative brightness of the red and green panels may also affect the end-point. Ageing or inappropriate light sources, dirty or fading filters and the reflectance of the chart surface may all have some slight affect on the end-result. With increasing age, a yellowish pigment accumulates in the crystalline lens. The lens becomes optically denser to shorter wavelengths so that the red background always appears brighter. This means that the duochrome test may over-minus (or under-plus) the elderly patient. In addition, the chromatic aberration of the eye reduces after age 55 and the pupil tends to become smaller. These effects lead to a reduction in the distance between the red and green foci and therefore a reduction in the difference between the blur of the targets on the two colours of the duochrome. The duochrome is unreliable if the defocus blur is more than 1.00 D from the patient's correct spectacle refraction as the targets on the red and green are so blurred as to make a comparison meaningless. The duochrome is of no real value until the ametropia has been corrected to the point where the acuity is at least 6/12.

At any age, if the targets on the green background are clearer, the result is unambiguous. There is uncorrected hypermetropia and more plus power needs to be added. If the targets on the red panel are clearer the situation is more complicated. Either there is uncorrected myopia, or the patient has yellowing of the crystalline lens or is accommodating. Therefore, if the patient is favouring the red background rather more than expected, it is worth checking the result with an alternative method. Fletcher (1991) also recommends that, in young patients, once equality has been achieved between red and green, an extra +0.50 DS be placed before the eye for a few seconds. This is then removed, and the patient asked on which colour the targets are clearer. It is not unusual to find a little extra plus when this is done.

Given the number of factors that may influence the result, it is perhaps surprising that the duochrome test is a popular as it is. However, practitioners tend to adapt their methods and prescribing preferences to the consulting room in which they are working. As with many clinical methods, accurate and consistent results are obtainable provided that the practitioner knows when to adapt. However, working in an unfamiliar environment may produce the odd interesting result.

The duochrome test: summary

- +0.25/−0.25 spheres are used, usually mounted on a handle
- Targets equal: add +0.25 DS – targets on red clearer
- Targets equal: add −0.25 DS – targets on green clearer

Technical information

- The duochrome is illuminated by tungsten lamps
- The filters used are Courtoid Red 15 and Green 16
- The filters are 0.25 mm thick
- The maximum transmittances are red 15 at 620 nm 15.9% LTF and green 16 at 535 nm 18.6% LTF

Patients with poor VA

As patients with poor VA may have difficulty in determining small changes in lens power and clarity, it is often necessary to make large power changes. A bracketing technique can be employed, e.g. use a +2.00 DS trial lens and compare this with a −2.00 DS trial lens. If the patient is able to differentiate between the two, the lens giving the better vision can be added and the technique repeated using, say, a +1.00 DS and a −1.00 DS lens, etc. It may be necessary to refract the patient with the chart placed at 3 m or less. If this is done the result is over-plussed (the chart at 3 m acts as a near object and therefore adds a vergence of about −0.30 D in the plane of the trial lens) and a correction should be made. The use of a pinhole disc is essential when vision is poor, in order to attempt to differentiate a refractive error from active pathology.

The Scheiner disc

The earliest instrument employed to assess refractive error was based on a principle described by Christopher Scheiner in 1619 (Bennett 1998). This trial case accessory is an opaque disc (the Scheiner disc) with two small circular apertures each 0.75 mm in diameter, 2–3 mm apart and equidistant from the centre in opposite directions along a common meridian (Figure 10.10). These dimensions allow light through both holes to enter the eye's pupil. The Scheiner disc is effectively a subjective optometer that can be used to detect and measure spherical ametropia. It is a somewhat outdated technique that is rarely used in optometric practice. However, the Scheiner disc can be used to demonstrate the existence of accommodation and is also used as the focusing mechanism in the one-position keratometer and

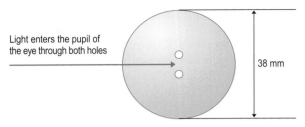

Figure 10.10 The Scheiner disc.

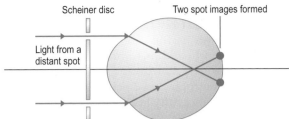

Figure 10.12 Scheiner disc: myopia.

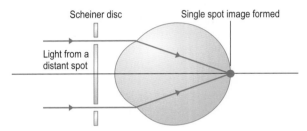

Figure 10.11 Scheiner disc: emmetropia.

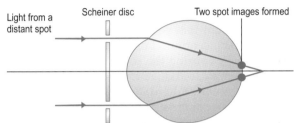

Figure 10.13 Scheiner disc: hypermetropia.

some autorefractors. To detect and measure spherical ametropia, the Scheiner disc is used with a spotlight at 6 m and is placed before one eye only, the other eye being occluded. If a point source of light at 6 m is viewed using the disc, the image is formed through two different portions of the pupil.

If the viewing eye is emmetropic, the two images thus formed are coincident and focused. The individual sees a single spot of light (**Figure 10.11**). Simple myopia causes the target to be imaged in crossed diplopia (**Figure 10.12**), whereas hypermetropia without accommodation produces uncrossed diplopia (**Figure 10.13**). The separation of the two retinal images depends upon the degree of ametropia present and the type of ametropia may be found by covering one of the two pinholes. In Figure 10.12, if the *upper* pinhole is occluded, the *lower* retinal image disappears in a myopic eye. As a result of retinal inversion, this is perceived by the patient as the disappearance of the *upper* image, whereas someone with hypermetropia reports the absence of the *lower* one. **Figure 10.14** compares image formation in hypermetropia, emmetropia and myopia. Scheiner also

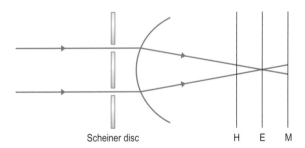

Figure 10.14 Image formation with the Scheiner disc in hypermetropia (H), emmetropia (E) and myopia (M).

described a version of the disc with three pinholes distributed in an equilateral triangle. The type of refractive error could then be determined by whether the observer saw an erect or inverted triangle of images. Spherical lenses may be interposed between the target and the eye to bring the diplopic images to coincidence and thereby determine the ametropia of the eye along the selected meridian. The Scheiner disc is rarely employed in subjective refraction in the consulting room, although it may be useful with those patients who are unable to cope with the more commonly employed methods.

Problems encountered when using the Scheiner disc include the following:

- The disc is difficult to centre correctly.
- The central part of the eye's optical system is not used during the test.
- The measurement of astigmatism is difficult.
- The patient's accommodation may not be relaxed during the test.

Concluding points for Chapter 10

Refraction consists of three phases, the first of which is the correction of the spherical element of the refractive error. There are a number of methods available to achieve this and each has inherent advantages and disadvantages. Experience helps the practitioner to select the best method for a given patient. It also informs the practitioner of those intuitive corrections that need to be made to allow for variations when using particular methods in a particular consulting room.

References

Bedford R E Wyrszecki G (1957) Axial chromatic aberration of the human eye. *Journal of the Optic Society of America* **47**:464–565

Bennett A G (1986) An historical review of optometric principles and techniques. *Ophthalmic and Physiological Optics* **6**:3–21

Brown (1927) Cited in Borish IM, Benjamin WJ (1998) *Clinical Refraction*. WB Saunders, Philadelphia

Fletcher R (1991) Subjective techniques. In: Allen RJ, Fletcher R, Still DC (eds), *Eye Examination and Refraction*. Blackwell Scientific Publications, Oxford: 93

Freeman H (1955) Working method – subjective refraction. *British Journal of Physiological Optics* **12**:20–30

Ivanof A (1949) Focusing wavelength for white light. *Journal of the Optic Society of America* **39**:718

Jenkins T C A (1962) Aberrations of the eye and their effects on vision: Part I. *British Journal of Physiological Optics* **20**:59–91

Jennings J A M, Charman W N (1973) A comparison of errors in some methods of subjective refraction. *Ophthalmic Optics* **13**:8

Lebensohn J E (1950) The pinhole test. *American Journal of Ophthalmology* **33**:1612–14

Mandell R B, Allen M J (1960) The causes of bichrome test failure. *Journal of the American Optometric Association* **31**:531

Millodot M, Sivak J (1973) Influence of accommodation on the chromatic aberration of the eye. *British Journal of Physiological Optics* **28**:169–74

Polasky M (1991) Monocular subjective refraction. In: Eskridge J B Amos J B, Bartlett J D (eds) *Clinical Procedures in Optometry*. JB Lippincott, Philadelphia: 174–88

Further recommended reading

Elliott D B (2003) *Clinical Procedures in Primary Eye Care*. Butterworth-Heinemann, Oxford

Michaels D M (1985) Subjective methods of refraction. In: Michaels D M (ed) *Visual Optics and Refraction*, 3rd edn. CV Mosby, St Louis, MO

Rabbetts R B (1998) *Bennett & Rabbetts' Clinical Visual Optics*. Butterworth Heinemann, Oxford

Tunnacliffe A H (1993) *Introduction to Visual Optics*. Association of the British Dispensing Opticians, London

Subjective Refraction: Principles and Techniques for the Correction of Astigmatic Ametropia

Andrew Franklin

Introduction

It is usual for a practitioner to attempt to find an astigmatic component of a patient's refractive error even if the visual acuity (VA) with the best vision sphere (BVS) in place is 6/6 or better. This is because most patients have some degree of astigmatism and only one in three displays no significant astigmatic error. About a third display astigmatism of between 0.25 DC and 0.50 DC, and a further third have astigmatism greater than this. The incidence of astigmatism is shown in Table 11.1.

The astigmatic component may be determined by two main methods. The most popular, the crossed-cylinder technique, allows cylindrical corrections found by objective means (retinoscopy) to be adjusted simply and rapidly. The alternative, the fan-and-block technique, is thought to be a little less accurate for small cylinders, although it probably controls accommodation better as a result of the use of a 'fogging' method, and may be better for larger degrees of astigmatism. It is also a useful back-up technique for those patients who do not respond well to the crossed-cylinder method. Although the crossed-cylinder is undoubtedly the most popular method, the fan-and-block method is sometimes very useful in certain cases. Practitioners therefore need to be familiar with both methods.

The usual starting point for either method is the spherocylindrical result of an objective refraction, refined to give the BVS. However, where retinoscopy is difficult and high astigmatism is present, the **stenopaic slit** (Figure 11.1) may be employed to give a starting point. This trial case accessory is an elongated pinhole that is used to find an approximate correction of astigmatism in cases where retinoscopy does not give an accurate result and high astigmatism is suspected. It is placed before the eye being tested with the BVS in place. The slit is rotated slowly and the position that gives the patient the best acuity is noted. The orientation of the slit approximates to one of the principal meridians of the eye. With the slit still in place, plus and minus spheres are added to give a BVS for this meridian. The slit is then rotated through 90° and the BVS for the second meridian found. If the astigmatism present is irregular, the second principal meridian may not be at 90° to the first. However, if spectacle correction is envisaged, it is not possible to prescribe lenses that reflect this. The powers found using the stenopaic slit are then converted to spherocylindrical form and the result placed in the trial frame, where it can be refined using the subjective techniques described below, if appropriate. It must be remembered that the power of a cylinder is always at right angles to its axis.

Chapter content

The crossed-cylinder technique

The Jackson crossed-cylinder is a lens that has a positive cylinder worked on one surface and a numerically equal negative cylinder on the other. The axes of the two cylinders are at right

Figure 11.1 A stenopaic slit: this is found in most trial cases.

Table 11.1 The incidence of astigmatism

Ocular astigmatism (DC)	Percentage of total population
0.00	32.0
0.25–0.50	34.6
0.75–1.00	17.7
1.25–2.00	9.8
2.25–3.00	3.8
3.25–4.00	1.5
>4.00	0.6

From Rabbetts (1989).

angles to each other. Thus the actual power of a ±0.25 crossed-cylinder is equivalent to a spherocylindrical lens of power +0.25 DS/ −0.50 DC. In general, the axes are marked with + and − signs and, historically, the plus axis is marked in red and the minus axis in white. However, some manufacturers now use opposite colours, so it is always worth checking before using a new set of 'crossed-cyls' to ascertain how the two meridians are marked.

In an eye with uncorrected astigmatism, an astigmatic pencil is formed inside the eye and, as described in Chapter 7, a line image, an out-of-focus disc or an out-of-focus ellipse is formed on the retina in the uncorrected astigmatic eye. The starting point for the crossed-cylinder technique is to place the out-of-focus disc, known as the **disc of least confusion** (DLC), on the retina. If the human visual system had no depth of focus, we could use the BVS without modification and proceed to investigate the cylinder. However, there is a measurable depth of focus even in young patients with large diameter pupils. On older patients with smaller pupils, and on those with severe astigmatism, whose principal foci are widely separated, this depth of focus is larger. Theoretically, the best way to ensure that the DLC is placed precisely on the retina is to allow patients to put it there themselves with the aid of their own accommodation, assuming of course that they have any. There is actually no real scientific proof that patients do this.

The total amount of blur is higher at the DLC than at other points, so there may be little incentive to focus it on the retina, although proceeding on this assumption seems to work in practice. The aim is, therefore, to allow the patient to accommodate *minimally* by adding slightly less plus or slightly more minus to the BVS. For patients who are likely to have a small depth of focus (young patients with low degrees of astigmatism) modifying the BVS by −0.25 DS is appropriate. For patients with higher degrees of astigmatism or larger depths of focus resulting from a small diameter pupil, a larger initial modification may be required. However, it should be borne in mind that a patient with late presbyopia has no accommodation with which to place the DLC on the retina. Some practitioners use the duochrome to assist with modification of the BVS. If the DLC is formed on the retina, the patient should report that the targets on the red and green backgrounds should be equally clear (or blurred). With the addition of −0.25 DS, preference should be given to the target on a green background. This is the situation preferred by most practitioners before starting the crossed-cylinder routine. The crossed-cylinder technique does need the DLC to be on the retina at all times and the positioning of the DLC can be checked using the duochrome at any stage of the crossed-cylinder routine once the patient can see the targets on the duochrome.

The usual starting point would be the BVS determined by the duochrome or Simultan

Table 11.2 Expected vision for any uncorrected cylinder

Best sphere VA	Astigmatic error with the best vision sphere in place
6/5	0.25 DC
6/6	0.50–0.75 DC
6/9	1.00–1.25 DC
6/12	1.50–1.75 DC
6/18	2.00–2.25 DC
6/24	2.50–3.00 DC
6/36	3.25–4.00 DC

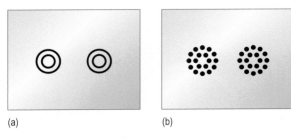

(a)　　　　　　　　(b)

Figure 11.2 Targets for use with the crossed-cylinder technique: (a) double black rings; (b) series of black dots.

techniques discussed in Chapter 10. In general, at this stage, the trial frame probably also holds a cylindrical lens close to the right correction (the 'working cylinder'), which has been determined by objective means (retinoscopy). However, where the astigmatic error appears small on retinoscopy, it may save time to leave out the cylinder entirely and check initially with a crossed-cylinder of a power roughly equal to the estimated astigmatic error over a purely spherical correction. The complete (or residual) astigmatic error may be estimated from Table 11.2.

The crossed-cylinder technique requires the patient to make decisions about the 'roundness' of a target. This could be a pair of rings or a series of dots. Figure 11.2 shows the common targets that are used with the crossed-cylinder technique. Generally, the target should be circular and a little bigger than the smallest letters that can be seen, because targets containing linear elements may prejudice the result if the circle of least confusion is not quite on the retina (O'Leary 1988). Most practitioners use the double black rings (**Figure 11.2**) on a white background as targets. The circles subtend about 6/12 and 6/5 so most patients can resolve the larger circles with their best sphere acuity. Sometimes the letter 'O' from a letter chart is used if the patient has BVS acuity poorer than 6/12. This is particularly useful when refracting a low-vision patient. The letter size must be just better than the patient's best sphere acuity for each eye. Some practitioners have advocated using several letters at a time, but this can

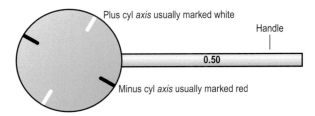

Figure 11.3 The crossed-cylinder.

confuse the patient (not to mention the practitioner) when some letters are better with position one, and some with position two. Circles or Landolt Cs of the appropriate size, or targets consisting of a pattern of dots, are preferable to single letters.

Remember that the crossed-cylinder is a sph–cyl lens mounted in a holder with a handle. The lens is effectively a pair of crossed plano cylindrical lenses of numerically equal powers but opposite sign, set with their axes at plus and –45° to the handle. The crossed-cylinder is illustrated in Figures 11.3 and 11.4. The number written on the handle indicates the astigmatic interval between the two meridians. A crossed-cylinder marked 1.00 D has meridians of power +0.50 D and –0.50 D, and the lens markings indicate the *axes* and *not* the power meridians. These axes are at ±45 degrees to the handle. Crossed-cylinders are available in ±0.25, 0.50, 0.75 and 1.00 D powers, which result in astigmatic intervals of twice these amounts. During a subjective refraction, the crossed-cylinder lens chosen should have an astigmatic interval equal to or less than the value of the estimated cylinder.

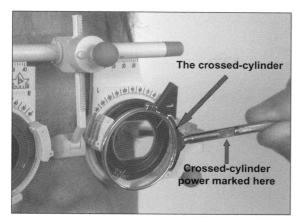

The crossed-cylinder

Crossed-cylinder power marked here

Figure 11.4 The crossed-cylinder in use. Here the minus axis is marked in white. It is therefore important to inspect the lens carefully before use.

The crossed-cylinder method follows the determination of the BVS and is divided into two parts:

1. To find and refine the cylinder axis
2. To refine the cylinder power

Determination of the axis

With the BVS only in the trial frame

The crossed-cylinder is presented to the trial frame so the handle lies at 45 degrees and the principal meridians lie along 90 and 180. We will assume that the minus axis lies along 180, and this is referred to as 'position 1'. The crossed-cylinder is then twirled or spun as quickly as possible so that the minus axis lies along 90 – 'position 2'. The patient is asked to select which of the two positions gives the clearest and roundest image. The crossed-cylinder is then presented with the handle along the 180 meridian and the principal meridians along 45 and 135. We assume that the minus axis first lies along 135 ('position 3'). The crossed-cylinder is twirled so that the minus axis now lies along 45 (position 4). Again, the patient is asked to select which of the two positions gives the clearest and roundest image. Let us assume that the patient chooses 'position 1' and 'position 3'. If this patient is astigmatic, the axis of the correcting *negative* cylinder lies somewhere between 180 and 135. With the BVS in place, the astigmatic error can be estimated

using Table 11.1. We assume that this is –1.00 DC. A –1.00 DC trial lens is now placed in the trial frame with its axis somewhere between 180 and 135, e.g. 160 degrees. The –1.00 DC trial lens is referred to as the 'working cylinder'.

With the working cylinder in place, the crossed-cylinder is presented so that its handle is *parallel* to the axis of the working cylinder, and the plus and minus axes lie at 45 degrees either side of the axis of the working cylinder. The crossed-cylinder should be presented for at least 1 second in each meridian, and spun as quickly as possible between meridians. The two 'spins' are usually referred to as 'position 1' and 'position 2'. Some crossed-cylinders have flat areas on the handle to assist this. Inexperienced practitioners often rush the crossed-cylinder, showing the first target for a fraction of a second before flipping to the second. Under these circumstances, extroverted patients tend to guess, and introverts dither, and in either case the result is needless repetition, chasing an elusive end-point! The patient is asked to select which of the two positions gives the clearest and roundest image. The axis of the working cylinder is then moved towards the minus cylinder axis of the crossed-cylinder in the *favoured position* (assuming that we are using minus cylinders). Initially, changes to the axis of the working cylinder should be generous (10 degrees), with the aim of bracketing the eventual end-point. Thus, if the working cylinder has its axis at 160 degrees and the crossed-cylinder indicates that this needs to be set more towards the vertical meridian, it should be changed to 140 degrees. If the crossed-cylinder now indicates that the axis should be more horizontal, it is then moved back to 155 degrees, which should produce equality between the two positions of the crossed-cylinder.

When the two positions give equal clarity, the axis of the working cylinder is correct. The bracketing technique is useful, because equality between the two positions of the crossed-cylinder does not always indicate a correct axis. The patients may be a poor observer, indecisive, bored or impatient. If we have an axis position that gives equality, plus one either side of it that favours the correct position of the crossed-cylinder, we can be fairly sure of the result.

After retinoscopy with the BVS and working cylinder in the trial frame

With the working cylinder in place, the crossed-cylinder is presented so that its handle is *parallel* to the axis of the working cylinder. In this position the plus and minus axes are at 45 degrees either side of the axis of the working cylinder. The crossed-cylinder should be presented for at least 1 second in each meridian, and spun as quickly as possible between meridians. Thus, if the working cylinder has its axis at 30 degrees and the crossed-cylinder indicates that this needs to be set more towards the vertical meridian, it should be changed to 40 degrees. If the crossed-cylinder now indicates that the axis should be more horizontal, it is then moved back to 35 degrees, which should produce equality between the two positions of the crossed-cylinder. When the two positions give equal clarity, the axis of the working cylinder is correct.

Determination of cylinder power

To determine the power of the cylinder, the crossed-cylinder is used with one of its principal meridians aligned parallel with the axis of the working cylinder. As before, the crossed-cylinder is shown in 'position 1' for at least 1 second and then flipped so that its other principal meridian aligns with the axis of the working cylinder ('position 2'). The patient is asked to compare the clarity of the two positions. A bracketing technique is also useful here, so changes to the power should initially be ±0.50 D rather than 0.25 D. Once the end-point has been 'over-shot', a 0.25 D change in the opposite direction should achieve equality between the two positions. It is important to modify the spherical component when making changes to the cylinder power in order to keep the disc of least confusion on the retina. As a general rule, every cylinder change should be matched with a change of sphere by half the amount and with the opposite sign, e.g. for each −0.50 DC, change the sphere by +0.25 DS. Alternatively, you can check the duochrome or the letter acuity with ±0.25 DS twirls at regular intervals. Automated refractor heads modify the sphere automatically, so practitio-

ners used to employing such devices occasionally forget to modify the sphere when faced with a trial frame. When the final cylinder power has been determined, the axis of the correcting cylinder should be checked using the method described above.

Summary of the crossed-cylinder method of refraction

As a stand-alone method

1. Obtain best sphere using letter chart and duochrome. Record the visual acuity.
2. Adjust best sphere so that the targets on the duochrome are just clearer on the green.
3. Estimate the amount of astigmatic ametropia using the best sphere visual acuity.
4. Use the crossed-cylinder to estimate the cylinder axis.
5. Place the estimated cylinder in the trial frame. Adjust the power of the best sphere so that the targets on the duochrome are just clearer on the green.
6. Use the crossed-cylinder to refine the cylinder axis.
7. Refine the cylinder power using the crossed-cylinder. Remember that the targets on the duochrome should be just clear on the green at all times so adjust the power of the sphere as required. When the power has been determined, double-check the cylinder axis.
8. Once the cylinder axis and power have been determined, check the final spherical component of the prescription using the duochrome and letter chart. Remember to maximise the plus.

After retinoscopy

1. With the objective result in the trial frame, adjust the sphere so that the targets on the duochrome are just clearer on the green.
2. Use the crossed-cylinder to refine the cylinder axis.
3. Refine the cylinder power using the crossed-cylinder. Remember that the targets on the duochrome should be just

clearer on the green at all times, so practitioners need to adjust the power of the sphere as required. Double-check the cylinder axis.

Once the cylinder axis and power have been determined, the final spherical component of the prescription is checked using the duochrome and letter chart. Most practitioners aim to give the most positive or the least negative prescription that a patient will tolerate. The fellow eye then needs to be refracted and the patient 'balanced'.

The fan-and-block technique

This technique is an alternative, but less popular, method of measuring the magnitude of a patient's astigmatic ametropia. Although subjective refraction generally follows retinoscopy, the fan-and-block method of refraction can be used alone without an objective result being available to the practitioner. Most, although not all, conventional tests charts have a fan and block at the top. It is a useful test to have, especially for those patients who do not respond well to the crossed-cylinder. As some modern charts do not always include a fan-and-block chart, practitioners must ensure that they are capable of performing alternative techniques (crossed-cylinder) should the chart be unavailable.

In the absence of an objective refraction (retinoscopy), there are two methods of determining the amount of astigmatic ametropia present using the fan-and-block technique: the **best sphere method** and the **blur method**. The fan and block is illustrated in Figures 11.5 and 11.6.

The fan-and-block test consists of three 'elements' (Figure 11.6):

1. The fan: to determine approximately the axis of the correcting cylinder
2. The Maddox V: to confirm the axis of the correcting cylinder
3. The blocks: to determine the power of the correcting cylinder.

The fan chart consists of radial lines at 10° intervals, each subtending the equivalent of a 6/18

Figure 11.5 The fan-and-block test incorporated into a typical sight testing unit.

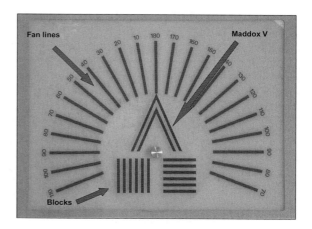

Figure 11.6 A close-up of the fan-and-block test.

limb at a distance of 6 m. The lines are arranged around a central rotation disc marked with an arrowhead, known as the **Maddox V**, with two blocks of parallel lines set at 90° to each other. This disc is designed so that, when the Maddox V is rotated towards the clearest line of the fan, the blocks align with the principal meridians of the eye. The larger numbers around the fan indicate the axis for the correcting negative cylinder. These numbers are at 90 degrees to the actual orientation of the lines and are for indirect (with a mirror) use. Some charts also have a series of smaller numbers, which are used when using the chart directly (without a mirror). The fan and block routine is performed *after* the best sphere has been determined and using minus cylinders only.

The starting point is the BVS, as for the crossed-cylinder, but then the techniques diverge, because we do not want to place the circle of least confusion on the retina. To understand why, let us consider an eye with simple myopic 'with-the-rule' astigmatism. In the uncorrected eye the astigmatic pencil consists of horizontal and vertical line foci, with the circle of least confusion separating and equidistant from them. If the DLC is focused on the retina, horizontal and vertical lines appear to be equally blurred. If, on the other hand, we position the posterior focal line on the retina, which in this case of simple myopic with-the-rule astigmatism will be the vertical one, the situation changes. Now, vertical lines are focused sharply, oblique ones less sharply and horizontal ones are the most blurred. Thus we have an indication of the two principal meridians of the eye. The degree of astigmatism present is indicated by the power of the cylindrical lens required to render the horizontal lines as clear as the vertical ones.

Figure 11.7 shows the position of the focal lines formed in an eye with an uncorrected ocular refraction of plano/−2.00 DC × 180 (simple myopic astigmatism). Figure 11.8 illustrates the appearance of the fan-and-block chart when viewed by this eye. In Figure 11.7, the horizontal focal line is formed in front of the retina and the vertical focal line is formed on the retina. This is reflected in the appearance of the fan chart shown in Figure 11.8, because the vertical line is clear and the horizontal line is the most blurred.

The fan-and-block best sphere method

1. The best sphere is determined using a letter chart and the duochrome test, as

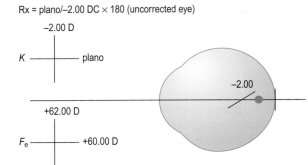

Figure 11.7 The focal lines formed in an eye with an uncorrected ocular refraction of plano/−2.00 DC × 180.

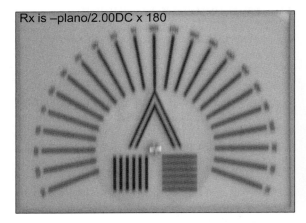

Figure 11.8 The appearance of the fan-and-block chart when viewed by the eye in Figure 11.7.

outlined in Chapter 10. It may not be possible to use the duochrome test at this stage if there is a large amount of uncorrected astigmatism and the vision is poor (<6/12).

2. Record the vision obtained using the best sphere and estimate the amount of astigmatic ametropia using Table 11.3.

3. As the disc of least confusion should be on or near the retina with the best sphere in place, half the value of the estimated cylinder is added as a positive sphere. This should bring the posterior line focus on to or near the retina, e.g. if the best vision sphere is +1.50 DS and the VA with this in place 6/18, the estimated cylinder would be about 2.00 DC. We therefore add half of this estimated cylinder power (+1.00 D) to

Table 11.3 Best sphere visual acuity with astigmatism

Best sphere VA	Astigmatic error (DC)
6/5	0.25
6/6	0.50–0.75
6/9	1.00–1.25
6/12	1.50–1.75
6/18	2.00–2.25
6/24	2.50–3.00
6/36	3.25–4.00

VA, visual acuity.

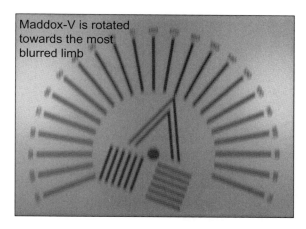

Maddox-V is rotated towards the most blurred limb

Figure 11.9 To obtain the correct axis, the Maddox V is rotated towards the most blurred limb of the 'V'.

the BVS, leaving us with +2.50 D in the trial frame. Elliott (1997) recommends adding plus power in 0.25 D steps until the VA has dropped by a line and the targets on the duochrome are clearer on the red.

4. The patient is asked to locate the position(s) of the clear line(s) on the fan chart. More lines appear clear in small degrees of astigmatism – say up to about 1.00 D, whereas only one line appears clear with 3.00 D or more. The purpose of this stage in the routine is to make an approximate determination of the axis of the correcting cylinder. The axis selection can be aided by asking: 'If this was a clock face, what time would the clearest line be pointing to?' If all of the lines in the fan seem equally clear to the patient, the spherical lens is increased by a further +0.50 DS and the fan is checked again. If the lines are still all equally clear, no significant astigmatism is present.

5. The Maddox V is now used to 'firm up' the axis direction. It needs to be equally blurred. When the limbs of the V are equally blurred, the point of the V is in the principal meridian of the posterior line focus. Rotate the V in the direction of the more blurred limb in order to equalise the appearance of the limbs (**Figure 11.9**).

6. The next step is to determine the power of the correcting cylinder. This involves using the blocks. After step 5, the blocks are in the orientations of the principal meridians of the eye. We need to confirm the position of the clear block to make sure

that the focal line is on the retina, which done by small adjustments of the sphere until the highest positive or least negative sphere is obtained, using the clear block as a target.

7. A further +0.50 D sphere is added to move the posterior line focus just in front of the retina. This forms the basis of a **check test** discussed below.

8. Add a minus cylinder (start by adding a value slightly less than the estimated cylinder), axis parallel to the blurred block. Each increment 'pushes' the anterior line focus back towards the retina. When the blocks appear equally clear the astigmatism has been corrected. Addition of a further −0.25 DC should reverse the blocks, i.e. the patient reports that the formally blurred block now appears clearer. This is the **check test**, which indicates that both focal lines were initially in front of the retina. The extra −0.25 DC is removed and the sphere adjusted using the letter chart and the duochrome test. The check test prevents over-correction of the astigmatism, but, to be able to perform it, the posterior focal line must be in front of the retina, so this accounts for the +0.50 DS addition in step 7. Reversal must be demonstrated, otherwise the posterior line was not in front of the retina and accommodation was not controlled.

Summary of the fan-and-block routine

- With the best sphere lens placed in the trial frame, measure the patient's letter acuity and estimate the astigmatic error. Add half of this estimated cylinder as a positive sphere.
- Add an additional +0.50 D sphere.
- Direct the patient's attention to the fan chart and ask the patient to locate the clearest line (or group of lines).
- Use the Maddox V to 'firm up' the cylinder axis. The two limbs of the V should be equally blurred.
- Direct the patient's attention to the two blocks. One block should be clearer than the other. The orientation of the clearest block should match the clear line on the fan chart.
- Place minus cylinders in the trial frame. The axis of the cylinder should correspond to the axis found using the fan chart and Maddox V until the two blocks appear equal.
- Increase the cylinder power by −0.25 DC. The blocks should now reverse, e.g. if the vertical block has been clearer all the way through the routine, the horizontal block should now be clearer for the first time. This is a check test. If the blocks do not reverse, abort the routine and start again.
- Remove the +0.50 D sphere from step 2.
- Adjust the power of the best sphere using the letter chart and the duochrome test.
- Record the final visual acuity.

The fellow eye then needs to be refracted and the patient 'balanced'.

As a further illustration of the fan-and-block technique, the prescription −2.00 DS/−2.00 DC × 180 will be found using the method described above (**Figures 11.10–11.17**):

- Using the given prescription, the power of the reduced surface and the ocular refraction is determined (**Figure 11.10**).
- Once the power of the reduced surface has been found, the positions of the focal lines in the uncorrected eye can be determined. The general appearance of the

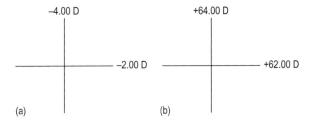

Figure 11.10 The power of the reduced surface: (a) principal powers of correcting lens will be −4.00 D along 90 and −2.00 D along 180; (b) power of the reduced surface will be +64.00 D along 90 and +62.00 D along 180.

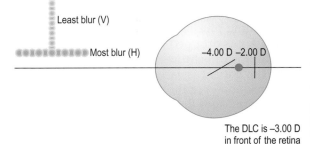

The DLC is −3.00 D in front of the retina

The astigmatic interval between the two line images is 2.00 D

Figure 11.11 The positions of the focal lines in the uncorrected eye. The astigmatic interval between the two line images is 2.00 D.

corresponding fan lines should also be considered (**Figure 11.11**).

- The power of the BVS is determined using a letter chart and, if possible, the duochrome test. The purpose of the BVS is to place the disc of least confusion on the retina. With the disc of least confusion in this position, the fan lines now appear equally blurred (**Figure 11.12**).
- Half the estimated cylinder is added as a positive sphere. In this example, we have been given the value of the cylinder but in practice it can be estimated from the VA achieved with the BVS in place. This positive sphere places the vertical focal line on the retina. The patient reports that the vertical line of the fan chart is now clear (**Figure 11.13**).
- An extra +0.50 DS is now added (the check test). This moves both focal lines to the left

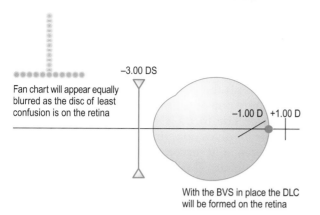

Figure 11.12 The positions of the focal lines with the best vision sphere in place (−3.00 DS). The fan chart will appear equally blurred as the disc of least confusion is on the retina. With the BVS in place the DLC will be formed on the retina.

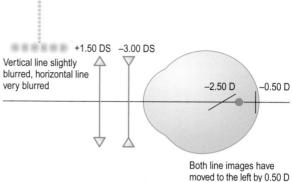

Figure 11.14 The positions of the focal lines with a +1.50 DS lens in place: vertical line slightly blurred, horizontal line very blurred. Both line images have moved to the left by 0.50 D.

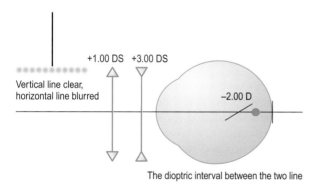

Figure 11.13 The positions of the focal lines with a +1.00 DS lens in place: vertical line clear, horizontal line blurred. The dioptric interval between the two line images is still 2.00 D (the cylinder value).

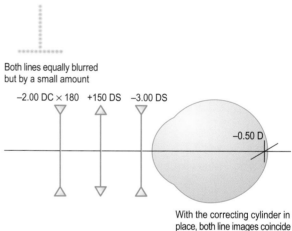

Figure 11.15 The position of the focal line with the correcting cylindrical lens (−2.00 DC × 180) and the check test lens (+0.50 DS) in place. Both lines equally blurred but by a small amount and with the correcting cylinder in place, both line images coincide.

by 0.50 D. The patient still reports that the vertical line is the clearest (Figure 11.14).

- In Figure 1.14, the horizontal focal line is furthest from the retina. If the astigmatism is to be corrected, both focal lines must coincide. To move the *horizontal* focal line, the *vertical* power of the lens must be adjusted. Remember that, in astigmatism, the orientation of the image formed is always at 90 degrees to the power meridian! As the power of a cylinder is always at 90 degrees to its axis, the axis of the correcting cylinder in this example is

180. The reader should be sure that this is understood! So, a minus cylinder, axis 180, is added until the two focal lines coincide. The patient reports that the fan lines look the same, slightly blurred, but the same (Figure 11.15).

- The power of the cylinder is increased by an extra −0.25 DC. If the procedure has worked the block will reverse and for the first time in the routine the patient reports

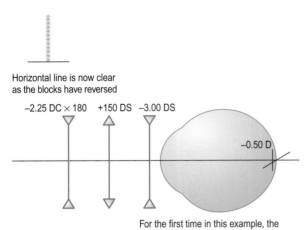

Horizontal line is now clear
as the blocks have reversed

−2.25 DC × 180   +150 DS −3.00 DS

−0.50 D

For the first time in this example, the
horizontal line image is closer to the retina

Figure 11.16 The positions of the focal lines with a
−2.25 DC × 180 cylindrical lens in place. The horizontal
line is now clear as the blocks have reversed and for the
first time in this example, the horizontal line image is
closer to the retina.

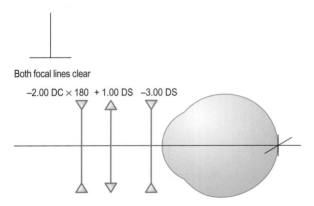

Both focal lines clear

−2.00 DC × 180 + 1.00 DS −3.00 DS

Figure 11.17 The positions of the focal lines with the
+0.50 DS lens removed and the correct cylinder in place.
Both focal lines clear. The final correction is −2.00 DS/
−2.00 DC × 180.

that the horizontal line is the clearest
(Figure 11.16).
- The +0.50 DS lens is now removed. Both
line foci should be formed on, or close to,
the retina. The power of the sphere can be
confirmed using the letter chart and the
duochrome test (Figure 11.17).

The fan-and-block blur method

The blur method involves starting with a high
positive sphere (+20.00 D) so that every patient
commences with both line foci in front of the
retina. With the patient viewing the fan chart,
the sphere power is gradually reduced until the
posterior line focus is on the retina. The blur
routine involves no estimation of the astigmatic
ametropia and accommodation is controlled
from the start. This is a very practical method
to use if a consulting room is equipped with a
refractor head because the sphere power can be
reduced very rapidly. Once the posterior line
focus is on the retina, the procedure follows on
from the routine described above.

Advantages and disadvantages of the different methods

Advantages and disadvantages of the fan-and-block method of subjective refraction

- An excellent method for very high cylinder
powers because one fan line is very clear.
- Not so reliable for very low cylinder
powers.
- Can be used for patients with a low VA.
- Some patients find this test less
stressful than other tests for astigmatism
because they are seeing the same target
all times.
- Not all sight-testing units have a fan-and-
block chart.
- The fan-and-block method can be useful
with patients who provide poor responses
to the crossed-cylinder method.
- The fan-and-block method provides a fast
check for the presence of astigmatism.

Advantages of the crossed-cylinder over the fan-and-block method

- Although the theory of this method may
appear complex, with practice it is a very
simple and quick method to use.
- Both positive and negative cylinders can
be used (fan and block uses negative
cylinders only).

- It is relatively unaffected by head tilt.
- Many modern charts do not include a fan-and-block component but all have a circular target of some sort. This point is particularly relevant for domiciliary visits.
- It is relatively unaffected by cortical lens changes.
- It can be performed for both distance and near vision.
- It can be very accurate (to 2.5 degrees) if the patient is discerning.
- This method is more accurate for small cylinders than the fan-and-block method.

Disadvantages of the crossed-cylinder over the fan-and-block method

- The spherical power must be correct otherwise errors occur.
- It requires the patient to have a good VA.
- The fan-and-block method can be more accurate for high cylinders.
- Some patients find the crossed-cylinder test confusing and stressful.
- Accommodation is better controlled with the fan-and-block method.

Concluding points for Chapter 11

The widely used crossed-cylinder method and the perhaps seldom-used fan-and-block technique both have advantages and disadvantages. For most patients either technique produces satisfactory results, but neither technique works well on every patient, so familiarity with both is essential. Both techniques require good control of accommodation before and during the application of the technique. The position of the patient's head should also be controlled during the examination.

References

Elliott D B (2003) *Clinical Procedures in Primary Eye Care*. Butterworth-Heinemann, Oxford

O'Leary D (1988) Subjective refraction. In: Edwards K, Llewellyn R (eds), *Optometry*. Butterworth-Heinemann, Oxford: 135

Further recommended reading

Rabbetts R B (1998) *Bennett & Rabbetts' Clinical Visual Optics*. Butterworth-Heinemann, Oxford

Tunnacliffe A H (1993) *Introduction to Visual Optics*. Association of the British Dispensing Opticians, London

Binocular Balancing and Binocular Refraction

Andrew Franklin

Introduction

Once any spherical and astigmatic ametropia has been corrected, final adjustments may be made to the distance refractive correction. At this stage, it is worth considering what ends we hope to achieve before looking in detail at ways to achieve them. We should perhaps look at this question from the point of view of the patient. What they would reasonably expect from an optical correction is clear comfortable vision. As most of our patients have two eyes that work together in some way, the ultimate aim of the refractive process should be best visual acuity and greatest comfort when both eyes are being used. If we consider a single eye in isolation, the monocular correction required is simply deduced using subjective means. If too much positive power is placed before the eye, distance vision is fogged and therefore acuity reduced. If too little plus power is applied and the patient is over-minused a pre-presbyopic eye is able to accommodate to achieve maximum acuity, but a presbyopic eye is not. It would therefore seem that the ideal correction would be the most positive or least negative lens that allows the best visual acuity to be achieved with the least effort.

Under binocular conditions things, however, become a little more complicated. Uncorrected hypermetropia stimulates the patient to accommodate, and in doing so a certain amount of accommodative convergence is induced. Should the hypermetropia be fully corrected, this accommodative convergence is not induced. The effect is to make the patient more exophoric. If the patient is already exophoric, greater fusional convergence is required to compensate the ocular motor balance. If the patient is esophoric, the extra plus power may help to reduce the esophoria. Thus, the best spherical correction for patient comfort may depend on ocular motor balance, and full correction of the hypermetropia may not always give the most comfortable vision. For this reason, we should discount any method of balancing that does not allow some degree of binocular fusion to occur. There are a number of these methods still in use, in which the eyes are completely dissociated by alternate occlusion, or by prisms, and all are equally unlikely to provide a meaningful result.

It is also pertinent to consider what it is that we wish to balance. There is little point in trying to equalise the visual acuity of the two eyes. In many cases, the quality of vision of the two eyes, even with an optimal refractive correction, is simply not the same even where the letter acuity is equal, and the only way to make it the same is to penalise the better of the two. There have been some authorities that have advocated this very course of action, although most have conceded that patients may be less keen on the idea. What most balancing techniques aim to do is to equalise the accommodative demand in the two eyes, because the accommodative response is generally the same in both eyes, even when the accommodative stimulus to each eye is different (Campbell 1960). On average only about 0.12 DS difference in accommodative response can be achieved between the two eyes, although certain individuals may achieve 0.50 DS. If the spherical corrections are unbalanced, the accommodative demand for a given target is also unequal. If the accommodative response is equal, it means that one eye or the other is blurred, unless they compromise, which would result in both being blurred.

Chapter content

Binocular balancing techniques

Binocular balancing after monocular refraction

If a subjective refraction has been performed under monocular conditions, the process of 'binocular balancing' must now occur. Binocular balancing is necessary in order to equalise the accommodative effort between the two eyes, because they do not necessarily accommodate by the same amount under monocular conditions as they do under binocular conditions. The eyes, however, always accommodate by the same amount under binocular conditions. If the right and left prescriptions are not 'balanced', the eyes accommodate by the least amount possible to produce a clear retinal image in one eye. The other eye may have a slightly blurred image and asthenopia could result.

If we are attempting to achieve a 'binocular balance', there are two main requirements:

1. There must be some way of providing a specific target for each individual eye.
2. There must be a target that both eyes can see, to act as a binocular lock and ensure that fusion takes place.

The monocular targets can be separated by the following methods.

Septum methods

The Turville infinity balance (TIB) technique was described by Turville in 1946, although he and Esdaile had been developing the technique since 1927. Originally the apparatus consisted of an indirect chart with two vertical columns of letters and a mirror that was divided into equal sections by a vertical septum 30 mm wide. The septum variously comprised a vertical bar or a rectangle 30 mm wide and 50 mm high, which had a surrounding frame that was intended to act as a binocular lock. The effect of the 30 mm septum is to hide a band of the test chart 60 mm wide from each eye, and also to act as a binocular target (although perhaps not a very good one).

Polarisation

Polarisation allows some of the test chart to be seen monocularly without the use of a septum, the positioning of which can be time-consuming. Furthermore, it is a more suitable method to use with projector charts. According to Bannon (1965) polarisation was first used at the Dartmouth Eye Institute in 1939. The left and right sides of the chart were covered with sheets of polarising film. Analysis lenses placed before the eyes allowed each eye to see only half of the chart, so it was necessary to place a central unpolarised fusional lock. The drawback with this method was that an unwanted effect of the polarisers was to cut down the light levels rather drastically. The contralateral half of the chart would be totally dark, and the background of the ipsilateral half would have half of its light cut out by the polarising sheet. The binocular target would also be rather dim. This problem was partly addressed by Cowen (1955) where the letters were polarised, but the background was not. Various polarised charts based on this idea were produced, with the aim of improving the binocular lock, and a design by Freeman (1955) added duochrome targets. A number of charts have been produced that feature polarised letters on an unpolarised coloured background, designed to give an impression of distance and thereby discourage accommodation and give a detailed fusional lock.

Grolman (1966) described vectographic slides in which a high-resolution printing process deposited dichroic ink on to a stretched film of polyvinyl alcohol. This provided an economically feasible way of providing polarised slides for projection, although the contrast obtainable was only about 80% of that of a polarised target on a conventional test chart. Polarising duochrome tests are featured in the design of some test charts. Such charts incorporate polarised duochrome test units so that patients can wear a polarising visor that allows them to view one set of red/green backgrounds with the right eye and the other set with the left eye. Under these more nearly normal binocular vision conditions, the right and left levels of accommodation can be equalised for distance vision.

Image degradation

Viewing part of the test chart using monocular vision only may be achieved by degrading one image with a neutral density filter or fogging lens. Humphriss (1984) has made use of this method for a full binocular refraction and this is discussed later in this chapter. When using such techniques the patient may view letters or duochrome targets.

Letter targets

These may be used in two different ways. O'Leary (1988) describes fogging both eyes by equal plus spheres and comparing the clarity of the letters seen with each eye. If one eye is seeing clearer, the fogging lens is increased on that side until the acuities are equalised. The fog is then reduced in 0.25 DS steps binocularly, checking that acuity is equal at each stage. Fletcher (1991) suggests adding +0.25 D to each eye in turn. This should in theory blur the vision but, if it is accepted by the patient, it may be added to the spherical correction. This approach is effectively comparing the vision of the eye with and without the extra correction, rather than comparing the clarity of the two eyes. As discussed earlier, the two eyes may not necessarily have inherently equal clarity of vision, and this second type of comparison may be the more logical. It is the basic type of comparison used in the Humphriss immediate contrast (HIC) technique of binocular refraction.

Duochrome targets

Here, the correction is adjusted to the same endpoint in both eyes. However, it may be impossible for the patient to equalise the red and green targets in one or both eyes. If this is the case, the targets on the other colour should be equally blurred for both eyes, and an equal amount of spherical power should reverse the appearance of the duochrome.

The Humphriss immediate contrast technique (HIC) test

This technique first described by Humphriss in 1962 employs a fogging lens before the eye that is not being refracted, which reduces the visual acuity (VA) of that eye by three or four lines. Humphriss recommended a +0.75 DS lens, although most practitioners use a +1.00 DS and some rather more. In younger patients with large pupils +1.00 D usually achieves the necessary fogging, although older patients may require a little more plus because their pupils tend to be smaller. The idea behind this technique is to reduce the acuity in this eye to about 6/12, at which point central vision is inhibited and the 'psychological septum' is established.

The HIC test is a binocular spherical lens balancing method that may be used after right and left monocular refractions. One eye is 'fogged' using a +0.75 or +1.00 spherical lens whereas the spherical component of the other eye's prescription is adjusted for the most comfortable vision. With the left eye fogged and the right eye looking at a row of letters, the procedure is to first hold a +0.25 DS lens in front of the right eye, followed after 2–3 seconds by a –0.25 DS. If the prescription is already balanced, the +0.25 DS blurs the right eye and the –0.25 DS has no effect so no adjustment to the right sphere is necessary. The fogging sphere is now placed in front of the right eye and the ±0.25 spherical lenses presented to the left eye. If the prescription is already balanced, the +0.25 DS blurs the left eye and the –0.25 DS has no effect. No adjustment to the left sphere is necessary. As with the duochrome, the ±0.25 DS lenses are usually mounted on a 'flip' or 'twirl'. If the vision is more comfortable with the +0.25 DS lens, this should be added to the test eye and the *same amount* added to the fogged eye because it is important that the amount of fogging be maintained during the procedure. The procedure is repeated, stopping when +0.25 DS blurs the vision. If the vision is more comfortable with the –0.25 DS lens, this should be added to the test eye and the fogging lens be reduced by 0.25 DS. The end-point is reached when adding +0.25 DS blurs the vision and adding −0.25 DS has no effect. When performing the HIC test it is important that the fogging lens is placed in front of the right eye *before* the left fogging lens is removed.

Binocular refraction

When an occluder is placed before an eye, the accommodation–convergence relationship is disrupted, and the pupils dilate. If the refractive error is at all irregular, as is the case in a patient with incipient cataract, the pupil dilation may produce a change in refractive error. With fusion suspended, the correction that has the most beneficial effect on binocular vision, which is the normal state for most patients, cannot be accurately determined. It is, however, possible to perform the entire refraction under binocular conditions and there is a considerable amount of published evidence that refractive corrections obtained monocularly do not necessarily correlate with those found under binocular conditions.

Morgan (1949) reported that 20% of individuals showed a difference of 0.25 DS in the end-points when monocular and binocular refractions were compared, and in 2% of cases the difference was more than 0.50 DS. Norman (1953) found that 35% had a difference in end-points of 0.25 DS and 12% a difference >0.50 DS. In addition 28% of his patients showed a difference of 0.25 DS in binocular sphere balance and 5.6% showed a difference of 0.50 DS or more. It is also known that different astigmatic axes may be found under binocular conditions, as a result of the presence of cyclophoria. The prevalence of significant cyclophoria is indeterminate, but some authorities have estimated that it could be as high as 15% of the population. It would appear to be high enough for some refractive surgery clinics to insist on binocular refraction for any patient about to undergo surgery.

Turville (1946) reported finding higher VAs in eyes with anisometropic amblyopia with the refractive results from binocular testing, and this has been confirmed a number of times since. Grolman (1966) found differences in visual acuities of as much as 1.5 lines on a letter chart. A full binocular refraction may be achieved by the use of polarisers or by modifying the TIB technique. However, polarised targets may suffer from poor contrast, and the TIB from an inadequate binocular lock. The method most commonly employed by British practitioners is a modification of the HIC technique, which was the result of a long programme of research that was summarised by Humphriss in 1984. The following were the essential conclusions:

- If one eye is fogged slightly with a plus sphere of +0.75 DS or +1.00 DS, the central fovea of that eye is inhibited so that, although there is binocular vision, the perception of edge sharpness depends on the unfogged eye alone, i.e. a 'psychological septum' is established.
- The inhibited area could be as small as 0.5 degrees, which is the area of the fovea necessary to recognise 6/9 letters.
- A +1.00 DS lens placed before one eye has very little effect on the positive fusional reserve. However, increasing the sphere to +1.50 DS may cause binocular vision to begin to collapse.

The third point is of interest because, although most practitioners use +0.75 DS or +1.00 DS to establish the psychological septum, a number employ +1.50 DS fogging lenses by the simple expedient of leaving their retinoscopic working distance correction in place when they proceed to subjective refraction. Corroboration on the effect of excessive fogging comes from research into monovision correction of presbyopia, where one eye is corrected for distance while the other is corrected for near by over-correction with a plus sphere. Pardhan and Gilchrist (1990) found that at a point between 1.00 and 1.50 D the eyes crossed over from binocular summation to binocular inhibition. When binocular summation is occurring, the binocular contrast sensitivity is about 40% higher than the monocular value. With binocular inhibition the binocular sensitivity is lower than the monocular one. This also correlates well with anecdotal evidence from contact lens practitioners and refractive surgeons that reading additions below +1.50 work better in

127

monovision. There are other ways in which the psychological septum may be established. Mallett (1964) recommended using a ±0.50 cross-cylinder and Lyons (1962) a neutral density filter of about 30%.

Refraction by HIC appears to have a number of clear advantages:

- Accommodation in the fogged eye is suspended with a plus sphere and tied to convergence. This helps to reduce fluctuations in accommodation in the eye undergoing refraction. Binocular refraction is generally less likely to under-correct patients with latent hypermetropia.
- No separate binocular balancing is needed. This saves quite a lot of time, an advantage to both the practitioner and the patient. Contrary to the belief of some eye care professionals, surveys have shown that patients are largely unimpressed by prolonged testing.
- Where latent nystagmus is present, it is reduced or eliminated under binocular conditions, allowing better VA and more precise subjective responses. Even in normal individuals, binocular acuity is usually better than monocular.
- Rotational heterophorias (cyclophorias), if present, do not reduce the final binocular VA. This is now known to be an important factor in refractive surgery patients, and binocular refraction is required in the clinical protocols for pre-treatment refractions in some refractive surgery clinics. Cyclophoria is present in a surprising number of patients and may be a significant factor both visually and in terms of patient comfort where the astigmatic element is high.
- Occlusion of the other eye is usually performed with an opaque black occluder. This may have a slight effect on pupil diameter. In older patients with lens changes this may, in turn, affect the final prescription obtained.

In summary, binocular refraction is quicker and probably more accurate on most patients, although there are some for whom binocular refraction is inappropriate. The plus sphere

used blurs the 'fogged' eye back to 6/12 or thereabouts, so if the eye being refracted has similar or worse acuity the potential for confusion is considerable. Some patients have a markedly 'dominant' eye which, when fogged, continues to be expressed in the binocular perception of the patient, leading to blurring or diplopia when refraction is attempted on the non-dominant eye. The solution in these cases is simple. The weaker or non-dominant eye is refracted first with the other eye occluded. The stronger eye may then be refracted under binocular conditions. The weaker eye is fogged if its VA requires it, or left unfogged if the acuity is already at 6/12 or worse.

'Hack Humphriss' technique

This is essentially a hybrid method employing both monocular and binocular elements, and it is probably at least as commonly employed as the 'proper' modified HIC technique because it is simple, quick and reliable in practice:

1. To refract the right eye, occlude the left eye.
2. Refract the right eye monocularly.
3. Apply the +1.00 blur test. If the VA is reduced to 6/12 or so leave the +1.00 in. If not, adjust the fogging lens until it does.
4. Refract the left eye with the right eye fogged, i.e. binocularly.
5. Apply the +1.00 blur test to the left eye.
6. Check the final sphere balance of the right eye binocularly with the +1.00 still in place before the left eye.

The +1.00D blur test

The +1.00D blur test is a commonly used procedure for checking the spherical component of a prescription at the conclusion of a subjective refraction. Essentially, a +1.00DS lens is added to the final distance correction in the trial frame and the distance VA is reassessed. The addition of a +1.00DS lens should result in a drop in VA of about four lines, e.g. 6/5 to 6/12 or 6/6 to 6/18. If a +1.00DS blurs a patient with a 'normal' acuity of, say, 6/5 to only 6/9, the patient may have been under-plussed (or over-minused) and the prescription should be re-checked. An insufficient relaxation of accommodation during the subjective routine may be the reason for under-plussing. It is for this reason that the +1.00 D blur test is mainly used with pre-presbyopic patients.

A clinical assessment of the +1.00 D blur test has been carried out by Elliott and Cox (2004). In pre-presbyopic patients the authors found that the +1.00 D blur test reduced VA from between one and seven lines on a logMAR chart, with the average value being approximately 0.40 logMAR (four lines on a logMAR chart). The authors concluded that this large range of normal values is not a result of unreliable measurements, but probably of either variability in the ocular optics or perceptual interpretation of reduced quality retinal images. In addition, larger reductions in VA with the +1.00 D blur test appeared to be partly caused by large pupils, whereas smaller reductions in VA were associated with smaller pupils. It is also important to note that VA values with the +1.00 D lens also depend on the original VA. If the original VA is about 6/4 then it blurs to about 6/12$^+$ with the +1.00 D lens. However, a VA of 6/9 or 6/6 with the +1.00 D lens may not necessarily indicate that the patient has been under-plussed (or over-minused). However, such values suggest that the spherical component of the correction should be re-visited.

The +1.00 DS blur test should be applied to each eye in turn, with the fellow eye being occluded. Some practitioners then leave the +1.00 DS lens in place in order to apply the Humphriss technique.

Binocular refraction at near

It is possible to perform a binocular refraction at near, although this is rarely attempted in practice except as a last resort. However, there are some patients whose spherical balance alters for near fixation as they accommodate and the pupil constricts. Furthermore, Scobee (1952) found that 77% of his patients showed a shift in astigmatic axis when accommodating. Although these changes are rarely significant visually, there are exceptions, particularly when the astigmatic element is high. It can be difficult to arrange appropriate fogging for a near target using plus spheres, and Mallet's (1964) suggestion of a plus or minus cross-cylinder for fogging may be helpful in these cases.

How accurate is subjective refraction?

There are a number of questions to consider when attempting to quantify the accuracy of subjective refraction. First, how likely is a single practitioner, refracting the same patient using the same methods, to obtain the same result twice? Humphriss (1958) re-refracted 132 subjects within 1–90 days and found some change in 78% of them. The standard deviation was 0.254 DS for the sphere and 0.214 DC for the cylinder. The deviation was slightly higher for those with pre-presbyopia (0.275 DS) than for those with presbyopia (0.208 DS).

Borish and Benjamin (1998) concluded that repeatability by the same examiner should be within ±0.25 D for sphere and cylinder power, and 5 degrees of the cylinder axis for routine cases. For higher powers, repeatability for power would be expected to go down, but it should improve when considering the cylinder axis. It would also be wise to consider whether two different practitioners using the same techniques are likely to obtain the same result.

Perrigen et al. (1982) investigated consistency between three optometrists. For spherical power no significant difference was found in 27% of patients. In 86% of the patients there were differences of 0.25 DS and in 98% differences of 0.50 DS. For the cylindrical power, the corresponding figures were 51%, 93% and 99%. Goss and Grosvenor (1996) reviewed the literature and found that 80% of refractions by two or three different practitioners were reproducible to within ±0.25 D and 95% to within ±0.50 D. The use of different techniques might be expected to affect the end-product, but Jennings and Charman (1973) found no significant difference in reproducibility between duochrome and Simultan techniques, and Safir et al. (1970) reported similar findings. Johnson et al. (1970) found that different methods of axis determination gave similar results. Some 73–85% of results were within 5 degrees and 93–98% within 10 degrees. It would appear, then, that subjective refraction could be expected to be reproducible to within 0.25 D most of the time. Adamson and Fincham (1953) found that **physiological ocular tolerance**, the range of vergence that does not induce a change of accommodation, and **perceptual tolerance**, the amount of defocus tolerated without noticeable blur, were also about ±0.25 DS. However, clinical experience suggests that tolerance of defocus is subject to some individual variation, and some patients are unusually sensitive, which may be a result of physiological factors, such as a large pupil, causing a reduced depth of focus or to psychological traits. In most patients a ±0.25 D change in power or less than 5 degrees in axis is potentially an artefact of testing, so unless there are correlating symptoms or a need for new spectacles a change in prescription might be hard to justify. Changes over these amounts are probably genuine, although if the VA is low, or the refraction inherently changeable (as in those patients with metabolic disease), some circumspection would still be appropriate. Perhaps the last word should go to Michaels (1985): 'Measuring the ametropia is one thing, prescribing for it may be something else.'

Concluding points for Chapter 12

For a patient with normal binocular vision, it is desirable to balance the accommodative effort of the two eyes under binocular conditions. This may be done as a separate operation after monocular refraction, and many practitioners achieve satisfactory results with this strategy. However, there are persuasive arguments in favour of full binocular refraction in most cases. The accuracy achieved in subjective refraction has finite limits, which are influenced by technique, experience and the patients themselves, and the decision whether or not to prescribe should always take account of these points. Prescribing itself is as much an art as a science, and even with experience surprises may occur from time to time.

References

Adamson J, Fincham E F (1953) Visual tolerances and their effect on the measurement of refraction. *Refractionist* 405–11

Bannon R E (1965) Binocular refraction – a survey of various techniques. *Optometry Weekly* **56**(31):25–31

Borish I M, Benjamin W J (1998) *Borish's Clinical Refraction*. W B Saunders Co., Philadelphia: 693

Campbell F W (1960) Correlation of accommodation between the two eyes. *J Opt Soc Am* **50**:738

Cowen L (1955) Binocular refraction, a simplified clinical routine. *British Journal of Physiological Optics* **16**:60–82

Elliott D B, Cox M J (2004) A clinical assessment of the +1.00 blur test. *Optometry in Practice* **5**:189–93

Fletcher R (1991) Subjective techniques. In: *Eye Examination and Refraction*. Blackwell Scientific Publications, Oxford

Freeman H (1955) Working method – subjective refraction. *British Journal of Physiological Optics* **12**:20–30

Goss D A, Grosvenor T (1996) Reliability of refraction – a literature review. *Journal of the American Optometric Association* **67**:619–30

Grolman B E (1966) Binocular refraction – A new system. *New England Journal of Optometry* **17**:118–30

Humphriss D (1958) Periodic refractive fluctuations in the healthy eye. *British Journal of Physiological Optics* **15**:30

Humphriss D (1962) Binocular vision technique – the psychological septum. *Review of Optometry* **99**:19–21

Humphriss D (1984) *Refraction Science and Psychology*. Juta & Co., Cape Town

Jennings J A M, Charman W N (1973) A comparison of errors in some methods of subjective refraction. *Ophthalmic Optics* **13**:8

Johnson B l, Edwards J S, Goss D A (1970) A comparison of three subjective tests for astigmatism and their inter-examiner reliabilities. *Journal of the American Optometric Association* **67**:590–7

Lyons J G (1962) Refraction and the binoculus. *Optician* July: 663–6

Mallett R F J (1964) The investigation of heterophoria at near and a new fixation disparity technique. *Optician* **148**:574–81

Michaels D M (1985) Subjective methods of refraction. In: Michaels D M (ed.), *Visual Optics and Refraction*, 3rd edn. CV Mosby, St Louis: 316–34

Morgan M W (1949) The Turville infinity binocular balance test. *American Journal of Optometry and* the *Archives of the American Academay of Optometry* **26**:231–9

Norman S L (1953) Plus acceptance in binocular refraction. *Optometry Weekly* **44**:45–6

O'Leary D (1988) Subjective refraction. In: Edwards K, Llewellyn R (eds), *Optometry*. Butterworth-Heinemann, Oxford: 135

Pardhan S, Gilchrist J (1990) The effect of monocular defocus on binocular contrast sensitivity. *Ophthalmic and Physiological Optics* **10**(1):33–6

Perrigen J, Perrigen D M, Grosvenor T (1982) A comparison of clinical refractive date obtained by three examiners. *American Journal of Optometry and Physiological Optics* **59**:515–19

Safir A, Hyams L, Philpott J (1970) Studies in refraction. I. The precision of retinoscopy. *Archives of Ophthalmology* **84**:49–61

Scobee R G P (1952) *The Oculorotatory Muscles.* Henry Kimpton, London

Turville A E (1946) *Outline of Infinity Balance.* Raphaels Ltd, London

Further recommended reading

Elliott D B (2003) *Clinical Procedures in Primary Eyecare.* Butterworth-Heinemann, Oxford

Rabbetts R B (1998) *Bennett & Rabbetts' Clinical Visual Optics.* Butterworth-Heinemann, Oxford

Tunnacliffe A H (1993) *Introduction to Visual Optics.* Association of the British Dispensing Opticians, London

Accommodation and Presbyopia

Andrew Keirl

Introduction

When compared with myopia, hypermetropia and anomalies of binocular vision, presbyopia is often considered the 'poor relation' of vision problems. However, for the person with life-long emmetropia, presbyopia is usually the first-time encounter with a refractive difficulty. Such a patient is often concerned, from an optical, lifestyle and ageing perspective, so it is unfortunate that some colleagues dismiss presbyopia as relatively insignificant. If patients are being given the impression that presbyopia is a minor hindrance, it is not surprising that some choose the ready-made option whereas others choose to purchase prescription reading spectacles via the internet or another retail outlet.

Chapter content

- Accommodation
- The clinical measurement of the amplitude of accommodation
- Definitions relevant to accommodation
- Presbyopia and the near addition
- Hypermetropia and accommodation
- Calculations relevant to the study of accommodation
- Depth of field

Accommodation

Accommodation is the eye's ability to change in power in order to focus on objects at different distances. To focus on a near object, it is necessary for the eye to increase its dioptric power. This change is referred to as **positive accommodation**. When an eye changes its focus from near to distance, a decrease in dioptric power is required, referred to as **negative accommodation**. In practice, we generally use the word **accommodation** when referring to an increase in the eye's dioptric power. The total amount by which an eye can change in power is known as the **amplitude of accommodation**, which is simply the eye's maximum power minus the eye's minimum power. The purpose of accommodation is to 'neutralise negative vergence from a near object'. Someone with emmetropia, for example, looking at a distant object can focus the image of the distant object on the retina with an unaccommodated eye because the incident vergence at the cornea is zero (**Figure 13.1**). If the eye now looks at a near object placed at a third of a metre from the eye (**Figure 13.2**) the incident vergence at the cornea is −3.00 D. This must be neutralised if a clear image is to be formed on the retina of this emmetropic eye. Positive power must be added to the eye's refracting system to ensure that a clear image is formed on the retina, and this addition is brought about by the eye's ability to accommodate. In this particular case, the eye must accommodate to add +3.00 D to the eye's dioptric power, which neutralises the 3 D negative vergence arising from the near object.

The components of the eye involved in the process of accommodation are:

- the third cranial nerve (N III)
- the ciliary muscle and ciliary body
- the zonules

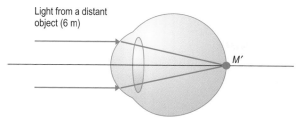

Light from a distant
object (6 m)

M′

Figure 13.1 Unaccommodated emmetropic eye.

- the crystalline lens
- the elastic capsule of the crystalline lens
- the vitreous

The gross structure of the human eye is shown in Figure 13.2a.

The traditional and most recognised theory of the process of accommodation is that provided by Helmholtz, who suggested that when looking into the distance the crystalline lens is pulled and stretched into a thinner and flatter form by the zonules, which are themselves pulled by movement of the ciliary muscle. In this unaccommodated state, the ciliary muscle is said to be relaxed. During near vision, the ciliary muscle contracts, causing it to move forward; this in turn relaxes the tension of the zonules, resulting in an increase in curvature of the crystalline lens. Some years later, Fincham (1937) suggested that the substance of the lens did not possess the necessary elasticity to 'relax' into a naturally accommodated state. He believed that, during accommodation, the crystalline lens is moulded into a more curved shape by the elastic properties of the capsule. Recent data has, however, shown that the crystalline lens is elastic and there is, almost without doubt, some involvement of the vitreous. Some authorities believe that there are two types of zonules: main zonules and tension zonules, the latter being placed under tension during accommodation. Tunnacliffe (1993) summarises the theory of accommodation as follows:

When the ciliary muscle is relaxed, the elastic tissues in the ciliary body hold the zonule under tension to focus the eye for distance vision. During accommodation for near, the ciliary muscle contracts, moves the ciliary body forwards and inwards, and reduces the tension in the zonule. The elastic capsule compresses the lens and aided by the pressure from the vitreous, the anterior lens surface bulges into a steeper shape at the pole. On returning to a distance focus, the ciliary body relaxes, the elastic tissue in the ciliary body restores the tension in the zonule and, aided by the elastic nature of the lens, the lens is pulled into a flatter, weaker shape necessary for distance focus.

When unaccommodated, the anterior and posterior crystalline lens radii are +10.0 mm and −6.00 mm, respectively. When fully accommodated they become +5.50 mm and −5.50 mm. It is interesting to note that most change in curvature occurs with respect to the anterior surface.

Physical evidence of positive accommodation occurring includes:

- on accommodation, the outer surface of the ciliary body moves forwards by approximately 0.50 mm
- on accommodation, the lens becomes tremulous, having lost the support of the zonule
- fine wrinkles, present when unaccommodated, disappear during accommodation, suggesting that the zonule is under tension when unaccommodated

There appears to be little work on possible changes in other ocular components during accommodation. However, Pierscionek *et al.* (2001) and He *et al.* (2003) found changes in corneal curvature with accommodation. Drexler and co-workers (1998) interestingly found that, as the lens accommodates, the length of the globe elongates. For an in-depth review of accommodation the reader is referred to Pierscionek (2005).

Classification of accommodation

Accommodation may be divided into various functional groups as follows:

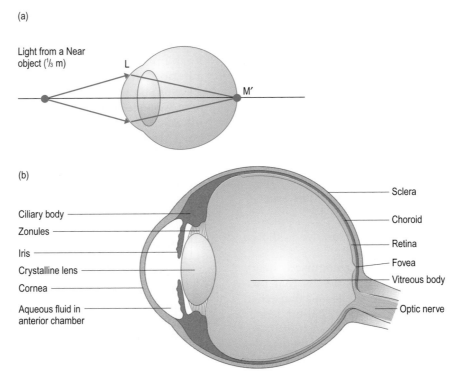

(a)

Light from a Near
object (⅓ m)

L

M′

(b)

Ciliary body

Zonules

Iris

Crystalline lens

Cornea

Aqueous fluid in
anterior chamber

Sclera

Choroid

Retina

Fovea

Vitreous body

Optic nerve

Figure 13.2 (a) Accommodated eye: light from a near object placed at ⅓ m produces a vergence at the eye of −3.00 D. The eye needs to accommodate by +3.00 D in order to neutralise this negative vergence. (b) The human eye.

- Tonic accommodation
- Convergence accommodation
- Proximal accommodation
- Reflex accommodation
- Voluntary accommodation.

Tonic accommodation represents the resting state of accommodation, being that amount of accommodation present in the absence of a stimulus. An eye is never totally unaccommodated which is why we describe an unaccommodated eye as an eye in its *weakest* dioptric state. This is also referred to as **accommodative tonus**, which is described in Chapter 8. **Convergence accommodation** is that amount of accommodation stimulated by convergence. In youth, the accommodative response follows convergence of the eye. The reaction time for convergence is about 0.2 seconds, which is almost twice as fast as that for accommodation. Accommodation lags behind and takes its

cue from convergence. Convergence is often said to 'drive' accommodation. **Proximal accommodation** is the amount of accommodation induced by the individual's awareness of the proximity of an object. **Reflex accommodation** is the normal involuntary response to blur in order to maintain a clear retinal image. **Voluntary accommodation** does not depend on the presence of a stimulus. The ability to relax accommodation from some near focus position is easily learned.

Spectacle and ocular accommodation

Accommodation can be 'measured' at two positions. Accommodation measured in the plane of the spectacle lens is known as **spectacle accommodation** and is defined as 'the accommodation required to neutralise negative vergence arising from a near object, measured in the plane of the spectacle lens'. The symbol for spectacle accommodation is A_{spec}. If accommodation is measured at the reduced surface, we have determined the **ocular accommodation**. This second term is defined as 'the accommodation required to neutralise negative vergence arising from a near object, measured at the reduced surface'. The symbol for ocular accommodation is A_{oc}. The general symbol for the amplitude of accommodation is A.

The clinical measurement of the amplitude of accommodation

The usual method of measuring the amplitude of accommodation for a subject involves a technique known as the **blur technique**, and requires the use of the RAF near point rule (Figure 13.3), which possesses a centimetre scale, a dioptre scale and an age scale! The amplitude of accommodation is measured with the distance correction in place, so any spherical and/or astigmatic ametropia is corrected and the individual has a real artificial near point (i.e. in front of the eye). The RAF rule should be placed below the patient's nose in a slightly depressed position to mimic the action of the eyes when reading. This method measures the spectacle amplitude of accommodation (the amplitude of accommo-

dation measured in the spectacle plane), which is the value required when calculating the power of the near spectacle correction. Monocular amplitudes are measured first (to screen for anomalies of the third cranial nerve) and should be approximately equal. The binocular amplitude is usually greater than either monocular value because convergence drives accommodation. Either the N5 letters or the telephone numbers are used as a target. The amplitude of accommodation is read off the scale from the back of the slider. There are two variations on the blur technique.

Method 1

The individual moves the near print along the sliding scale from the remote end of the rule until the print just blurs. The amplitude of accommodation is read off at this point. This is called the 'push-up-to-blur' method.

Method 2

The individual starts with the print at the end of the scale closest to the face and moves the print away until it is just legible. The amplitude of accommodation is read off at this point. This is called the 'push-down-to-clear' method.

If both techniques are used, the recorded amplitude of accommodation should be the average of the readings obtained with both methods. Possible subjective variations in recording the clinical amplitude of accommodation may be caused by the following:

- The depth of field: when accommodation occurs, the pupil constricts, so increasing the depth of field. This means that the patient can bring the target closer than the conjugate focus with the macula when fully accommodated.
- Target luminance: the individual may have trouble reading a certain size print in low or poor illumination, but may read it easily in improved illumination. This may also constrict the pupil, causing a further increase in the depth of field.

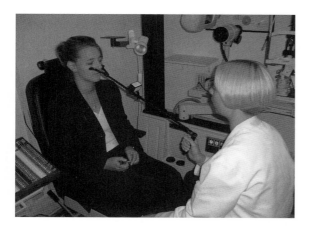

Figure 13.3 The RAF Near Point Rule (Haag-Streit UK Ltd).

- Variation in tolerance to blur: the individual tolerates blur on a large-sized letter better than on a small-sized letter, if the size of the blur disc is assumed to be the same for both letters. The blur ratio (disc size : letter size) means that a large letter may still be legible whereas a small letter may not with the same amount of blur.

Definitions relevant to accommodation

Accommodation, amplitude of accommodation, spectacle accommodation and ocular accommodation have all been defined in the preceding text. The following are other terms of interest.

The true far point M_R

That point conjugate with the centre of the macula by refraction at the uncorrected and unaccommodated eye.

In other words, how far an uncorrected and unaccommodated eye can see.

The true near point M_P

That point conjugate with the centre of the macula by refraction at the uncorrected but fully accommodated eye.

How close an uncorrected and fully accommodated eye can see.

The range of accommodation

The distance between the true far point and the true near point.

As this a distance between two true points, the eye is uncorrected.

The artificial far point Art_{MR}

That point conjugate with the centre of the macula by refraction at the corrected and unaccommodated eye.

This translates into how far a corrected and unaccommodated eye can see. With regard to reading spectacles, the position of the artificial far point is found by taking the reciprocal of the reading addition.

The artificial near point Art_{MP}

That point conjugate with the centre of the macula by refraction at the corrected but fully accommodated eye.

How close a corrected and fully accommodated eye can see.

The range of clear vision

The distance between the artificial far point and the artificial near point.

The use of the word artificial assumes that the eye is corrected.
Notice how these definitions vary by only one or two words.

Presbyopia and the near addition

The amplitude of accommodation decreases gradually with advancing age. This is well known and is demonstrated in Table 13.1. It is the reduction in amplitude of accommodation with age that causes a patient to require near vision spectacles.
Causes of the drop in amplitude with age include the following:

Table 13.1 The variation of amplitude of accommodation with age

Age (Years)	Amplitude of Accommodation (D)
10	14
20	10
30	8
40	5–6
45	3–4
50	2
60	1
70	<1

- The lens capsule gradually loses elasticity with continued growth of the lens.
- From about 30 years of age the lens nucleus begins to stiffen.
- A thickening of the ciliary muscle and growth of the lens reduce the tension in the zonules.

At around 45 years of age for someone with emmetropia or corrected ametropia, the amplitude of accommodation has reduced to such an extent that near vision becomes difficult or impossible to maintain comfortably. At this point, some additional plus power is required to make up for the eye's inability to accommodate sufficiently. This additional plus power is called the **near vision addition**. **Presbyopia** is said to be present when the amplitude of accommodation has dropped to around 3.00 D. Clinical evidence taken from innumerable case histories suggests that patients are able to maintain comfortable near vision if they exert no more that two-thirds of their amplitude of accommodation. Any sustained focusing requirement that exceeds this accommodative demand is likely to produce symptoms, which include:

- near blur
- distance blur after prolonged close work
- eyestrain and headache caused by excessive contraction of the ciliary muscle
- headache around the temple and forehead
- reading difficulty in poor light

The reading addition or 'Add' for a particular patient depends on the required working distance and the amplitude of accommodation available, and can be estimated using the equation:

$$\text{Add} = |L| - 2/3\,\text{Amp}$$

where:

$$L = \frac{1}{l}$$

and is the vergence of the incident light arriving at the lens. The symbol l is the working,

or near object, distance in metres. The vertical lines either side of L mean that the sign is ignored. If a patient habitually works at a distance of, say, 20 cm, the onset of presbyopia is earlier than for a patient who habitually works at, say, 50 cm. As the patient's amplitude of accommodation becomes insufficient, the patient must either increase the working distance or wear a corrective lens to neutralise the negative vergence that can no longer be overcome by accommodation.

All near vision tests should be performed after the distance prescription has been ascertained and the trial frame adjusted to represent the correct near centration distance for the required working distance. The first task is to estimate the addition using the equation given above or the values quoted in Table 13.2. The power of the addition must then be refined and there are various techniques available to do this. Determination and refinement of the near addition are described in Chapter 14. The estimated addition for various working distances and amplitudes of accommodation are given in Table 13.2.

Near test charts

When assessing near vision in practice, most near charts are constructed using a Times New Roman font, e.g. the reading test types approved by the Faculty of Ophthalmologists (**Figure 13.4**). The numbers used within such a chart refer to the point size and are prefixed by the letter N, e.g. N5, N6, N8, N10 and N12. In modern digital typography, a letter is 'imagined' to be enclosed in a rectangular boundary box and the height of this box is called the point size. The point size refers to the vertical space occupied by a letter, but does not allow us to state the letter heights, which can be determined only by measurement. Font is the term used for a typestyle, e.g. Times New Roman, Arial. Knowing the point size gives no indication of the letter heights because different fonts in the same point size generally have slightly different letter heights. In addition, as the letters are not the same design as Snellen chart letters, it is not possible to obtain a genuine Snellen equivalent with near vision test types in Times New Roman.

Table 13.2 The estimated addition for various working distances and amplitudes of accommodation

Amplitude (D)	Age (years)	Working distances (cm)						
		25	30	35	40	45	50	55
		Estimated addition (D)						
5.00	40	0.75	0.00	0.00	0.00	0.00	0.00	0.00
3.00	45	2.00	1.25	1.00	0.50	0.25	0.00	0.00
2.00	50	2.75	2.00	1.50	1.25	1.00	0.75	0.50
1.50	55	3.00	2.25	2.00	1.50	1.25	1.00	0.75
1.00	60	3.25	2.75	2.25	1.75	1.50	1.25	1.00
0.75	65	3.50	2.75	2.50	2.00	1.75	1.50	1.25
0.50	70	3.75	3.00	2.50	2.25	2.00	1.75	1.50

After Tunnacliffe (1993).

READING TEST TYPES
as approved by
THE FACULTY OF OPHTHALMOLOGISTS,
LONDON, ENGLAND

N. 5

He moved forward a few steps: the house was so dark behind him, the world so dim and uncertain in front of him, that for a moment his heart failed him. He might have to search the whole garden for the dog. Then he heard a sniff, felt something wet against his leg — he had almost stepped upon the animal. He bent down and stroked its wet coat. The dog stood quite still, then moved forward towards the house, sniffed at the steps, at last walked calmly through the open door as though the house belonged to him. Jeremy followed, closed the door behind them; then there they were in the little dark passage with the boy's heart beating like a drum, his teeth chattering, and a terrible temptation to sneeze hovering around him. Let him reach the nursery and establish the animal there and all might be well, but let them be discovered, cold and shivering, in the passage, and out the dog would be flung. He knew so exactly what would happen.
(From "Jeremy" by Hugh Walpole).

wire sons vain error unwise cream remove

N. 6

The camp stood where, until quite lately, has been pasture and ploughland; the farm house still stood in a fold of the hill and had served us for battalion offices; ivy still supported part of what had once been the walls of a fruit garden; half an acre of mutilated old trees behind the washhouses survived of an orchard. The place had been marked for destruction before the army came to it. Had there been another year of peace, there would have been no farmhouse, no wall, no apple trees. Already half a mile of concrete road lay between bare clay banks, and on either side a chequer of open ditches showed where the municipal contractors had designed a system of drainage. Another year of peace would have made the place part of the neighbouring suburb. Now the huts where we had wintered waited their turn for destruction.
(From "Brideshead Revisited" by Evelyn Waugh)

nervous manner immune over unanimous wear

N. 8

And another image came to me, of an arctic hut and a trapper alone with his furs and oil lamp and log fire; the remains of supper on the table, a few books, skis in the corner; everything dry and neat and warm inside and outside the last blizzard of winter raging and the snow piling up against the door. Quite silently a great weight forming against the timber; the bolt straining in its socket; minute by minute in the darkness outside the white heap sealing the door, until quite soon when the wind dropped and the sun came out on the ice slopes and the thaw set in a block would move, slide and tumble high above, gather way, gather weight, till the whole hillside seemed to be falling, and the little lighted place would crash open and splinter and disappear, rolling with the avalanche into the ravine. *(From "Brideshead Revisited" by Evelyn Waugh)*

immense snow came near arrow use.

Figure 13.4 The Faculty of Ophthalmologists' Reading Test Types (Keeler Ltd).

When assessing near vision, practitioners should also note reading fluency along with the size of print resolved.

Uses of near test charts in practice include:

- to determine near acuities at a reading distance
- to present a suitable target to view while determining a near refractive correction
- to verify acuity when spectacles are being collected
- to investigate the effect of a correction on reading fluency
- to compare acuity and reading fluency with different optical corrections.

The following are ideal design features of a near test chart:

- The reading task of a near chart should be the same at each print size level.
- A near chart should be suitable for use by those with good and poor vision.
- The progression of size levels should be logarithmic.
- The separation between lines should relate to the acuity size of the print.
- The number of words and letters on each line should be kept constant.

The following are problems with existing designs:

- The most common charts are not useful if only a rapid assessment is required.
- Sometimes it is more appropriate to obtain an estimate of the Snellen equivalent.
- Some cards are not suitable for young children.
- Context can be used to predict words.
- Sometimes, patients are reluctant to stop reading until they have 'finished the story'.
- Passages, and words within a passage, are not matched for difficulty.
- The changes in size between successive passages may not be regularly scaled.

As mentioned above, most near test charts used in practice are Printers Point System Times New Roman style charts. Alternative and superior near test charts include:

- Bailey–Lovie Word Reading Chart (**Figure 13.5**)
- Practical Near Acuity Chart (**Figure 13.6**)
- Institute of Optometry Near Test Chart (**Figure 13.7**)

These are all logarithmic in design.

The Bailey–Lovie Reading Chart is a word-reading chart designed to test the resolution components of the ability to see to read and not the ability to read. It does not use continuous text but extends from N80 to N2 to measure resolution threshold. It is a logarithmic design with the scale 80, 64, 48, 40, 32, 24, 20, 16, 12, 10, 8, 6, 5, 4, 3, 2.5, 2. The Bailey–Lovie Reading Chart is very useful in low vision practice because it gives clues to distance change, size change, dioptric power or magnification in

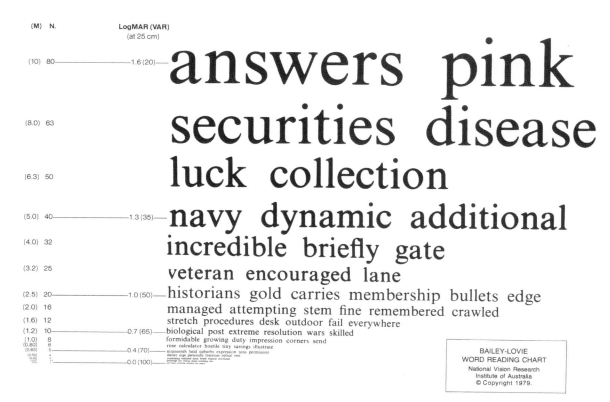

Figure 13.5 The Bailey–Lovie Word Reading Chart (Haag-Streit UK Ltd).

Figure 13.6 The Practical Near Acuity Chart (IOO Marketing Ltd).

order to achieve specific gains in resolution. For example, a patient achieves N24 with a +4.00 D Add at 25 cm but an acuity of N6 is desired. Remember that the key to using any logarithmic design is in the number of steps. Using the above scale we can see that the patient requires a six-step gain in resolution (N24 to N6 is six steps). To achieve N6 acuity we can:

1. Move the chart six steps closer (25 cm → 6 cm) or
2. Increase the reading addition by six steps (+4.00 D → +16.00 D)

Interestingly, a +16.00 D Add gives a working distance of 100/16 = +6.25 cm which, in practical terms, is the same distance suggested in (1) above. The traditional way to determine the required magnification is to divide the achieved acuity by the desired acuity. In this case N24/N6 = 4× magnification. Nominal magnification is given by F/4 which in this case is +16.00/4 = 4×. It all makes sense!

The Practical Near Acuity Chart uses print sizes from N80 to N5. The designers did not incorporate print smaller than N5 to avoid demoralisation and also to address the practical irrelevance of very small print. The chart design is a logarithmic degression of print size and spacing (0.1 logMAR). It has an equal number of related words on each line and employs one three-letter word, one four-letter word and one five-letter word. The designers suggest that the five-letter word scores 0.04 logMAR units and the four- and three-letter words score 0.03 logMAR units (total 0.1 logMAR units). This degree of accuracy is useful if very precise acuity monitoring is required.

The Institute of Optometry Near Test Chart incorporates groups of words in a pseudo-random order from N36 to N4.5 and also includes a column of single words from N36 to N4.5. This single column is useful because it provides a rapid indication of near acuity. The groups of words can then be used to determine threshold. The words used have been selected from 15 of those that are most commonly encountered by young people, so even poor readers should be able to cope with the task.

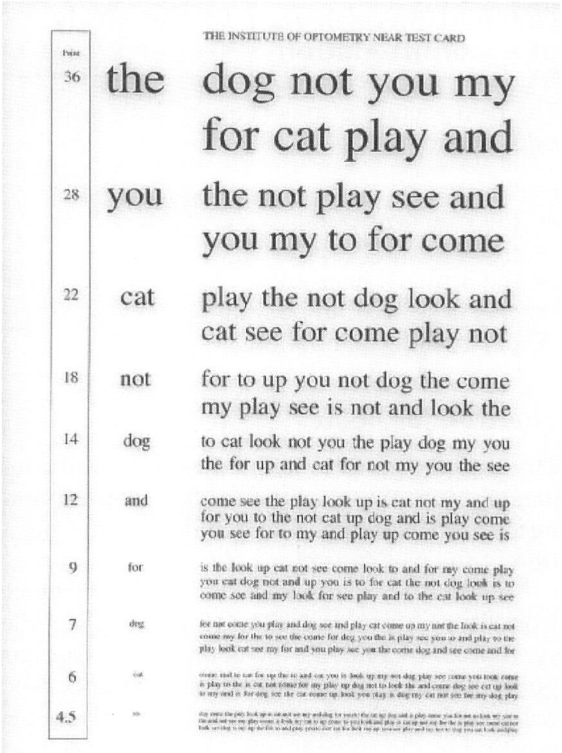

Figure 13.7 The Institute of Optometry Near Test Card (IOO Marketing Ltd).

Suggested viewing distances in centimetres for the chart are given in Table 13.3.

These distances form a logarithmic scale, which can be used to overcome the 'ceiling' and 'floor' effects experienced with non-logarithmic designs. The 'ceiling' effects occurs when the patient reads the bottom line before the threshold is reached and the 'floor' effect occurs when the patient cannot read the top line of the chart. For example, a patient easily reads N4.5 at the recommended distance of 40 cm. The practitioner moves the chart away to 64 cm. This is two 'steps' further out on the above scale (Table 13.3). The patient can now just read N4.5. Moving two 'steps' down (Table 13.4) shows that the threshold near acuity is N3, logMAR −0.1 and Snellen 6/5.

In another example using the Institute of Optometry Near Test Chart, a patient can just read the top line (N36) at a distance of 20 cm. As 20 cm is three 'steps' closer than 40 cm (Table 13.3) we must move three 'steps' up (Table 13.4). The patient's near acuity at 40 cm is N70, logMAR 1.3 and Snellen 6/120. For a comprehensive review of the Institute of Optometry Near Test Chart the reader is directed to Evans and Wilkins (2001).

Hypermetropia and accommodation

Facultative and absolute hypermetropia

For a hypermetropic patient the ocular refraction (K) is always greater than zero (K is always plus). Therefore accommodation may be used to give the extra positive power required by the eye. For hypermetropia we have three possibilities:

1. $K > A$
2. $K = A$
3. $K < A$.

Table 13.3 Suggested viewing distances for use with the Institute of Optometry Near Test Chart

Distance (cm)	80	64	48	40	32	24	20	16	12	10	9

Table 13.4 Conversion table for use with the Institute of Optometry Near Test Chart

Point size	UK Snellen	Decimal	LogMAR
	Equivalents of different scales of visual acuities		
N70	6/120	0.05	1.3
N60	6/100	0.06	1.2
N48	6/80	0.07	1.1
N36	6/60	0.10	1.0
N28	6/48	0.12	0.9
N22	6/38	0.16	0.8
N18	6/30	0.20	0.7
N14	6/24	0.25	0.6
N12	6/18	0.30	0.5
N9	6/15	0.40	0.4
N7	6/12	0.50	0.3
N6	6/9	0.60	0.2
N4.5	6/7.5	0.80	0.1
N3.5	6/6	1.0	0.0
N3	6/4.8	1.2	−0.1

For (1), it is *not* possible for the patient to obtain a clear retinal image by accommodation. The amount of hypermetropia that can be overcome by accommodation is called **facultative hyper-metropia** (H_F). That which cannot be overcome by accommodation is called **absolute hyper-metropia** (H_A).

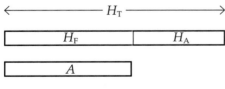

$H_F = A$
$H_F + H_A = H_T$.

As the amplitude of accommodation (A) reduces with age so does H_F; therefore, most of the **total hypermetropia** (H_T) becomes **absolute**.

Latent and manifest hypermetropia

In young people with hypermetropia, it is not unusual to find that the individual does not accept the full refractive error as found by an objective method, e.g. retinoscopy. The maximum positive sphere accepted by the individual while maintaining clear distance vision

is called the **manifest hypermetropia** (H_M). The remaining hypermetropia is called the **latent hypermetropia** (H_L).

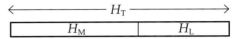

Latent hypermetropia is the result of an over-accommodated state, so that, when the manifest correction is worn, the eyes are still accommodating to some extent. In youth, latent hypermetropia may be as high as half the total hypermetropia, but it usually reduces to zero in middle age.

Calculations relevant to the study of accommodation

The next section of this chapter deals with the calculation of:

- the true far point
- the true near point
- the artificial far point
- the artificial near point
- spectacle accommodation
- ocular accommodation.

The true far point

The **true far point** is defined as the point con-jugate with the macula in the *unaided* and *unaccommodated* eye. The symbol for the posi-tion of the true far point is M_R.

The true far point distance is the distance from the reduced surface to M_R and is given the symbol k. To find the position of the true far point, the ocular refraction is required. The reciprocal of the ocular refraction gives the position of the true far point in metres:

$$k = \frac{1}{K}$$

The true far point in the unaccommodated emmetropic eye is at *infinity*. The true far point in the unaccommodated myopic eye is at some distance *in front* of the reduced surface. The true far point in the unaccommodated hypermetro-pic eye is at some distance *to the right* of the reduced surface (**Figures 13.8–13.10**).

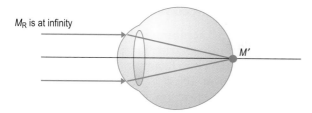

Figure 13.8 The true far point in emmetropia (unaccommodated eye): M_R is at infinity.

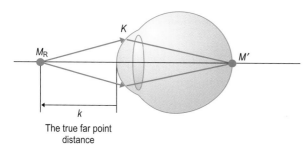

Figure 13.9 The true far point in myopia (unaccommodated eye): M_R is to the left of the reduced surface.

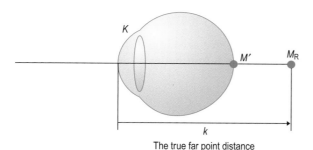

Figure 13.10 The true far point in hypermetropia (unaccommodated eye): M_R is to the right of the reduced surface.

Example 13.1
Find the position of the true far point in a myopic eye where $K = -4.00\,D$.

This is very straightforward! As we have been given the ocular refraction, all we have to do is take the reciprocal. We then obtain the position of the far point in metres:

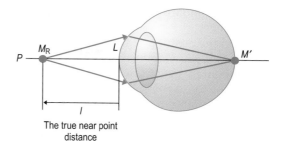

The true near point distance

Figure 13.11 The true near point: M_P is the true near point. L is the vergence at the fully accommodated eye: $L = K + (-A)$ and $l = 1/L$.

$$k = \frac{1}{K} \qquad k = \frac{1}{-4.00} = -0.25\,m$$

The true far point is therefore 25 cm in front (to the left) of the reduced surface.

Example 13.2
Find the position of the true far point in a hypermetropic eye where $F_{sp} = +5.00\,D$ at 14 mm.

This example is slightly more involved as we have not been given K directly. It can of course be easily found using F_{sp} and the given vertex distance. We need to use the expression:

$$K = \frac{F_{sp}}{1 - (dF_{sp})}$$

$$K = \frac{+5.00}{1 - (0.014 \times +5.00)} = +5.376\,D$$

so:

$$K = \frac{1}{K} \qquad k = \frac{1}{+5.376} = +0.186\,m$$

The true far point is therefore 18.6 cm behind (to the right of) the reduced surface.

The true near point

The **true near point** (Figure 13.11) is the point conjugate with the macula in the *unaided* but *fully* accommodated eye. The symbol for the position of the true near point is M_P. The true near point distance is the distance from the reduced surface to M_P and can be **143** given the standard symbol for an object distance, l.

To find the position of the true near point we need to find the vergence arriving at the fully accommodated eye from an object placed as close to the eye as possible, while at the same time maintaining a clear retinal image. Put another way, this is the maximum negative vergence that the amplitude of accommodation of the eye can cope with. The symbol for this value is simply L_{eye} and is given by:

$$L_{eye} = K + (-A)$$

where K is the ocular refraction and A is the amplitude of accommodation. As always, l is given by:

$$l = \frac{1}{L}$$

The above expressions can be used to find the position of the true near in people with emmetropia, myopia and hypermetropia.

Example 13.3

A myopic eye has an ocular refraction of −8.00 D and an ocular amplitude of accommodation of 4.00 D. Find the positions of the true far and near points.

As the ocular refraction is given, the position of the true far point can be found immediately from:

$$k = \frac{1}{K} \qquad k = \frac{1}{-8.00} = -0.125\,m$$

The vergence arriving at the fully accommodated eye is found using:

$$L_{eye} = K + (-A)$$

$$L_{eye} = -8.00 + (-4.00) = -12.00\,D$$

Hence, the position of the true near point can be found using:

$$l = \frac{1}{L_{eye}} \qquad l = \frac{1}{-12.00} = -0.0833\,m$$

The true far point is located 12.5 cm in front of the eye and the true near point is located 8.33 cm in front of the eye.

The range of accommodation

The **range of accommodation** is simply the distance from the true far point to the true near point, which in Example 13.3 is 4.17 cm. As long as the near object is placed between these two points a clear image is formed on the retina.

Example 13.4

A hypermetropic eye with a spectacle refraction F_{sp} of +3.817 D at 12 mm has an ocular amplitude of accommodation of 6.00 D. Find the positions of the true far and near points.

The position of the true near point is given by the distance k. We therefore need to find the ocular refraction using the given F_{sp} and vertex distance:

$$K = \frac{F_{sp}}{1 - (dF_{sp})}$$

$$K = \frac{+3.817}{1 - (0.012 \times +3.871)} = +4.00\,D$$

and:

$$k = \frac{1}{K} \qquad k = \frac{1}{+4.00} = +0.25\,m$$

The vergence arriving at the fully accommodated eye is found using:

$$L_{eye} = K + (-A)$$

$$L_{eye} = +4.00 + (-6.00) = -2.00\,D$$

Again, the position of the true near point can be found using:

$$l = \frac{1}{L_{eye}} \qquad l = \frac{1}{-2.00} = -0.50\,m$$

The true far point is located 25 cm behind the eye and the true near point is located 50 cm in front of the eye.

The artificial far point

The **artificial far point** (Art$_{MR}$) is the point conjugate with the macula in the *corrected* but *unaccommodated* eye. When discussing the artificial far point, two possible clinical situations need to be considered.

1. If the eye is fully corrected by a spectacle lens, the artificial far point is at infinity. This is what we usually want to happen when we correct a patient for distance vision.
2. If the eye is not fully corrected, the artificial far point is at some distance from the spectacle lens. The most practical example of this is when a patient is given a reading addition because distance vision is blurred if a reading correction is used.

Example 13.5

An emmetropic eye is given an addition of +1.00 D for reading. What is the furthest point from the spectacle lens that clear vision is possible?

To find the *furthest* point at which clear vision is possible when a reading addition is worn, simply take the reciprocal of the reading addition. This gives the position of the artificial far point in metres. A +1.00 D Add therefores produce an artificial far point of 1 m. A +2.00 D Add gives an artificial far point of 0.50 m, etc.

Example 13.6

A person with −10.00 D myopia wears a −8.00 D lens for reading. What is the position of the artificial far point?

The above tells us that this patient requires a −10.00 D lens to see clearly at infinity. If the patient wears a −8.00 D lens, −2.00 D is missing and distance vision blurred. It is important here to remember that a near object generates *negative vergence*. Placing an object at the appropriate distance from the lens produces the missing −2.00 D and provides clear vision. To find the appropriate object distance, we can simply use:

$$l = \frac{1}{L}$$

where L is the 'missing vergence' and l the object distance required to generate this 'missing vergence'. In this example, the missing vergence is −2.00 D. The required object distance l is therefore:

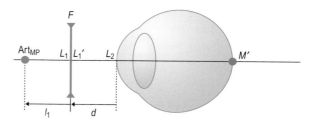

Figure 13.12 The artificial near point (fully accommodated eye): L_2 is the vergence at the fully accommodated eye: $L_2 = K + (-A)$ and $l_1 = \dfrac{1}{L_1}$.

$$l = \frac{1}{-2.00} = -0.50 \, \text{m}$$

So, if an object is placed at a distance −50 cm from the lens, the vergence arriving at the lens L is −2.00 D. As the power of the lens is −8.00 D, the vergence leaving the lens L' is:

$$L' = L + F = -2.00 + (-8.00) = -10.00 \, \text{D}$$

As this eye needs a spectacle correction of −10.00 D to see, a clear image is formed on the retina if an object placed is placed at a distance of −50 cm to the left of the lens.

The artificial near point

The **artificial near point** (Art$_{MP}$) is the point conjugate with the macula in the *corrected* but *fully* accommodated eye (**Figure 13.12**). To calculate the position of the artificial near point, the vergence at the fully accommodated eye (L_2) is required. This is given by:

$$L_2 = K + (-A)$$

where K is the ocular refraction and A the amplitude of accommodation. In calculations involving the artificial near point, step-back ray-tracing is used to find L_1', L_1 and hence l_1. The distance l_1 gives the position of the artificial near point Art$_{MP}$. As usual:

$$l_1 = \frac{1}{L_1}$$

which gives the position of the artificial near point from the spectacle lens in metres. **145**

Step-back ray-tracing is revised in Chapter 1. Remember that the ocular refraction must always be found using the *distance* spectacle correction.

Example 13.7

A patient is corrected for distance vision using a −4.00 D at 14 mm. The amplitude of accommodation A is 8.00 D. Find the position of the artificial near point.

First find the ocular refraction using the *distance* spectacle correction and vertex distance:

$$K = \frac{F_{sp}}{1-(dF_{sp})}$$

$$K = \frac{-4.00}{1-(0.014 \times -4.00)} = -3.79\,\text{D}$$

The vergence at the fully accommodated eye (L_2) is given by:

$$L_2 = K + (-A)$$
$$L_2 = -3.79 + (-8.00) = -11.79\,\text{D}.$$

Now use a step-back ray-trace to find L_1', L_1 and l_1. As always, work in two columns. The required vergences are shown in Figure 13.12. The computation is given below:

Vergences (D)   Distances (m)

$$L_2 = -11.97\,\text{D} \quad \rightarrow \quad l_2 = \frac{n}{L_2}$$

$$l_2 = \frac{1}{-11.79} = -0.0848\,\text{m}$$

$$l_1' = l_2 + d$$

$$L_1' = \frac{n}{l_1'} \quad \leftarrow \quad l_1' = -0.0848 + 0.014 = -0.0708\,\text{m}$$

$$L_1' = \frac{1}{-0.0708} = -14.12\,\text{D}$$

$$L_1' = L_1 + F$$

which rearranges to become:

$$L_1 = L_1' - F$$

$$L_1 = -14.12 - (-4.00) = -10.12\,\text{D} \quad \rightarrow \quad l_1 = \frac{1}{L_1}$$

$$l_1 = \frac{1}{-10.12} = -0.0988\,\text{m}$$

The artificial near point is therefore located 9.88 cm to the left of the lens.

Example 13.8

An individual with a distance spectacle correction of +6.00 D at 14 mm is given a +1.00 D addition for reading. Find the position of the near point when wearing the reading lens if the patient exerts 2.50 D of accommodation.

If the distance correction is +6.00 D and the reading addition +1.00 D, the power of the reading lens is +6.00 + (+1.00) = +7.00 D.

The question asks us to find the position of the near point when wearing the reading lens, which is of course the artificial near point Art$_{MP}$.

First of all we must use the *distance* correction, to find the ocular refraction:

$$K = \frac{F_{sp}}{1-(dF_{sp})}$$

$$K = \frac{+6.00}{1-(0.014 \times +6.00)} = +6.55\,\text{D}$$

The vergence at the fully accommodated eye (L_2) is given by:

$$L_2 = K + (-A)$$
$$L_2 = +6.55 + (-2.50) = +4.05\,\text{D}$$

Now use a step-back ray-trace to find L_1', L_1 and l_1. The required vergences are shown in (Figure 13.12) and the computation is given below:

Vergences (D)   Distances (m)

$$L_2 = +4.05\,\text{D} \quad \rightarrow \quad l_2 = \frac{n}{L_2}$$

$$l_2 = \frac{1}{+4.05} = +0.247\,\text{m}$$

$$l_1' = l_2 + d$$

$$L_1' = \frac{n}{l_1'} \quad \leftarrow \quad l_1' = +0.247 + 0.014 = +0.261\,\text{m}$$

$$L_1' = \frac{1}{+0.261} = +3.83\,\text{D}$$

$$L_1' = L_1 + F$$

which rearranges to become:

$$L_1 = L_1' - F$$

$$L_1 = +3.83 - (+7.00) = -3.17\,\text{D} \qquad \rightarrow \qquad l_1 = \frac{1}{L_1}$$

$$l_1 = -\frac{1}{-3.17} = -0.316\,\text{m}$$

The artificial near point is therefore located 31.6 cm to the left of the +7.00 D reading lens.

The range of clear vision

The range of clear vision is the distance from the **artificial far point to the artificial near point.** This translates into how far (distance to near) an individual can see when wearing a spectacle lens.

Example 13.8

An individual is corrected for distance vision using a −8.00 D lens at 10 mm. He is given a reading addition of +2.00 D and has an amplitude of accommodation of +1.50 D. Find the range of clear vision obtained when wearing the reading lens if the individual exerts two-thirds of his available amplitude of accommodation.

The range of clear vision is the distance from the artificial far point to the artificial near point. To find the position of the Art_{MR}, all we have to do is take the reciprocal of the reading addition:

$$\text{Art}_{MR} = \frac{1}{\text{Add}} \qquad \text{Art}_{MR} = \frac{1}{2.00} = 0.50\,\text{m}$$

The artificial far point is therefore −50 cm from the lens. The next task is to find the ocular refraction using the *distance* correction and vertex distance:

$$K = \frac{F_{sp}}{1 - (dF_{sp})} \qquad K = \frac{-8.00}{1 - (0.010 \times -8.00)} = -7.41\,\text{D}$$

The individual has to exert only two-thirds of his available amplitude of accommodation. Two-thirds of 1.50 is 1.00 D, so A is taken to be 1.00 D. The vergence at the fully accommodated eye (L_2) is therefore:

$$L_2 = K + (-A) \qquad L_2 = -7.41 + (-1.00) = -8.41\,\text{D}$$

We can now use a step-back ray-trace to find L_1', L_1 and l_1. The required vergences are again illustrated in Figure 13.12 and the computation is given below.

If the distance correction is −8.00 D and the reading addition +2.00 D, the power of the reading lens is $-8.00 + (+2.00) = -6.00\,\text{D}$. The question states that the patient is wearing the reading lens so, in our ray-trace, $F = -6.00\,\text{D}$ at 10 mm. The ray-trace begins at L_2.

Vergences (D)   Distances (m)

$$L_2 = -8.41\,\text{D} \qquad \rightarrow \qquad l_2 = \frac{n}{L_2}$$

$$l_2 = \frac{1}{-8.41} = -0.119\,\text{m}$$

$$l_1' = l_2 + d$$

$$L_1' = \frac{n}{l_1'} \qquad \leftarrow \qquad l_1' = -0.119 + 0.010 = -0.109\,\text{m}$$

$$L_1' = \frac{1}{-0.109} = -9.18\,\text{D}$$

As always:

$$L_1' = L_1 + F$$

which rearranges to become:

$$L_1 = L_1' - F$$

$$L_1 = -9.18 - (-6.00) = -3.18\,\text{D} \qquad \rightarrow \qquad l_1 = \frac{1}{L_1}$$

$$l_1 = \frac{1}{-3.18} = -0.314\,\text{m}$$

The artificial near point is therefore located 31.4 cm to the left of the −6.00 D reading lens. When wearing the reading lens, the individual will achieve a range of clear vision from 50 cm to the left of the lens to 31.4 cm to the left of the lens. The actual distance between Art_{MR} and Art_{MP} is therefore 18.6 cm.

Example 13.9

A reduced eye has an ocular refraction of −7.299 D and an ocular amplitude of accommodation of 1.50 D. Find the positions of the true far and near points.

For the same eye, assume that the eye can comfortably use 1.00 D of accommodation to view an object 35 cm in front of a reading lens at 12 mm. Find the vergence at the eye, the vergence leaving the lens, the incident vergence at the lens and, hence, the power of the reading lens.

The position of the true far point is found by stating the distance k. As the ocular refraction is given in the question, k is given by:

$$k = \frac{1}{K} \qquad k = \frac{1}{-7.299} = -0.137\,\text{m}$$

To find the position of the true near point, we must first determine the vergence at the fully accommodated eye L_{eye}. This is found using:

$$L_{eye} = K + (-A)$$
$$L_{eye} = -7.299 + (-1.50) = -8.799\,\text{D}$$

Again, the position of the true near point can be found using:

$$l = \frac{1}{L_{eye}} \qquad l = \frac{1}{-8.799} = -0.114\,\text{m}$$

The true far point is therefore located 13.7 cm in front of the eye and the true near point is located 11.4 cm in front of the eye.

In the second part of this example we are told the following:

- The amplitude of accommodation actually used is 1.00 D.
- The object viewed is 35 cm in front of the reading lens.
- The vertex distance of the lens is 12 mm.

With reference to Figure 13.12, in order to find the power of the reading lens we need to know L_1 and L_1'. We can then find F using:

$$F = L_1' - L_1$$

L_1 can be found as the object distance l_1 has been given (−35 cm). We know the ocular refraction and the amplitude of accommodation, so the vergence at the eye can be calculated using:

$$L_2 = K + (-A)$$

From L_2 we can step-back to find L_1' and hence F, the power of the reading lens. However, the first task is to determine L_1:

$$L_1 = \frac{n}{l_1}$$

$$L_1 = \frac{1}{-0.350} = -2.86\,\text{D}$$

Now $K = -7.299$ and $A = +1.00$ so:

$$L_2 = K + (-A)$$
$$L_2 = -7.299 + (-1.00) = -8.299\,\text{D}$$

We can now use a step-back ray trace to find L_1'. The required vergences are again illustrated in Figure 13.12. Working in two columns:

Vergences (D)   Distances (m)

$$L = -8.299\,\text{D} \quad \rightarrow \quad l_2 = \frac{n}{L_2}$$

$$l_2 = \frac{1}{-8.299} = -0.120\,\text{m}$$

$$l_1' = l_2 + d$$

$$L_1' = \frac{n}{l_1'} \quad \leftarrow \quad l_1' = -0.120 + 0.012 = -0.108\,\text{m}$$

$$L_1' = \frac{1}{-0.108} = -9.259\,\text{D}$$

Now:

$$F = L_1' - L_1$$
$$F = -9.259 - (-2.86) = -6.399\,\text{D}$$

The power of the reading lens is therefore −6.399 D at 12 mm.

Spectacle accommodation

The accommodation required to neutralise negative vergence arising from a near object measured in the plane of the spectacle lens is known as **spectacle accommodation**. The symbol for spectacle accommodation is A_{spec} and its value

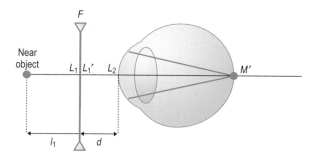

Figure 13.13 The spectacle and ocular accommodations: $A_{spec} = L_1$ with the sign changed. $A_{oc} = K - L_2$.

is simply the object vergence L with its sign changed.

Ocular accommodation

Ocular accommodation is the accommodation required to neutralise negative vergence from a near object measured in the plane of the eye. A_{oc} is the symbol for ocular accommodation. A_{oc} is given by:

$$A_{oc} = K - L_2$$

where K is the ocular refraction and L_2 the vergence arriving at the eye from the near object. The spectacle and ocular accommodations are illustrated in Figure 13.13.

Example 13.10

A thin spectacle lens of power −6.00 D is placed as a vertex distance of 14 mm. An object is placed ⅓ m from the spectacle plane. Calculate the spectacle and ocular accommodations.

The spectacle accommodation is found by determining the vergence arriving at the spectacle lens from the near object. If the object distance l is −⅓ m then:

$$L = \frac{1}{l} \qquad L = \frac{1}{-1/3} = -3.00\,\text{D}$$

The vergence arriving at the spectacle lens from the near object is therefore −3.00 D. Spectacle accommodation is the accommodation required to neutralise negative vergence from a near object. In this example, the spectacle accommodation is therefore

+3.00 D. The determination of the ocular accommodation involves a step-along ray-trace from the near object to the eye. We need to find the vergence arriving at the eye, L_2, and then compare this with the ocular refraction. The difference between the two is the ocular accommodation.

The step-along is performed using two columns, one for vergences (in dioptres) and one for distances (in metres). The vergences are illustrated in Figure 13.13.

Vergences (D)		Distances (m)

$$L_1 = \frac{1}{l} \qquad \leftarrow \qquad l_1 = -1/3\,\text{m}$$

$$L = \frac{1}{-1/3} = -3.00\,\text{D}$$

$$L_1 = -3.00\,\text{D}$$

$$F_1 = -6.00\,\text{D}$$

$$L_1' = L_1 + F_1$$

$$L_1' = -3.00 + (-6.00) = -9.00\,\text{D} \;\rightarrow\; l_1' = \frac{n}{L_1'} = \frac{1}{-9.00}$$

$$= -0.11\dot{1}\,\text{m}$$

$$l_2 = l_1' - d$$

$$L_2 = \frac{n}{l_2} = \frac{1}{-0.125} = -7.99\,\text{D} \;\leftarrow\; l_2 = -0.11\dot{1} - 0.014$$

$$= -0.125\,\text{m}$$

The vergence arriving at the eye from the near object is −7.99 D. We must now compare this with the ocular refraction to determine the ocular accommodation.

So, using the *distance* correction and vertex distance:

$$K = \frac{F_{sp}}{1 - (dF_{sp})} \qquad K = \frac{-6.00}{1 - (0.014 \times -6.00)} = -5.53\,\text{D}$$

Now:

$$A_{oc} = K - L_2$$
$$A_{oc} = -5.53 - (-7.99) = +2.64\,\text{D}$$

Essentially, when viewing the near object, the eye is receiving more negative vergence than it needs (L_2 is what it is receiving and K is what it **149**

$$L = \frac{1}{-0.347} = -2.88\,\text{D}$$

As an emmetropic individual has an ocular refraction of zero, the emmetropic eye must accommodate in order to neutralise the above negative vergence. The ocular accommodation A_{oc} is therefore +2.88 D.

Ocular accommodation: those with emmetropia versus those with ametropia

The values from the above ocular accommodation calculations show that the myopic patient accommodates *less* than one who is emmetropic when viewing the same near object and the hypermetropic patient accommodates *more* than one who is emmetropic when viewing the near same object. Thus hypermetropic patients may require a reading addition slightly earlier than myopic patients.

Depth of field

The depth of field may be defined as 'the distance over which an object may be moved along the axis of an optical system without causing its image to move away from the correct image plane by more than a stated tolerable amount'. If an optical system did not have any depth of field, an object not placed exactly in the correct object plane would be out of focus in the system's image plane. As far as the eye is concerned, this would mean that an object not situated at the far point, near point or within the area between theses two points would be blurred. A practical advantage of depth of field is that the eye does not need to adjust its focus for small variations in object distance, e.g. when reading a book where the various points on the page do not differ in distance from the eye by more than a few centimetres. The depth of field depends on pupil diameter, visual acuity, contrast and object distance.

Depth of field is of little interest in the prepresbyopic patient because the amplitude of accommodation is usually sufficient to deal with any problematic blurring of the retinal image. However, depth of field is useful to the presbyopic patient. As previously discussed it is expected that the amplitude of accommodation reduces with age. It is actually unproductive to try to measure the amplitude of accommodation of patients in those aged over 55, so it is rarely attempted in practice. The normal amplitude of accommodation declines with age until around the age of 55 years, when all that is left is depth of field. This tends to increase as a result of the increasing miosis (reduction in the pupil diameter) associated with ageing. It is interesting to note that between the ages of 42 and 60 years about 1.75 D of the clinical amplitude of accommodation is actually caused by depth of field and tolerance to blur. It is for this reason that depth of field has been termed the 'fictitious amplitude of accommodation'.

Concluding points for Chapter 13

This lengthy chapter has included:

- the purpose and function of accommodation
- definitions relevant to the study of accommodation
- the anatomical components involved in accommodation and accommodation theory
- the clinical measurement of the amplitude of accommodation
- presbyopia and the near addition
- hypermetropia and accommodation
- the true far point
- the true near point
- the artificial far point
- the artificial far point
- the artificial near point
- spectacle accommodation
- ocular accommodation
- calculations relevant to accommodation.

References

Drexler W, Findl O, Schmetterer L, Hirzenberger C K, Fercher A F (1998) Eye elongation during accommodation in humans – Differences between emmetropes and myopes. *Investigative Ophthalmology and Visual Science* **39**:2140–7

He J C, Gwiazda J, Thorn F, Held R, Huang W (2003) Change in shape and corneal wavefront aberrations with accommodation. *Journal of Vision* 3:456–63

Fincham E F (1937) The mechanism of accommodation *British Journal of Ophthalmology* 21(suppl VIII):1–80

Pierscionek B K (2005) Accommodation revisited. *Points de Vue* 52:20–6

Pierscionek B K, Popiolek-Masajada A, Kasprzak H (2001) Corneal shape change during accommodation. *Eye* 15:766–9

Tunnacliffe A H (1993) *Introduction to Visual Optics.* Association of the British Dispensing Opticians, London: Chapter 4

Further recommended reading

Elliot D (2003) *Clinical Procedures in Primary Eye Care.* Oxford: Butterworth-Heinemann

Bailey I L, Lovie J E (1980) The design and use of a new near vision chart. *American Journal of Optometry and Physiological Optics* 57:378–87

Evans B J W, Wilkins A J (2001) A new near vision test card. *Optometry Today* December 15

Rabbetts R B B (1998) *Bennett & Rabbetts' Clinical Visual Optics.* Oxford: Butterworth Heinemann

Tunnacliffe A H (1993) *Introduction to Visual Optics.* London: Association of the British Dispensing Opticians

Wolffsohn J S, Cochrane A L (2000) The practical near acuity chart and prediction of visual disability at near *Ophthalmic and Physiological Optics* 20(2):90–7

Zadnik K (1997) *The Ocular Examination – Measurements and findings.* Philadelphia: Saunders

Determination of the Near Addition

Andrew Franklin

Introduction

Once we have established a correction for distance vision, we can turn our attention to near vision. Clinically, 'normal' patients may be divided into three groups:

1. Those with pre-presbyopia
2. Those with early presbyopia
3. Those with late presbyopia

Pre-presbyopia

Patients in this group have enough accommodation and accommodative stamina to focus at near to their satisfaction. In these cases the amplitude of accommodation is measured and the two eyes compared because significant differences between them may indicate pathology. Provided that the ocular motor balance (OMB) for near is compensated, no extra correction needs to be made for near. It should be noted that, for this group of patients when testing accommodation, convergence and the OMB, the optical centres of the trial frame should remain at the distance setting.

Early presbyopia

Patients with early presbyopia are those who no longer have sufficient accommodation to provide adequate near vision without the aid of a correction. The age of onset of presbyopia varies with the level of visual task undertaken and with the fixation distance required, i.e. if the print is big and bold and you need to look at it only for a short time every day, you won't need reading glasses so soon. Furthermore, a patient who is 5 feet (1.5 m) tall probably needs a reading addition some years before another who is 6 feet 4 inches (1.9 m), because the latter normally has longer arms, which allows a longer working distance. People with myopia who need a medium-to-high correction often cheat a bit by looking obliquely through the

lower part of their spectacles. This is easier when deep frames are in fashion and may delay the need for a near correction for some years. However, those with low myopia who habitually read without spectacles may enfeeble their accommodation and require an earlier near correction. Most patients in this group are aged between 40 and 55 years but some younger patients who have abnormally low accommodative tonus may also be encountered.

The near addition required by someone with early presbyopia depends on both the working distance and the accommodation is still available, so the amplitude of accommodation may still be usefully measured. However, in 'real' optometric practice it is rare for a patient who already has a reading addition to have accommodative amplitude measured unless there are other clinical indications, e.g. a suspected third nerve problem. With that in mind, the early presbyopic group might consist of those patients about to have their first reading addition prescribed. Those with pre-existing reading additions are dealt with as having late presbyopia.

Late presbyopia

Patients with late presbyopia have no remaining accommodation of their own, so their reading addition is largely determined by the working distance required. Depth of focus remains and those with small pupil diameters may need a lower reading addition when compared with those with larger pupils. Tall people usually need lower reading additions as a result of their greater working distance. However, the ability to resolve fine detail, especially in low-contrast situations, may decline with both age and ocular disease. In patients with ocular disease such as age-related maculopathy, there may be a need to increase the magnification by increasing the addition, although this incurs a shorter working distance and smaller depth of focus. It is unproductive to try to measure the amplitude of accommodation of patients in the over-55 group, so it is rarely attempted. The normal amplitude of accommodation declines with age until around the age of 55 years, when all that is left is depth of focus. This tends to increase as a result of the increasing miosis associated with ageing. Even for asymptomatic patients there is considerable individual variation in measured amplitudes of accommodation. Reduced accommodation may also be associated with the following:

- Latent or inadequately corrected hypermetropia.
- Poor health (e.g. Graves' disease, alcoholism), drug treatment (e.g. for asthma, antidepressants) or drug abuse.
- Hysteria and stress, usually associated with over-stimulation of the sympathetic nervous system. Children whose parents have divorced recently, or who are attending a new school, are prone to this.
- Students (more usually female) around the age of 12–14 may experience a temporary accommodative palsy that usually resolves

spontaneously after a short time. Help with reading may be necessary for a while.
- Ocular disease (e.g. glaucoma, anterior uveitis, Adie's pupil).
- People with myopia often read without spectacles and their accommodation may become feeble through disuse. They also tend to have larger pupils, which reduces the depth of focus.
- People who have lived in sunnier climates often struggle at near when they come to Britain. This may be associated with lower levels of light at the blue end of the spectrum. There is also a correlation between ambient temperature and the age of onset of presbyopia (Weale 1981).

Higher than expected amplitudes of accommodation may be recorded in older patients or in those with small pupils caused by enhanced depth of focus. Examples include the following:

- Patients using pilocarpine eyedrops have small pupils and ciliary spasm.
- Some older patients (again more often female) develop spasm of the near reflex as a response to excessive demands on accommodation or convergence. Often cycloplegia is necessary to rule out myopia.

Chapter content

- Measurement of near vision function
- Determination of the reading addition
- VDU users

Measurement of near vision function

Dynamic retinoscopy

In Chapter 8 (discussing **static** methods of retinoscopy) much emphasis was given to the prevention of accommodation. In contrast, during **dynamic** retinoscopy, the patient is encouraged to focus on an accommodative target set at their habitual working distance. The target may consist of letters or an array of dots, and may be fixed on the retinoscope below the sight hole or be separate from the retinoscope. If the correct distance prescription is in place, we should in

theory see reversal at the habitual working distance, or indeed any other distance. In practice, however, it is usual to see a 'with' movement, which has been termed the 'dynamic' or 'accommodative lag'. In normal patients this is between 0.25 and 1.00 DS. It can be measured by either of the methods described below, no allowance for the working distance being needed with any method:

1. If a separate target is being used, it is kept at a constant distance while the working distance of the retinoscope is increased to the point at which 'reversal' occurs. The difference between the positions of the retinoscope and the target are measured with a centimetre rule and converted into dioptres.
2. More usually, plus spheres are placed before the eyes and the power increased until reversal is obtained. The most convenient way to do this is by using a

binocular rack of lenses, which allows rapid measurement and comparison between the eyes. The amount of plus sphere required is termed the 'low neutral'.

3. If we continue to add plus power, reversal is usually still seen, presumably because some negative relative accommodation is present. Typically this continues up to a value of about +1.50 DS in those aged under 30, and rather less in older patients. At this point an 'against' reflex is seen (the 'high neutral').

In those with pre-presbyopia, an excessive accommodative lag may indicate uncorrected hypermetropia or poor accommodative tonus. A small accommodative lag may indicate that the patient is 'calling up' extra innervation to overcome exophoria at near, which is in turn inducing more accommodation than usual. People with presbyopia should be tested wearing an estimated near addition over their distance prescription. Those with late presbyopia probably require the 'high neutral' value on top of the estimated addition, but those with early presbyopia probably require about 0.50 D less. In either case, subjective confirmation would be wise if practicable.

Amplitude of accommodation

There is a surprising lack of unanimity, among those who have written on the subject, on the precise method used to measure accommodation in a clinical setting. While acknowledging that dynamic retinoscopy can, with practice, give a useful objective measurement of the accommodative abilities of the patient, most people use one of the many variations of the 'push-up' method of Donders. The trouble is that no two sources seem to agree exactly what this method is. For a start, we need to consider target size because larger print allows a greater depth of focus. Strictly speaking accommodation is the ability to retain resolution at closer working distances, so the target used should equate to the distance acuity, yet some authorities recommend using a paragraph of N5 print, which is equivalent to a distance acuity of about 6/9. Even the point of reference for the measurement is in

dispute. Should we use the spectacle plane, the bridge of the nose or the lateral canthus, all of which have their advocates? Then there is the dilemma over whether to do a 'push-up' or a 'push-up and pull-down' technique, as described below. When using the RAF rule (see Chapter 13) to measure the amplitude of accommodation, a suitable and practical technique is as follows:

1. The target should be the smallest print that can be seen clearly when the sliding scale is at the remote end of the ruler, or at arm's length if using a budgie stick or separate card.
2. The target is brought slowly towards the patient along the ruler, with the eyes slightly depressed, until it becomes blurred ('push up').
3. Patients are then asked if they can bring the target back into focus. If they can, the target is moved slowly towards them until they can no longer regain clear focus. The accommodation can be read from the appropriate scale on the rule, or the distance of the target from the spectacle plane may be measured in centimetres and converted into dioptres.
4. The target is then moved back until the patient can see it clearly, and this distance is also noted ('pull-back' value).
5. The amplitude of accommodation is the average of the 'push-up' and 'pull-back' values.

The RAF rule is usually used, although a 'budgie stick' and measuring tape are also satisfactory. Accommodation should be measured both monocularly (to screen for third nerve anomalies) and binocularly. There are those who advocate repeating each measurement three times to assess the effects of fatigue. There may be value in this approach in some clinical situations, especially in the paediatric arena, but in a routine refraction this is rarely a constructive use of time.

Determination of the reading addition

The prescribing of a reading addition consists of two phases. Initially an addition is estimated by considering the age, accommodation and any existing correction that the patient may have. This is then refined subjectively, taking into account the occupational and lifestyle needs of the individual patient.

Estimating the near addition

The 'Add' for a particular patient depends on the required working distance and the amplitude of accommodation available, and can be estimated using the equation:

$$\text{Add} = |L| - 2/3\,\text{Amp}$$

which was explained in Chapter 13. Alternatively, Table 14.1 can be used to estimate the required reading addition.

The near correction needed depends on the amplitude of accommodation remaining and the patient's habitual working distance. For those with early presbyopia both need to be considered. Bennett and Francis (1962) suggest leaving a third of the amplitude of accommodation in reserve. Millodot and Millodot (1989) found that, although this was suitable for those with early presbyopia, an allowance of half of the amplitude was more suitable for women aged over 52 and men aged over 63. In routine practice what most practitioners use as the starting point is the old reading correction. In general, reducing the power (and hence the magnification) should be avoided unless the patient is having problems.

Equally, large increases often cause problems with adaptation. The higher the addition, the smaller the depth of focus and, in progressive power lenses, the progression characteristics and the near field of view may be affected. The historical technique of adding +0.25 DS to the addition if a progressive power lens is prescribed is both unnecessary and unhelpful.

The effects of a reduced accommodative stimulus on the ocular motor balance should also be considered because the phrase 'these new glasses, they draw my eyes' is often encountered. Also, those patients who are prone to decompensated exophoria need to be watched carefully. Unequal additions are rarely a good idea, unless there are clear reasons for prescribing, e.g. a unilateral pathological condition affecting ciliary tonus. If an unequal addition appears to be required for no good reason, the distance sphere balance may well be wrong. This should be checked before prescribing unequal additions, which rarely seem to work well, particularly if progressive power lenses are to be prescribed. Those with late presbyopia who are anisocoric may appear to need different reading additions, because the depth of focus is greater in the eye with the smaller pupil. However, this eye also has a greater tolerance to defocus, so prescribing the addition required by the eye with the larger pupil (this is generally higher than that required by the other eye) usually works.

Table 14.1 The estimated addition for various working distances and amplitudes of accommodation

Amplitude (D)	Age (years)	Working distances (cm)						
		25	30	35	40	45	50	55
		Estimated addition (D)						
5.00	40	0.75	0.00	0.00	0.00	0.00	0.00	0.00
3.00	45	2.00	1.25	1.00	0.50	0.25	0.00	0.00
2.00	50	2.75	2.00	1.50	1.25	1.00	0.75	0.50
1.50	55	3.00	2.25	2.00	1.50	1.25	1.00	0.75
1.00	60	3.25	2.75	2.25	1.75	1.50	1.25	1.00
0.75	65	3.50	2.75	2.50	2.00	1.75	1.50	1.25
0.50	70	3.75	3.00	2.50	2.25	2.00	1.75	1.50

After Tunnacliffe (1993).

Methods of refining the near addition

Given the singular lack of a consensus over the testing of amplitude of accommodation, it is probably no surprise that practitioners have found numerous ways to refine the near addition. Whichever is used, the aim is to arrive at an addition(s) that allows clear and comfortable vision at all the distances at which the patient habitually works. If too little plus is prescribed, the patient has to accommodate if he can, or tolerate blur if he cannot. Too much and the depth of focus and ocular motor balance are affected. This would be simple enough if all patients were alike in their tolerance levels to blur, depth of focus and, of course, change in general. All near vision tests should be performed after the distance prescription has been ascertained and the trial frame adjusted to represent the correct near centration distance for the required working distance. The first task is to estimate the addition using the equation given above or using the values quoted in Table 14.1. The power of the addition must then be refined and there are various techniques available to do this. The more common techniques are described below. Suitable near test charts are described in Chapter 13.

The range of clear vision

This is defined as the distance between the artificial far point and the artificial near point (see Chapter 13). The range of clear vision should 'straddle' the most usual reading distances in a way that is most useful to the individual, and a good range of clear vision is important for most patients. This should always be demonstrated, measured and recorded. Most non-tolerances to reading spectacles are a result of the addition being too strong, causing a reduced range of clear vision. When dispensing to fulfil occupational and vocational requirements, it is often necessary to modify the given addition to allow for a shorter or longer working distance. This is something that a registered optometrist or dispensing optician can do that an unqualified seller cannot.

Occupational or recreational factors may need to be considered, e.g. the patient may need to be able to focus on a computer screen as well as on written material. Alternatively, a patient who constructs scale models in his spare time may need a correction with a shorter than normal working distance and a certain amount of base-in prism. In some cases it is impossible to cover all a patient's requirements with one near addition and additional pairs of spectacles may be required to cover part of the range. Alternatively, progressive power lenses, enhanced reading lenses or occupational progressive power lenses may be required.

Trial lenses

With the estimated reading addition in place, the patient observes a reading card at the preferred reading distance. Plus and minus spheres of 0.25 D are added binocularly until neither improves the vision. At this point, extra plus upsets the accommodation/convergence relationship, whereas extra minus requires extra accommodative effort.

Near duochrome

With the estimated addition in place, the patient holds a near duochrome test (often a pattern of black dots or rings on the two colours) and is asked to state whether the black targets appear darker on the red or the green background, or

Figure 14.1 The Freeman–Archer near vision unit showing the duochrome test.

if they appear equally dark. If the targets on the red background appear darker, the size of the estimated near addition is reduced. The opposite is true if the dots are clearer on the green background. The addition is usually adjusted to give a sight clarity preference for the targets on the green background, or perhaps 'red equals green'. From this point, the detail on the red background should become clear if the target is moved away from the patient. The range is then checked by moving nearer and further away.

As accommodation is brought into play, the wavelength focused on the retina moves towards the blue end of the spectrum. If red–green equality is taken to be the end-point in duochrome near vision tests, the reading addition prescribed on that basis may be theoretically 0.25 or 0.50 D too strong. To overcome this, the use of yellow and blue filters has been suggested, which would shift the mid-point of the test by about 0.25 D compared with the normal red and green filters. However, in practice, the conventional red–green duochrome test provides an excellent procedure for the presbyopic patient. Age-related yellowing of the lens causes a red bias and consequent underestimation of positive sphere, but this may not be a bad thing in many cases. The use of plus and minus spheres along with a near test chart is the usual technique to finalise the result.

Near crossed-cylinder

A ±0.25 D cross-cylinder (+0.25 DS/−0.50 DC) is placed before each eye with the minus cylinder axis at 90°. The two principal powers of the lens are +0.25 D along 90 and −0.25 D along 180. The patient fixates a grid or cross target (**Figure 14.2**) aligned with the principal meridians of the cross-cylinder, placed at their customary working distance. The addition is adjusted to give equal clarity to the vertical and horizontal elements of the target. When 'equal clarity' of the vertical and horizontal lines is reported, the disc of least confusion (DLC) is formed on the patient's retina, the horizontal line 0.25 D in front of the retina and the vertical line 0.25 D behind (**Figure 14.3**). If the patient reports that the horizontal lines are sharper, plus lenses are added binocularly in 0.25 D steps. If the vertical

Figure 14.2 Determination of the near addition: target for use with the crossed-cylinder technique.

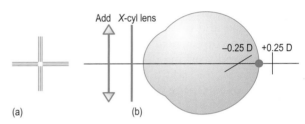

Figure 14.3 Positions of the focal lines with the near crossed-cylinder lens *and* the correct reading addition for the required working distance in place. The horizontal focal line is −0.25 D from the retina and the vertical line +0.25 D from the retina. The disc of least confusion is formed on the retina. If the addition is increased (more plus) the vertical lines should appear clearer. If the addition is reduced (more minus) the horizontal lines should appear clearer. (a) The vertical and horizontal lines of the target will have the same clarity; (b) with the best vision sphere in place the disc of least confusion will be formed on the retina.

lines are sharper then minus lenses are added, again binocularly and in 0.25 D steps. This method is often used with both manual and automated refractor heads but also works well with a trial frame. The author has a pair of +0.25 DS/−0.50 DC lenses edged to fit his trial frame and routinely uses this method for the refinement of the near addition. It is useful to allow the patient to see the target first without the cross-cylinder, then insert the cross-cylinder and (so that the patient has no time to adjust the accommodation) immediately ask which element of the target is clearer. This test is usually performed under binocular conditions.

Further considerations

It should be borne in mind that in some patients the astigmatic element of the refraction is not identical at all fixation distances. Both the power and axis may vary as a result of irregularities in the crystalline lens, which become relevant when the pupils constrict or as a result of cyclo-rotation of the eyes in near vision. Normally these differences are of little consequence except when high degrees of astigmatism are present. It should also be remembered that the cylinder axes obtained by binocular refraction might not be identical to those obtained monocularly. If monocular refraction is performed, checking the axes during the balancing phase may be appropriate if the cylindrical element is significant.

When high-powered spheres are prescribed, the form of the lens may influence its effectivity at near. For higher plus prescriptions, once the lens is made in the usual curved form as a spectacle correction, its effective power may be as much as 0.50 D less than it was in the trial frame. Some manufacturers include tables in their technical literature to enable correction factors for this **near vision effectivity error** to be applied to the final prescription to allow for this.

Along with the value of the reading addition, near acuity and testing distance optometrists should be encouraged to record the following information on the patient's record card:

- The required working distance for a particular task
- The position of the artificial near point
- The position of the artificial far point
- The range of clear vision produced by the reading correction.

VDU users

The rather archaic term 'visual display unit' (VDU) is still widely used to describe computer or electronic display screens, now a significant part of most people's visual environment and ocular demands.

The Association of Optometrists in the UK has issued recommendations on visual standards for VDU users. These are as follows:

- The ability to read N6 at a distance of ⅔ m to ½ m.
- Monocular vision or good binocular vision: near heterophorias over ½ prism dioptre (Δ) vertical or 2 Δ of esophoria and 8 Δ of exophoria **at the specified working distance** are contraindicated and should be corrected **unless well compensated** or deep suppression is present.
- No central (20°) field defects in the dominant eye.
- Near point of convergence normal.
- Clear ocular media checked by **ophthalmoscopy** and **slitlamp**.

The above recommendations are intended to increase the level of operator comfort, and therefore efficiency, but failure to achieve the standard does not exclude a person from working with a VDU. Common sense is required when testing VDU users. The recommendations on heterophoria should not be applied to Maddox rod or wing results, because most people fail. When measuring heterophoria in a VDU user, the prism bar cover test should be used. When prescribing for VDU users, bifocal or trifocal spectacles rarely seem to work well. A modern progressive power lens, enhanced reading lens or occupational progressive power lens often works better, although in some cases single vision correction may be needed, particularly if binocular fatigue requires prismatic correction.

Concluding points for Chapter 14

In practice, no one method is found to be reliable for all patients and all consulting rooms. Combinations of the above methods and consideration of the patients' needs are generally the best tactic, so it pays to be familiar with a number of methods, which allows a flexible approach. As most non-tolerances are caused by the reading addition being too strong for the required working distance, it is important to take care when discussing the patient's near vision requirements. Ask appropriate and clear questions and listen carefully to the patient's responses. Always demonstrate the range of clear vision attainable with a prescribed reading addition and record all distances carefully on the patient's record.

References

Bennett A G, Francis J L (1962) Ametropia and its correction. In: Davson H (ed.), *The Eye*, Vol IV. Academic Press, New York: 131–80

Millodot M, Millodot S (1989) Presbyopia correction and the accommodation in reserve. *Ophthalmic and Physiological Optics* **9**:126–32

Weale RA (1981) Human ocular ageing and ambient temperature. *British Journal of Ophthalmology* **65**:869–70

Further recommended reading

Elliott D B (2003) *Clinical Procedures in Primary Eye Care*. Butterworth-Heinemann, Oxford

Rabbetts R B (1998) *Bennett & Rabbetts' Clinical Visual Optics*. Butterworth-Heinemann, Oxford

Tunnacliffe A H (1993) *Introduction to Visual Optics*. Association of the British Dispensing Opticians, London

Convergence

Andrew Keirl

Introduction

This relatively short and straightforward chapter discusses the subject of convergence. The chapter is essentially a series of worked examples that the reader should attempt to follow. Historically, some students have found this topic a challenge. However, the key to success lies in the drawing of an accurate diagram to illustrate what is going on and also in the application of basic optical principles and simple trigonometry.

Chapter content

- Definition of convergence
- Convergence: emmetropia versus corrected ametropia

Definition of convergence

Convergence is defined as 'the movement (rotation) required from the primary position, for the eyes to fixate an object point on the mid-line.' This is illustrated in **Figure 15.1**.

There is an important and sometimes delicate relationship between convergence and accommodation. For patients with normal binocular vision, the amount of convergence and accommodation should be equal. Also, remember that convergence 'drives' accommodation. There are binocular vision problems that are caused by an excess of accommodation resulting in an excess of convergence, and there are also binocular vision problems that result from a convergence insufficiency. Convergence can be expressed in degrees, prism dioptres or the metre angle.

The 'base-in' effect of minus lenses centred for distance vision means that myopic subjects

hypermetropic subjects corrected with spectacle lenses converge *less* for near vision than those with emmetropia. On the other hand, the 'base-out' effect of plus lenses centred for distance vision means that hypermetropes corrected with spectacle lenses converge *more* for near vision than those with emmetropia. The above statements relating to convergence with minus and plus spectacle lenses are important and can be illustrated using a series of calculations.

Example 15.1

A patient is corrected using a pair of −8.00 D lenses positioned at a vertex distance of 14 mm and centred for distance vision. The distance interpupillary distance is 66 mm and the centres of rotation of the eyes lie 2 mm behind the spectacle plane. An object is placed on the midline and at a distance of ⅓ m from the spectacle plane.

We will calculate the convergence required for this individual to view the near object.

It is important that you refer to **Figure 15.2** when working through this example. Inspection of Figure 15.2 shows the following:

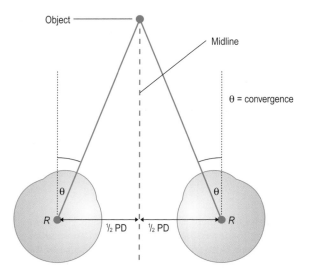

Figure 15.1 An illustration of the convergence required to view a near object placed on the midline.

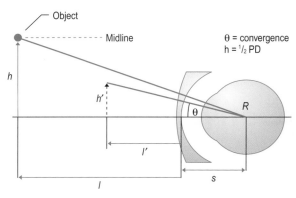

Figure 15.2 Convergence for a myopicsubject corrected with a spectacle lens.

- The minus lens is placed at a distance s from the eye. This is the distance from the back vertex of the lens to the centre of rotation of the eye, R, and is called the **fitting distance**. Do not confuse this with the vertex distance, which is measured from the back vertex of the spectacle lens to the corneal apex.
- We are told that an object is placed on the midline. This is a line that bisects the face. The distance from the principal axis passing through the centre of rotation of the eye to the midline is taken to be the object size h.
- In Figure 15.2, h is equivalent to the semi-interpupillary distance (½ PD) and is placed at a distance l from the lens.
- The image of an object formed by a minus lens is virtual, upright and diminished. Such an image, labelled h', has been added to Figure 15.2. The correct position of the image formed is important because the calculation of the convergence required hinges around this fact. The image h' is formed at a distance l' from the lens.
- When the eye rotates to view the near object, it does not converge to view the object directly but instead rotates towards the image. The angle between the principal

axis passing through the centre of rotation of the eye and the top of the image has been labelled θ. This angle represents the rotation or convergence required to view the image formed by the lens. We therefore need to find the angle θ.

From inspection of Figure 15.2:

$$\text{Tan}\,\theta = \frac{h'}{l' + s}$$

Note: when finding the angle θ, the sign of l' is ignored in order to produce a positive value for θ.

So, we need to find h' and l'. l' can be found using:

$$L' = L + F$$

and:

$$l' = \frac{1}{L'}$$

Similar triangles can be used to find h' as follows:

$$\frac{h'}{h} = \frac{l'}{l}$$

which becomes:

$$h' = h \times \frac{l'}{l}$$

We now have to substitute the values from the question into the above expressions. Be sure to refer to Figure 15.2. The given data is summarised as follows:

- $F = -8.00\,\text{D}$
- Interpupillary distance is 66 mm, so the ½ PD and therefore $h = 33\,\text{mm}$
- $s = 27\,\text{mm}$
- $l = -\frac{1}{3}\,\text{m}$ from the spectacle plane

$$L = \frac{1}{l} \qquad L = \frac{1}{-1/3} = -3.00\,\text{D}$$

$$L' = L + F \qquad L' = -3.00 + (-8.00) = -11.00\,\text{D}$$

$$l' = \frac{1}{L'} \qquad l' = \frac{1}{-11.00} = -0.091\,\text{m}$$

$$h' = h \times \frac{l'}{l} \qquad h' = 0.033 \times \frac{-0.091}{-1/3} = +0.009\,\text{m}$$

All values in the above equation are in metres. And finally:

$$\text{Tan}\,\theta = \frac{h'}{l' + s}$$

$$\text{Tan}\,\theta = \frac{+0.009}{|0.091| + 0.027} = 0.0763$$

All values in the above equation are in metres. In order to give a positive angle, the minus sign for l' has been ignored:

$$\text{Tan}^{-1} = 4.36°$$

The rotation or convergence required to view the near object placed ⅓ m from the lens is therefore 4.36°.

Convergence can also be expressed in prism dioptres. The prism dioptre is given by:

$$p = 100 \tan\theta$$

which in this example is:

$$p = 100 \tan 4.36 = 7.63\Delta$$

Example 15.2
Example 15.1 is repeated, replacing the −8.00 D lens with a +8.00 D lens. All the other parameters remain unchanged and the arrangement is shown in Figure 15.3. The basic principles for this calculation are the same as for the minus lens. Figure 15.3 shows the following:

- The plus lens is placed at a distance s from the eye. This is the distance from the back vertex of the lens to the centre of rotation of the eye and is called the fitting distance. Once again, do not confuse this with the vertex distance.
- An object is placed on the midline which is the line that bisects the face. The distance from the principal axis passing through the centre of rotation of the eye to the midline is taken to be the object size h.
- In Figure 15.3, h is equivalent to the semi-interpupillary distance (½ PD) which is placed at a distance l from the lens.
- The image of an object formed by a plus lens is real and inverted. An appropriate image h' is shown in Figure 15.3. As with the minus lens, the correct positioning of the image is important in the calculation of convergence. The image h' is formed at a distance l' from the lens.
- When the eye rotates to view the near object, it does not converge to view the object directly but instead rotates towards the image. The angle between the principal axis passing through the centre of rotation of the eye and the top of the image is θ, which represents the rotation or convergence required to view the image formed by the lens. So once again, we need to find the angle θ.

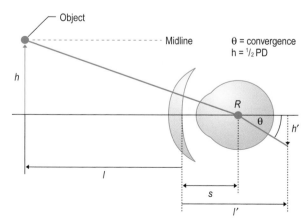

Figure 15.3 Convergence for a hypermetropic subject corrected using a spectacle lens.

The procedure is essentially identical to that used in Example 15.1. From inspection of Figure 15.3:

$$Tan\,\theta = \frac{h'}{l' - s}$$

When finding the angle θ, the sign of h' is ignored in order to produce a positive value for θ.

We need to find h' and l'. l' can be found from:

$$L' = L + F$$

and:

$$l' = \frac{1}{L'}$$

Using similar triangles:

$$\frac{h'}{h} = \frac{l'}{l}$$

and:

$$h' = h \times \frac{l'}{l}$$

The appropriate date is now substituted into the equations:

- $F = +8.00\,D$
- Interpupillary distance is 66 mm, so the ½ PD and $h = 33$ mm
- $s = 27$ mm
- $l = -\frac{1}{3}$ m from the spectacle plane

$$L = \frac{1}{l} \qquad L = \frac{1}{-1/3} = -3.00\,D$$

$$L' = L + F \qquad L' = -3.00 + (+8.00) = +5.00\,D$$

$$l' = \frac{1}{L'} \qquad l' = \frac{1}{+5.00} = +0.20\,m$$

$$h' = h \times \frac{l'}{l} \qquad h' = 0.033 \times \frac{+0.20}{-1/3} = -0.0198\,m$$

All values in the above equation are in metres:

$$Tan\,\theta = \frac{h'}{l' - s} \qquad Tan\,\theta = \frac{|0.0198|}{0.20 - 0.027} = 0.1144$$

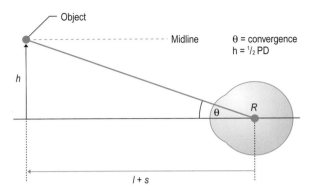

Figure 15.4 Convergence for someone with emmetropia.

To give a positive angle, the minus sign for h' has been ignored.

$$Tan^{-1} = 6.53°$$

The rotation or convergence required to view the near object is therefore 6.53°. The convergence in prism dioptres is:

$$p = 100\tan 6.53 = 11.44\,\Delta$$

Example 15.3
We now have to determine the convergence required if an emmetropic individual views the same near object. This is illustrated in Figure 15.4.

In the case of emmetropic subject, there is of course no image formed because there is no lens. The eye can simply rotate to view the object directly. From Figure 15.4 the distance between the centre of rotation of the eye and the object is given by:

$$l + s$$

and the convergence or rotation required to view the object is given by:

$$Tan\,\theta = \frac{h}{l + s}$$

Once again, h is equivalent to the ½ PD.
The data used in this calculation is:

- $l = -1/3$ m
- $s = 27$ mm
- $h = 33$ mm

All values are entered in metres and in order to give a positive angle, the minus sign for l is ignored:

$$\text{Tan}\,\theta = \frac{h}{l+s} \qquad \text{Tan}\,\theta = \frac{0.033}{|1/3|+0.027} = 0.0916$$

$$\text{Tan}^{-1} = 5.23°$$

The rotation or convergence required to view the near object is therefore 5.23°. The convergence in prism dioptres is:

$$p = 100\tan 5.23 = 9.16\,\Delta$$

Table 15.1 A summary of the results of Examples 15.1–15.3

	Convergence (°)	Convergence (Δ)
In myopia	4.36	7.63
In emmetropia	5.23	9.16
In hypermetropia	6.53	11.44

Convergence: emmetropia versus corrected ametropia

The results of the three examples above is summarised in Table 15.1. All three individuals are converging to view the *same* near object.

To view a near object, the emmetropic patient needs to converge *more* than the myopic patient who is corrected with spectacle lenses. However, the emmetropic patient needs to converge *less* than the hypermetropic patient who is corrected with spectacle lenses to view the same near object. Does this sound familiar? Exactly the same relationship applies to accommodation (see Chapter 13). When viewing the same near object, the person with corrected myopia is required to accommodate less than the person with corrected hypermetropia, with the person with emmetropia being somewhere in between. This has important practical significance in contact lens wear (Chapter 27).

Concluding points for Chapter 15

This chapter has covered:

- the definition of convergence
- the convergence demand for emmetropic subjects
- the convergence demand for myopic subjects
- the convergence demand for hypermetropic subjects

It is important that the reader understands and appreciates the optometric significance of the examples in this chapter.

Further recommended reading

Rabbetts R B B (1998) *Bennett & Rabbetts' Clinical Visual Optics*. Butterworth Heinemann, Oxford
Tunnacliffe A H (1993) *Introduction to Visual Optics*. Association of the British Dispensing Opticians, London

Automated Methods of Refraction

William Harvey

Introduction

Objective refraction includes retinoscopy and the use of autorefractors, which are appearing in optometric practice in increasing numbers. In general, objective methods are not required to give us a final prescription. They merely need to get us to a point from which subjective methods can take us to the end-point accurately and quickly. With an alert and compliant patient it is possible to get an accurate result using subjective methods alone, but it takes time. In general, excessively prolonged refraction of normal patients is usually indicative of poor technique rather than 'professionalism'. With practice, and if a previous prescription is known, objective refraction should take seconds rather than minutes. There are patients who are unable to participate in a subjective refraction, because of limitations of understanding or communication. Very young children, or people with Alzheimer's disease or a learning impairment may require a prescription to be arrived at purely from the objective findings.

The use of a machine to measure refractive error has a long history. The original optometers could use either subjective methods (the forerunners of modern phoropters) or objective methods, and it is the automated objective refraction instruments that are now described as autorefractors. Autorefractors use an infrared light source (around 800–900 nm), which allows good ocular transmission, but requires a −0.50 D adjustment to the final refraction as a result of error introduced by reflection from the choroid and sclera.

The source projects light via a beam splitter and a Badal lens system to form a slit image within the eye, the reflection of which passes out via the beam splitter to reach a light sensor. Throughout, the patient is encouraged to relax any available accommodation (a major source of error for autorefractor measurement) by use of a fixation target or, in some cases, an open view to allow fixation on a distant target. The calculation of refractive error is based on analysis of how the patient's eye influences the infrared radiation. The way in which this analysis is performed varies. Most of the original instruments used some form of image quality analysis, relying on positioning of the Badal lens system to achieve a maximum signal to the light sensor. Most modern autorefractors, of which there are many, rely on an adapted Scheiner disc principle (see Chapter 10). The original Scheiner disc consisted of two holes in a card placed before the eye. A myopic eye sees the two images from the holes swapped over or crossed, whereas the person with hypermetropia sees them uncrossed. This may be done in various meridians to give information about the nature of any astigmatic element to the refractive error. Autorefractors simulate this using LED (light-emitting diode) light sources, the images of which are detected by a light sensor or photodetector and the position of the LED needed to achieve a single image over the photodetector is related to the patient's refractive error. A further method employed by a few machines is an adaptation of retinoscopy, where the instrument analyses the speed of movement of a reflex of infrared light to measure the refractive error.

Most studies suggest that autorefractors are quick, simple, repeatable and accurate (with some qualification). With cycloplegia or good accommodative control the results are very accurate. Indeed the spherical aberration introduced by the dilation of someone with cycloplegia makes the method preferable to retinoscopy in many cases. Its ease of use makes it suitable to be carried out by ancillary staff, so reducing the burden on the optometrist. The machines may directly link to an automated phoropter head, again making routine refraction more fluid. It is useful to remember that even the most accurate objective measurement may not be the one preferred by the patient, so a subjective approach is always preferable to ensure a tolerable refractive result, even though this is sometimes modified away from the actual refractive error present.

The main errors in autorefraction are the result of poor fixation (very much dependent on the target of the instrument), accommodative fluctuation (proximal accommodation in young people invariably leads to over-minused

measurements) and media difficulties (which are also likely to reduce the effectiveness of retinoscopy). The lack of portability of instruments is less of an issue nowadays because there are several portable models, and some have found a use in child-screening programmes, where the main outcome is not precise error measurement but detection of large amounts of ametropia or anisometropia. Other new models incorporate some subjective assessment, with the patient responding to prompts to clarify a presented image.

Chapter content

- Advantages of autorefraction
- Disadvantages of autorefraction
- Design features common to all autorefractors
- Optical principles employed in autorefractors

Advantages of autorefraction

Autorefraction represents a reliable alternative to retinoscopy and, as such, removes much potential for human error. The reliability of autorefractors has obviously improved and many studies have been undertaken looking at the validity of results. In a large comparative trial in 1993, McCaghrey and Matthews looked at the validity of results of a wide range of instruments and found acceptable reliability. Some of the results for a number of instruments are summarised in Table 16.1. Walline *et al.* (1999) noted that the machines were particularly accurate at determining the size and orientation of astigmatism.

An autorefractor may provide exceptional accuracy after cycloplegia, assuming fixation is maintained. There is some evidence that, in such situations, accuracy may exceed that of retinoscopy, which may be affected by the difficulty in interpreting the larger reflex within the dilated pupil. In cases where subjective responses are questioned, e.g. in those with severe learning or cognitive impairment, the autorefractor offers a useful confirmation of

retinoscopy results. It also offers an alternative when the retinoscope is rejected by the patient, rare as this is. It is easy enough for non-professional staff to operate and may be used as a useful screening instrument. In private practice such pre-examination data allows the optometrist to refine the result and generally makes the examination more time efficient. Portable autorefractors allow large-scale refractive error screening, e.g. in a school environment, to detect cases of large ametropia or anisometropia, which may allow the screener to recommend a more thorough eye examination (**Figure 16.1**). Trials of such a method appear promising in detecting amblyopia at a stage where intervention might still prove beneficial.

Many autorefractors may be 'hard wired' and directly linked to an automated refractor head (phoropter) within the consulting room. They also measure the pupillary distance when the instrument is aligned before one eye, then the other. This information aids in the initial set-up of the phoropter. Patients increasingly expect a certain level of automation from modern eye care practices, and anecdotal and marketing research tends to reflect a public preference for the automated compared with the apparently more traditional methods. Research trials, such as those looking at myopia patterns within a population, benefit from the excellent repeatability and ease of use of autorefractors to acquire large amounts of refractive error data from large sample populations in a short time, often using teams of auxiliary staff with minimal training requirements.

Table 16.1 Validity of a selection of autorefractors compared with subjective refraction

Comparison	Individuals	Results
Subjective refraction versus Hoya AR550	100 consecutive eyes in practice No cycloplegia No details of sample	Mean spherical difference: −0.0150 Confidence limits: −0.69, 0.66
Subjective refraction versus Humphrey 550	100 consecutive eyes in practice No cycloplegia No details of sample	−0.0530 Confidence limits: −0.88, 0.78
Subjective refraction versus Inami GR12	100 consecutive eyes in practice No cycloplegia No details of sample	−0.2200 D Confidence limits: −1.08, 0.64
Subjective refraction versus Nidek AR1000	100 consecutive eyes in practice No cycloplegia No details of sample	−0.0450 Confidence limits: −0.91, 0.82
Subjective refraction versus Nikon NR5000	100 consecutive eyes in practice No cycloplegia No details of sample	0.0050 D Confidence limits: −0.51, 0.52
Subjective refraction versus Nikon NR5100	100 consecutive eyes in practice No cycloplegia No details of sample	0.0450 D Confidence limits: −0.92, 0.83
Subjective refraction versus Topcon RMA2000	100 consecutive eyes in practice No cycloplegia No details of sample	0.0230 D Confidence limits: −0.82, 0.87
Subjective refraction versus Takagi ARI	90 eyes No cycloplegia	−0.0056 D Confidence limits: −0.64, 0.63

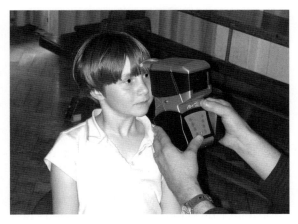

Figure 16.1 Portable autorefractors have some use in screening children for significant refractive error.

Disadvantages of autorefraction

Results tend towards the minus if a patient is unable to relax accommodation fully. Although possible with any pre-presbyope, it is a particular concern in those with latent hypermetropia, especially young children, who exhibit significant proximal accommodation. For this reason, cycloplegia is essential for this younger age group if autorefractor results are to be useful. As discussed later in this chapter, accuracy depends on accurate fixation and an interpretation of an adequate incident and reflected light (infrared) beam. Errors occur, therefore, as a result of a number of common factors, including media opacities (vitreous changes, such as asteroid hyalosis, as well as cataract and corneal changes), poor fixation (either attention related or nystagmus), smaller pupils, pseudophakia, and in some cases high ametropia, amblyopia or reduced central acuity, as with age-related macular degeneration.

Many retinoscopists argue that the retinoscope provides useful clinical information via the quality of the reflex. The effective retroillumination of the lens allows for a very useful assessment of lenticular integrity and this may often be the first clinical test to detect early lens changes, such as posterior subcapsular vacuoles. Also in some corneal conditions, typically keratoconus, the distortion of the reflex provides another indicator of severity of corneal compromise. For the former, some instruments (such as the RetinoMax – Figure 16.2) do include a retroillumination feature for cataract assessment, whereas the advent of topography has largely superseded the latter technique's usefulness. Expense is an issue, with the original investment usually being well over five times that for investing in a good retinoscope.

Design features common to all autorefractors

As discussed later, there are several methods employed by instrument designers that allow autorefractors to measure refractive error.

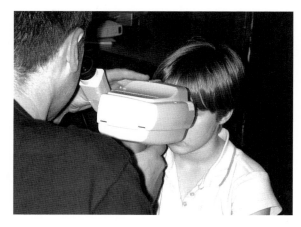

Figure 16.2 The RetinoMax in use.

However, there are many features common to all the instruments.

Infrared radiation

To measure the refractive nature of an eye, it is necessary to interpret a directed or diffuse reflection of light from an area of fundus on axis. This area acts as a secondary source and may be measured variously to reveal the refractive characteristics of the eye. To produce this secondary source, all autorefractors use infrared radiation at wavelengths between 780 nm and 950 nm (the near infrared or NIR). There are two reasons for this:

1. The pigment within the retina poorly absorbs NIR so that it is reflected more efficiently, e.g. the retina reflects just 1% of visible green light (550 nm) but up to 9% of the NIR (880 nm). It does, however, show excellent transmittance (around 90%) through clear media.
2. NIR is invisible to the human eye, so there is no effect on patient comfort, pupil diameter or accommodation response.

As most of the reflected NIR is scattered diffusely and does not exit the pupil for analysis, all autorefractors use a very intense primary NIR source, one that, were it to be using visible light, would be far too intense for safe usage. Furthermore, the systems are designed to offer as few sources of unwanted reflections as possible to the reflected light to improve accuracy, both by minimising reflections and by keeping optical components along the path of the light to a minimum. Accuracy is reduced also by the corneal reflex, and all use techniques to minimise corneal reflex effects, such as aperture stops and polarisers.

Null point versus open loop

Autorefractors use one of two basic principles to arrive at the refractive error of the test eye. As with retinoscopy, many instruments change their power until the **null point** of the reflected light path is reached. As the reflex is optimised near to the null point (an analogy here would be the observed increased brightness of the reflex seen through the retinoscope as neutralisation is neared), these instruments are able to function with higher signal/noise ratios. Some instruments analyse the reflected beam to arrive immediately at the refractive error without the use of any focusing components. This so-called **open loop** approach is generally much quicker and, because fewer moving parts are involved, more robust long term.

Vertex distance adjustments

Although the refraction obtained by an autorefractor is initially that at the plane of the anterior corneal surface, all autorefractors offer a default setting (or range of settings) to convert to a more usable spectacle plane result.

Accommodation and fixation control

As already mentioned, changes in accommodation or loss of fixation by the patient present the biggest challenge in accuracy for the autorefractor on a healthy patient. A few modern instruments now employ a binocular assessment (such as the Topcon BV-1000, the Shin Nippon NVISION K 5001 or the new PlusOptix Power Refractor), which minimise accommodative reflex. Most, however, use a monocular visible light fixation target brought into focus by a spherical focusing system and presented along the same axis as the NIR primary source. Focusing occurs once the secondary source allows the instrument to locate the far point of the eye. Patients with a large astigmatic error present a source of error to such machines where a spherical focusing system is employed.

Furthermore, when moving from one eye to the other, the state of accommodation of each eye is often different. This potential error resulting from proximal accommodation is often overcome by use of a photographic target containing prominent distance features of low spatial frequency within a view containing a wide range of spatial frequencies. Despite such efforts, accommodation error remains a significant problem for most modern monocular autorefractors used on young, actively accommodating patients.

Optical principles employed in autorefractors

The Scheiner disc principle

The Scheiner principle was first put forward by Thomas Scheiner in 1619 and gained prominence when used by Thomas Young in his research into the origins of refractive error. The principle is based on the fact that a disc with two holes in it appears different to someone with hypermetropia and someone with myopia. For the former, the pencils of light from the two holes cross behind the retinal plane, so the two holes appear 'uncrossed' whereas, for the latter, the light pencils cross anterior to the retinal plane and so the holes appear 'crossed'. The person with emmetropia perceives the holes as one and no diplopia is perceived (**Figure 16.3**).

This principle is the basis of the operation of most modern autorefractors (e.g. those of Nidek, Takagi or Topcon). The holes are in effect replaced by two LED light sources imaged in the pupillary plane. **Figure 16.4** shows a schematic breakdown of the principle within a Nidek autorefractor.

The NIR from the two sources passes through a moveable stop and is focused on to the retina

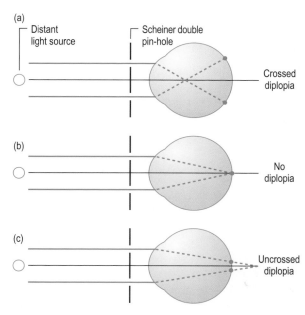

Figure 16.3 Principle of the Scheiner disc: (a) myopia; (b) emmetropia; (c) hypermetropia.

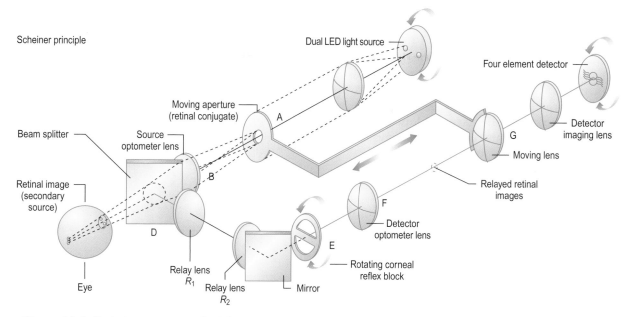

Scheiner principle

Dual LED light source

Four element detector

Moving aperture
(retinal conjugate) A

Beam splitter

Source
optometer lens

Detector
imaging lens

G

Moving lens

Retinal image
(secondary
source)

B

Relayed retinal
images

D

F

Detector
optometer lens

E

Eye

Relay lens
R_1

Relay lens
R_2

Mirror

Rotating corneal
reflex block

Figure 16.4 Optical components of a Scheiner principle autorefractor. (After Benjamin 2007, with permission of Elsevier Ltd.)

to provide the primary NIR source. The two LEDs are rotatable through 180° to allow for astigmatic error measurement along any specific axis. Only when focused on the retina is a single circular area achieved, and residual error is detected as either crossed or uncrossed images of the LEDs. The 'crossing' may be measured by the fact that the LEDs rapidly flicker alternately. For accuracy of measurement, it is essential that the fixation target be coaxial with the NIR; to ensure this many modern instruments include an auto-alignment feature that registers the measurement only when the correct alignment is achieved.

Retinoscopic principles

Some autorefractors, such as the Nikon NR series, the RetinoMax or the Tomey TR series, use a similar principle to that of the retinoscope. They may detect either the direction of movement of an image or the speed of movement. For the former, rectangular images are presented by means of a revolving drum and the direction of movement of the images is detected and 'neutralised'. The far point for any particular meridian is reached when the movement is

no longer seen. For the latter, the speed of movement of the secondary source is detected and this is seen to increase the closer it gets to the neutralisation point. For a detailed study of how these two end-points are achieved see Campbell et al. (2006).

Best focus principle

Originally used in the Dioptron instrument of the mid-1970s, this technique is rarely seen in modern instruments (the Canon R1 is an exception). The contrast of an image reduces as it defocuses. In principle, therefore, it is possible to produce a secondary source that, as a lens moves the image in and out of focus, is focused optimally when the contrast measured is maximal.

Knife-edge principle

Foucault famously used this principle to determine the refractive uniformity of mirrors and lenses. A knife edge may be moved along the optic axis of a lens and mirror placed behind the image, such that the edge and its image become conjugate. All light returning back

171

through an optical system should return to its source and so for a uniform lens no light should escape past the knife edge. An autorefractor could use the optical system of the eye as the test lens and the reflecting fundus as the mirror. In this way the machine could move a 'knife-edge' target to a position where total reflected light returns to the target. Only Carl Zeiss instruments use this technique in commercially available autorefractors.

Ray deflection principle

This technique is found more commonly in recent instruments because the ability to measure small changes to the reflected light path is the basis of the Hartmann–Shack wavefront sensor, originally used in astronomy but now found in many aberrometers. In essence, the instruments use a focusable incident light source that then reflects into a detector. This may comprise a number of linear element detectors (such as in the Canon R-30) or a number of lenslets (as in the Hartmann–Shack-based instruments, including the Bausch & Lomb Z-Wave system or the Welch-Allyn Sure-Sight), allowing analysis of light from an array of points throughout the pupil area. In each case, the degree of deviation from a point focus indicates the amount of refractive error, and focusing of the pencil of light to achieve the point focus indicates the total error. Most Hartmann–Shack systems are used to measure higher-order aberrations but, as with the Sure-Sight (Figure 16.5), may accurately determine sphero-cylindrical error. As these systems become commercially more viable outside refractive surgery centres, it is likely that optometrists will be able to measure higher-order aberrations as part of the autorefraction process.

Image size analysis principle

The size of the retinal image is of course related to the refractive error. Topcon were the first to incorporate this principle into an autorefractor (the RM-A7000 and KR-7000). An annular secondary source may be analysed in terms of both size and, for astigmatism, meridional changes by what is, in effect, a specialised fundus camera.

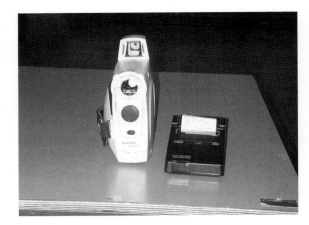

Figure 16.5 The Welch–Allyn Sure-Sight: an example of a Hartmann–Shack-based instrument.

A computer analyses the change in the image presented to the CCD camera detection chip surface.

Concluding points for Chapter 16

There is an array of techniques employed by manufacturers of autorefractors, although most modern instruments rely on the Scheiner disc principle. A number of 'human' factors, not least the need for steady fixation, relaxed accommodation and clear media, means that there are always inherent sources of inaccuracy. As such, autorefraction is used as an adjunct rather than a replacement for retinoscopy. It is, however, proving increasingly useful in screening and research programmes, something likely to continue as instruments become more portable.

References

Benjamin WJ (2007) *Borish's Clinical Refraction*, 2nd edn. Elsevier Science, Oxford

Campbell C E, Benjamin W J, Howland H C (2006) Objective refraction: retinoscopy, autorefraction and photorefraction. In: Benjamin W J (ed.), *Borish's Clinical Refraction*, 2nd edn. Elsevier Science, Oxford: Chapter 18

McCaghrey G E, Matthews F E (1993) Clinical evaluation of a range of autorefractors. *Ophthalmic and Physiological Optics* **13**:129–37

Walline J J, Kinney K A, Zadnik K, Mutti D O (1999) Repeatability and validity of astigmatism measurements. *Journal of Refractive Surgery* **15**:23–31

Further recommended reading

Elliott D B, Wilkes R (1989) A clinical evaluation of the Topcon RM-6000. *Clinical and Experimental Optometry* **72**:150–3

Wesemann W, Dick B (2000) Accuracy and accommodation of a handheld autorefractor. *Journal of Cataract and Refractive Surgery* **26**:62–70

Wong S C, Sampath R (2002) Erroneous automated refraction in a case of asteroid hyalosis. *Journal of Cataract and Refractive Surgery* **28**:1707–8

Part 2

Contact Lenses and Their Application in Vision Correction

The Correction of Refractive Errors Using Contact Lenses

Andrew Keirl and Caroline Christie

Introduction

This brief chapter is a simple introduction to the correction of refractive errors with contact lenses. Almost all types of refractive error can be corrected with spectacle lenses and 'special' spectacle lens forms are available for the correction of differential prism and to incorporate prism at distance but not at near (or vice versa). However, the incorporation of a prismatic correction into a contact lens is limited in terms of both lens availability and success.

It is important to realise that almost all refractive errors can be corrected using contact lenses. The correction of myopia, hypermetropia, astigmatism and presbyopia presents few challenges to the competent contact lens practitioner. Irregular astigmatism, including medical conditions such as keratoconus and keratoplasty, are also often best corrected using contact lenses. In addition patients who have experienced less than satisfactory results after refractive surgery are also often best corrected using contact lenses.

Chapter content

- Rigid and soft contact lenses
- The correction of vision: soft versus rigid gas permeable (RGP) contact lenses
- RGP lenses: selection of the best vision sphere
- The correction of vision using soft/silicone hydrogel contact lenses
- The correction of astigmatism
- Contact lenses for the correction of presbyopia
- Contact lenses versus spectacles

Rigid and soft contact lenses

Rigid and **soft** contact lenses are the principal contact lens types fitted today. The term 'rigid' refers to a rigid gas permeable or RGP contact lens, whereas the term 'soft' relates to traditional hydrogel and silicon hydrogel materials (see Chapter 21). The advantages and disadvantages of RGP and soft contact lenses are outlined in Tables 17.1 and 17.2.

RGPs are particularly indicated when a patient seeks:

- full-time contact lens wear
- optimal vision
- relief from previous problems with soft contact lenses
- accurate correction of astigmatism
- best possible vision in presbyopia

The correction of vision: soft versus RGP contact lenses

When a soft lens is fitted to an eye, the flexible nature of the lens material means that the geometry of the back surface of a soft lens conforms to that of the anterior corneal surface, i.e. the soft lens 'drapes' to fit the cornea and any corneal astigmatism is transmitted to the contact lens. This is particularly true for traditional hydrogel materials. A well-fitted single vision soft lens should centre properly on the cornea, produce only small amounts of lateral move-

Table 17.1 Advantages and disadvantages of spherical and aspheric rigid gas permeable (RGP) contact lenses

Advantages	Disadvantages
Provide good vision	Perceived comfort problems
Can correct up to 2.50 D of astigmatism	Lack of wearing schedule flexibility
Quality of vision does not depend on lens rotation	Replacement costs
Durability	Requires greater skill to fit
Ease of care	Not suitable for some sports
Good handling	
High oxygen transmissibility	
Low incidence of microbial infection	

Table 17.2 Advantages and disadvantages of spherical and aspheric soft contact lenses

Advantages	Disadvantages
Provide good vision if ametropia is spherical	Reduced visual acuity in astigmatism ≥0.75 D
Good initial comfort	Can be prone to surface deposits
Quality of vision independent of rotation	On-eye dehydration with some lens materials
Flexible wearing schedule	Low oxygen transmissibility with traditional hydrogel materials

ment and hence introduce insignificant amounts of additional aberration into the lens–eye system. The diameter of the optic zone of a soft lens normally exceeds that of the entrance pupil of the eye under all lighting conditions. This means that haloes around light sources (flare) are avoided. The refractive index of a traditional hydrogel lens is close to that of the cornea, so that reflection losses are comparable to those in the natural eye. Possible optical disadvantages

of soft lenses include lens flexure and hydration changes, which may result in on-eye power changes. When ordering a soft contact lens for a patient it is necessary to specify the patient's ocular refraction (K). With soft contact lenses, the required back vertex power (BVP) of the soft contact lens measured in air is assumed to be equal to the patient's ocular refraction. This applies to both spherical and astigmatic soft contact lenses.

The correction of ametropia with a RGP contact lens is very different to that of a soft contact lens. Unlike soft lenses, which drape to fit the cornea so that, when on the eye, the geometry of the back surface of the soft contact lens mimics the anterior corneal surface, the back surface of a rigid lens maintains its shape. This means that a tear lens of predictable form and power is formed between the back surface of the rigid contact lens and the anterior surface of the cornea. The **contact lens–tear lens system** formed when a RGP contact lens is placed on an eye means that there are three elements involved in the formation of the final retinal image:

1. The RGP contact lens
2. The tear lens (also known as the liquid lens or lacrimal lens)
3. The eye

The contact lens–tear lens system consists of two elements: the RGP contact lens and the liquid lens. When performing calculations with RGP contact lenses, it is convenient to imagine that each element – the contact lens and the tear lens – is separated by an infinitely thin 'air film' or 'air gap' (**Figure 17.1**). The power of the tear lens depends on the relative geometry of the back surface of the rigid lens (back optic zone radius or BOZR) and the anterior corneal surface. The tear lens may be positive (a steep fitting RGP lens – **Figure 17.2**), plano (an RGP lens fitted in alignment with the cornea – **Figure 17.3**) or negative (an RGP lens fitted flat – **Figure 17.4**). If a RGP contact lens is to correct a distance refractive error, the vergence leaving the posterior surface of the tear lens in an RGP contact lens–liquid lens system must be equal to the patient's ocular refraction. This value is also the BVP of the RGP contact lens–liquid lens

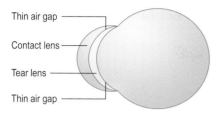

Figure 17.1 The contact lens-tear lens system.

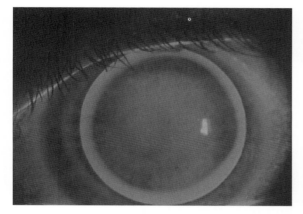

Figure 17.2 An RGP contact lens that has been fitted steep relative to the curvature of the cornea. This relationship will produce a positive tear (or liquid) lens. (See also colour plates.)

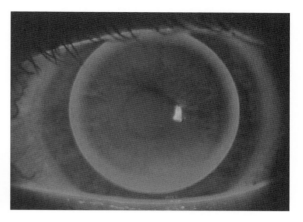

Figure 17.3 An RGP contact lens that has been fitted in alignment with the cornea. This relationship will produce an afocal or plano tear lens. (See also colour plates.)

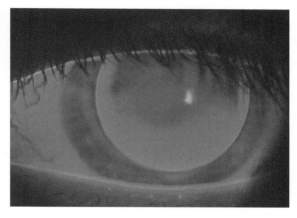

Figure 17.4 An RGP contact lens that has been fitted flat relative to the curvature of the cornea. This relationship will produce a negative tear lens. (See also colour plates.)

system, which is not the same as the BVP of an RGP contact lens measured in air. So, unlike a soft lens, the patient's ocular refraction is *not* equal to the BVP of an RGP contact lens measured in air. The only exception to this is when the shape of the back surface of the RGP contact lens is in exact alignment to the anterior corneal surface and the power of the resulting tear lens is zero.

RGP lenses: selection of the best vision sphere

In practice, the power of an RGP contact lens can be determined by using an empirical method based on the tear lens power and manifest refraction, or by performing an over-refraction with a correctly fitting diagnostic trial lens *in situ*. The empirical method is outlined below:

1. Transpose the spectacle prescription to minus cylinder form and convert to an ocular refraction (when >±4.00 D).
2. Ignore any cylinder.
3. Determine the tear lens power by calculating the dioptric difference between the BOZR and the flattest corneal radius.
4. Add together the spherical component of the ocular refraction and the tear lens power to determine the final RGP contact lens power.

Over-refraction of a trial lens

To avoid compensating for a change in vertex distance, the power of the trial contact lens should be as close as possible to the patient's

177

ocular refraction. The vertex distance should be taken into consideration if the over-refraction is greater than ±4.00 D.

The tear lens and the BOZR

The BOZR is the radius of curvature of the back surface of the contact lens measured within a specified central region (the back optic zone diameter) of the back surface of the contact lens. The BOZR is used to specify fit of an RGP contact lens. It is important to be able to determine the probable magnitude of the tear lens and how it varies as the BOZR is changed. This can help fit evaluation.

The following are rules of thumb for RGP contact lenses:

- The tear lens power increases by +0.25 D for each 0.05 mm that the BOZR is steeper than the cornea.
- The tear lens power increases by −0.25 D for each 0.05 mm that the BOZR is flatter than the cornea.

The correction of vision using soft/ silicone hydrogel contact lenses

It is important to appreciate that soft and silicone hydrogel contact lenses *do not* form a tear lens. For these lenses to correct a patient's refractive error, the BVP of the contact lens measured in air should be *equal* to the patient's ocular refraction. In practice, the power of the soft contact lens ordered is usually the sum of the initial trial contact lens and any over-refraction necessary to provide good visual acuity. Remember, there is no liquid lens to consider. Once again, to avoid compensating for vertex distance, the power of the trial contact lens should be as close as possible to the patient's ocular refraction. The vertex distance should be taken into consideration if the over-refraction is greater than ±4.00 D. As practices are often supplied with banks of soft contact lenses, it is usually possible to insert a trial lens that is very close to the patient's ocular refraction. This often eliminates the need for large amounts of over-refraction when fitting soft lenses. Over-refraction techniques in contact lens practice are discussed in Chapter 26.

The correction of astigmatism

Astigmatism can be corrected using the following.

Spectacles

There is no real limit for the correction of astigmatism with spectacle lenses.

Spherical and aspheric RGP contact lenses

These can correct up to 2.50 D of corneal astigmatism.

Soft spherical and aspheric contact lenses

Only small amounts of astigmatism (about 0.25–0.50 D) are effectively masked with spherical soft contact lenses. Soft aspheric designs, thicker monocurve and first-generation silicone hydrogel designs may mask up to 0.75 DC.

Soft toric contact lenses

Relatively high amounts of astigmatism (about 3.75 D) can be corrected using 'stock' toric soft lenses and higher levels with custom-made lenses.

RGP toric contact lenses

Relatively high levels of astigmatism can be corrected with custom-made RGP toric and bitoric contact lenses.

A high percentage of patients with *corneal* astigmatism can be successfully fitted with a spherical/aspheric RGP contact lens. However, as the level of astigmatism increases a significant percentage attains optimal vision and/or mechanical comfort only if fitted with a toric RGP lens (see Chapter 22).

The incidence of astigmatism is shown in Table 17.3.

Contact lenses for the correction of presbyopia

Presbyopia, as discussed in Chapter 13, is an admission of ageing and inconvenience; it requires a modification to one's normal lifestyle and is depressingly progressive! It does,

Table 17.3 Incidence of astigmatism

Ocular astigmatism (DC)	Percentage of total population
0.00	32.0
0.25–0.50	34.6
0.75–1.00	17.7
1.25–2.00	9.8
2.25–3.00	3.8
3.25–4.00	1.5
>4.00	0.6

After Rabbetts (1998).

however, offer opportunities to the eye-care practitioner. The percentage of the population aged over 45 is of course growing. An estimated 37.1% of the UK population was aged over 45 in 1991. It is projected that this figure will increase by 6.4% by 2011. Patients who were fitted with contact lenses some 20–25 years ago are now becoming presbyopic and these patients need advice on the correction of their presbyopia with both spectacles and contact lenses. We are often told that the one of the largest potential areas for growth in contact lens fitting is the presbyopic market and that the age range 45–60 years is probably the target group for multifocal contact lenses.

Relatively speaking, not that many presbyopic contact lenses are fitted and such fitting is sometimes described as a 'niche activity' carried out by 'specialists'. Presbyopic contact lens fitting can be profitable because it is regularly quoted in the press that 87% of the country's wealth is in the hands of the over-50s! Happy presbyopic patients rarely seek alternative sources of lens supply or professional guidance; they are loyal and are a good source of referrals.

It is undoubtedly easier to correct presbyopia with spectacles than with contact lenses. The spectacle correction of presbyopia is usually by means of:

- single vision lenses, including ready-made reading spectacles
- enhanced reading lenses
- bifocal lenses
- trifocal lenses
- progressive power lenses.

Methods of correcting presbyopia with contact lenses include the following:

- Modifying the distance prescription (less minus or more plus)
- Reading spectacles to wear together with distance contact lenses
- Monovision: generally fitting the distance lens to the dominant eye
- Bifocal contact lenses: alternating (translating) designs in rigid and soft forms
- Multifocal contact lens: simultaneous designs in rigid and soft forms
- Monovision in conjunction with multifocal lenses:
  - *modified*: adjusting the multifocal lenses to give the best vision compromise while compromising on true binocular vision
  - *enhanced*: fitting in general the dominant eye with a distance-only correction, leaving the multifocal lens in the non-dominant eye.

All of the above are a compromise in one way or another and are discussed in Chapter 25.

Contact lenses versus spectacles

When comparing the correction of refractive errors with spectacles and contact lenses the following points should be considered.

Vision

People with myopia tend to see better with contact lenses than spectacles. This is because the size of the retinal image formed with a contact lens is larger than the retinal image formed with a spectacle lens. A patient with a very high myopic correction may see significantly better when wearing a contact lens than a spectacle lens. The opposite is true for those with hypermetropia.

Accommodation and convergence

To view a near object, the person with myopia needs to accommodate and converge *more* when wearing contact lenses than the person with myopia who is corrected with spectacle lenses. However, the hypermetropic patient needs to accommodate and converge *less* when wearing contact lenses than when corrected with spectacle lenses. This has some practical significance in contact lens practice because, when a patient changes from contact lenses to spectacles (and vice versa), the accommodation convergence ratio is only minimally disturbed. Also, the myopic contact lens patient may require a reading addition slightly earlier than the person who is corrected with a spectacle lens.

The topics of accommodation and convergence are covered in Chapters 13 and 15. Binocular vision considerations in contact lens wear are discussed in Chapter 31.

Anisometropia

Anisometropia (meaning a difference in the right and left prescriptions) gives rise to several optical and mechanical problems when corrected with spectacle lenses. These problems include the following:

- Horizontal and vertical differential prismatic effects: these occur when the individual views through points away from the optical centres of the lenses.
- A difference in right and left spectacle magnifications resulting in aniseikonia (a difference in cortical image sizes). Unequal magnifications also give rise to poor cosmesis when wearing the lenses.
- Poor cosmesis caused by unequal edge or centre thicknesses.
- Unequal weights.

Most of the above problems do not occur with contact lenses. As a contact lens moves with the eye, prismatic effects are of no optical consequence. The poor cosmesis resulting from unequal spectacle magnifications and unequal edge thicknesses is negated with contact lenses. A difference in the weight of a pair of contact lenses is not usually an issue, although the density of a lens material may be an important factor when attempting to reduce the weight of an individual contact lens to improve centration on the eye. (Contact lens materials are discussed in Chapter 21.) The only optical problem that

may be an issue with the anisometropic contact lens wearer is that of aniseikonia, in the case of *axial* ametropia. In this situation, the patient experiences less aniseikonia and therefore more comfortable binocular vision, if *spectacles* rather than contact lenses are prescribed. This is discussed in more detail in Chapter 20.

Irregular refraction

This condition, described in Chapter 7, is initially corrected with spectacles and later with contact lenses. Where the irregularity is in the cornea, e.g. in keratoconus, keratoplasty or post-refractive surgery, a rigid contact lens is often prescribed because the tear layer that is trapped between the back surface of the contact lens and the cornea 'fills in' any surface defects. Significant improvements in visual acuity can be obtained with contact lenses compared with spectacle lenses.

Occlusion: partial and total

Indications for partial occlusion with a contact lens include albinism, pupil abnormalities, Marfan syndrome and pigmentary retinal degeneration. For any of these, there is less chance of light entering the eye from around the periphery of a contact lens and, if only certain wavelengths of light are to be admitted to the eye, a contact lens would again prove more efficient than the closest-fitting spectacles. Indications for use where total occlusion is required include intractable diplopia and amblyopia treatment.

Therapeutic

In therapeutic contact lens practice the objective is rarely to achieve an optimal visual result; rather it is to protect or assist in the healing process of the compromised cornea. It is rare that these lenses are fitted on the basis of a single indicating factor; multiple aims are more commonplace. Pain relief, mechanical protection and corneal wound healing are frequently required in combination. Soft lenses, both conventional and disposable, provide most therapeutic lens applications, but the latest silicone hydrogel, conventional silicone elastomers, scleral and large diameter corneal lenses also have a role to play.

Cosmetic

A dictionary definition of cosmetic is 'having no other function than to beautify'. Medical cosmetic lenses, however, may also perform a visual function if the eye is sighted although in general this is not their prime function. Common uses include:

- masking opaque hypermature and traumatic cataracts
- hiding corneal opacities and dystrophies
- covering a large angle squint where a surgical procedure is not indicated
- restoration of the iris and pupil to normal size, appearance and colour
- covering iris holes that give diplopia and polyopia
- albinism

The simplest solution is often the best as well as the easiest to reproduce when a replacement lens is required. An example is masking a white cataract. A soft lens with a black pupil is the simplest, least expensive and least time-consuming option.

Myopia control

Historically, polymethyl methacrylate (PMMA) and RGP lenses have been prescribed with the aim of reducing the progression of myopia in young patients. No plausible hypothesis has ever been given as to why this might occur. Furthermore, the studies that have been reported to date, which purport to show such an effect, are limited because of high rates of loss to follow-up, low rates of progression and poor experimental methodology (lack of appropriate controls, masking and randomisation). Recently the Contact Lens and Myopia Progression (CLAMP) study, one of the largest of this type, reported the results of a 2-year observer-masked randomised study of 158 children, aged 6–11 years, fitted with rigid lenses and 225 age- and sex-matched spectacle wearers (controls). The mean progression in myopia in the two groups

over the 2-year period was 1.30 D, and there was no difference in this rate of myopia progression between the two groups. Thus, the clinical argument that rigid lenses are beneficial because they arrest the progression of myopia must be regarded with some degree of scepticism.

Although orthokeratology has been around for centuries, a more recent development known as accelerated orthokeratology, which involves the wearing of specially designed rigid contact lenses overnight, temporarily to correct the patient's refractive error, has been gaining clinical and scientific credibility. Accelerated orthokeratology can be defined as 'the use of RGP lenses to gradually and systematically alter the shape of the cornea to cause a temporary and reversible reduction in myopia'. From the patient's perspective it can be described as a safe and reversible form of scientific, non-surgical, semi-permanent, vision correction therapy.

The essentials of accelerated orthokeratology are the reduction of myopia by the application of appropriately fitted high Dk (oxygen permeability) reverse-geometry RGP lenses, which reduce the corneal eccentricity and flatten the apical radius of the cornea. The lenses are removed during waking hours and re-inserted before sleep. These lenses contain the patient's correction, so they can be worn to enhance vision during the day if required. The maximum realistic change that is possible using contemporary designs is up to −5.00 D of myopia and 1.50 D of with-the-rule astigmatism. In some regions, particularly the Far East, orthokeratology is being promoted as a means of 'controlling' or preventing myopic creep. The jury in the west, however, is still out on this one, although young myopic teenagers do make good candidates, as do those with low myopia who are considering refractive surgery.

Off-axis aberrations

Aberrations are the result of inherent shortcomings of a lens and can occur with lenses manufactured using the best materials, and free from manufacturing and other defects. Aberrations that afflict spectacle lenses can be broadly divided into two types: image-deforming aberrations and image-degrading aberrations. In addition, some aberrations occur with monochromatic light, the **monochromatic aberrations** (1–5 in the list below) and other aberrations occur with light that contains at least two wavelengths (**chromatic aberrations,** 6). The six aberrations are:

1. Spherical aberration
2. Coma
3. Oblique astigmatism
4. Curvature of field
5. Distortion
6. Chromatic aberration

In addition, spherical aberration and chromatic aberration both have a longitudinal (axial) and a transverse (lateral) variety, each with its own causes and corrections.

The above aberrations occur when an individual looks through off-axis points on a spectacle lens, so the challenge for the spectacle lens designer is to minimise the effects of oblique astigmatism and curvature of field in order to provide good off-axis performance when the eye views through peripheral regions of a spectacle lens. The contact lens designer is faced with a different challenge. As a contact lens moves with the rotation of the eye in all directions of gaze, the troublesome aberrations for the lens designer are spherical aberration and coma, which occur when the contact lens decentres as a result of lens lag.

Field of view

As one would expect, the field of view obtained when wearing a contact lens is much greater that that provided by a spectacle correction. This is particularly true for the hypermetropic patient because, when compared with those with myopia, people with hypermetropia experience a reduced field of view.

Summary of the optical advantages of contact lenses

A contact lens *does not* give rise to:

- oblique astigmatism
- distortion
- chromatic aberration
- a limitation in the field of view
- a scotoma in the visual field caused by the spectacle frame
- differential prismatic effects in anisometropia
- unequal accommodative requirements in anisometropia
- large amounts of spectacle magnification in a corrected aphakia eye

Concluding points for Chapter 17

This chapter is designed to form an introduction to the second part of this text. All the points and issues raised in this chapter are expanded upon later in the text.

References

Rabbetts R B (1998) *Bennett & Rabbetts' Clinical Visual Optics*. Butterworth Heinemann, Oxford.

Further recommended reading

Douthwaite W A (2006) *Contact Lens Optics and Lens Design*. Elsevier, Oxford
Efron N (2002) *Contact Lens Practice*. Butterworth-Heinemann, Oxford
Mountford J, Ruston D, Dave T (2004) *Orthokeratology: Principles and Practice*. Elsevier, Oxford

Measurement of the Cornea: The Keratometer and Beyond

CHAPTER 18

Caroline Christie and Andrew Keirl

Introduction

'Keratometry' is the term used to describe the measurement of the principal radii of curvature of the anterior surface of the cornea (strictly speaking, the tear layer). During keratometry, the cornea is regarded as a convex mirror and in effect, it is the radius of curvature of this 'mirror' that is being measured. Essentially, keratometry is a method of measuring the size of an image of an object formed by reflection. Given the object size and the distance from the image to the object, the radius of curvature of the cornea can be calculated. In keratometry, the illuminated object, or **mire**, may be two separate mires or two points at distinct distances on individual mires. Reflection is from an area of about 3–4 mm in the central zone on the cornea, although the exact distance depends on the individual instrument and corneal size. The optical principles of keratometry employ the first Purkinje image (the reflection from the anterior surface of the cornea), which is formed just within the cornea. The size of the image formed depends on the size of the object (mires), the radius of curvature of the cornea and the distance of the mires from the cornea.

Chapter content

- Keratometry
- Practical use of the keratometer
- Keratoscopy
- Corneal topography

The keratometer equation

If, during keratometry, the cornea is regarded as a spherical mirror, with radius of curvature r, then an object of height h is imaged with a height h'. Using Newton's magnification relationship, the keratometer equation is

$$r = 2\frac{h'}{h}d$$

In theory, if h' could be read off against a measuring scale placed on the eyepiece graticule, the above equation could be used to find r. The formation of the reflected image is shown in Figure 18.1. The derivation of the keratometer equation is as follows: Newton's magnification equation is:

$$m = \frac{h'}{h} = -\frac{f}{x}$$

where h is the size if the object, h is the size of the reflected image, f is the focal length of the mirror and x (the extra-object focal distance) is the distance from F to h. The extra-image focal distance x' is also shown in Figure 18.1.

For a curved mirror:

$$f = \frac{r}{2}$$

which means that Newton's magnification equation becomes:

$$\frac{h'}{h} = -\frac{r}{2x}$$

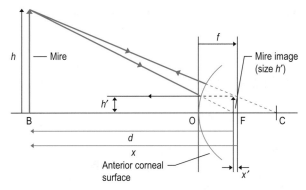

Figure 18.1 The formation of the mire image. (After Douthwaite 2006, with permission of Elsevier Ltd.)

With reference to Figure 18.1, the distance x is measured from F to B. This means that x is a negative value. If we assume that the distance d is approximately equal to $-x$ we obtain:

$$r = 2\frac{h'}{h}d$$

In the above equation, h' is the size of the reflected corneal image, h is the object (mire) size and d is the distance between the mires and the corneal image.

There are two problems associated with the keratometer equation. Although the equation is the basis of the keratometer design, it is a *paraxial* equation and therefore leads to errors in determining the corneal radius, because the angles involved in the reflection are not small. Remember that small angles are the basis of paraxial equations. It has been shown that the use of the keratometer equation leads to significant errors (up to 5%) in the recording of the corneal radius. This is almost 0.40 mm on a 7.90 mm radius. During the manufacturing process, keratometers are therefore calibrated against spherical surfaces of known radii. Second, because the eyes are constantly making small movements, the reflected image would never be still enough to allow its direct measurement. This second problem is overcome by doubling the image and the size of the reflected image is obtained by measuring the displacement of this doubled image. Although this is not a problem as such, it is important to note that the true refractive index of the cornea

(1.3760) is not used to calibrate keratometers. Instead, an index of 1.3375 is usually used, which allows the instrument to read the total corneal power (about 90% of the front surface power).

Image doubling

A doubling device (a biprism back to back with a planoconvex lens) produces two images of the reflected image within the cornea. This occurs within the viewing system of the keratometer and allows the observer to adjust the amount of doubling until the two images appear to touch top to bottom. As mentioned above, a doubling device is necessary because it is impossible to measure the size of the image directly because the patient's eye is never stationary. However, it is relatively easy to judge when the top and bottom of the doubled images just touch, even though they are moving. Keratometers are calibrated to obtain the correct anterior surface corneal radius when the doubled images are correctly set, i.e. 'just touch'. The image doubling equation is:

$$h' = \frac{dP}{100}$$

where d is the distance between the image and the mires and P is the power of the doubling device in prism dioptres.

Doubling principles

There are two ways of producing a doubled image. The first, as the amount of doubling produce by a biprism is dependent on the position of the prism with respect to the objective lens, is to vary the position of the biprism lens. If this distance is reduced, the extent of the doubling reduces and if it is increased, the extent of the doubling increases. Therefore, by varying the position of the biprism, the amount of doubling can be made to equal the size of the image. Keratometers that incorporate this principle are known as **variable doubling keratometers**. Alignment of the doubled images can also be obtained by altering the size of the mires, while the amount of doubling is kept constant. Keratometers that incorporate this principle are known as **variable mire keratometers**.

Types of keratometer

There are two types of instrument in common use. Each employs one of the doubling principles outlined above to measure the separation of the mire images. As previously mentioned, doubling is necessary in taking the corneal radius measurement, because rapid eye movements would otherwise make this task extremely difficult. Keratometers measure the curvature of the central cornea over an area of approximately 3–4 mm and can determine:

- the radii of curvature
- the directions of the principal meridians
- the degree of corneal astigmatism
- the presence of any corneal distortion

The same cornea measured with more than one instrument can result in a variety of readings because different keratometers employ:

- different mire separations, so that the area of cornea used for reflection varies
- different refractive indices for calibration, so that the same radius could give a variety of surface powers.

Variable doubling

With this type of instrument, the mires have a fixed separation. The separation of the mire images is found by varying the position and therefore the power of the doubling device. The Bausch & Lomb keratometer (**Figure 18.2**) has two variable doubling devices and two sets of fixed mires. Both principal meridians can therefore be measured simultaneously, and the design is called a **one-position** instrument. The mires used in this instrument are shown in **Figure 18.3(a)**. The two independently adjustable prisms, situated behind a special stop consisting of four apertures A, B, C and D (**Figure 18.4**), double the mire image along two perpendicular meridians. When the instrument is correctly aligned the practitioner sees three images of the instrument's mires. Aperture C relates to the vertically displacing prism whereas aperture D relates to the horizontally displacing

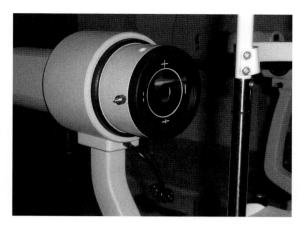

Figure 18.2 The mires of a Bausch and Lomb keratometer.

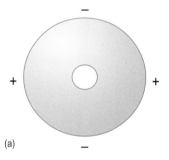

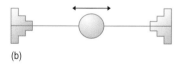

Figure 18.3 The illuminated mires used in the (a) Bausch & Lomb and (b) Javal–Schiötz keratometers. Keratometer mires: (a) fixed mire in one position and (b) variable mire in two position.

prism. Movement of the vertical prism results in movement of the vertically displaced image whereas movement of the horizontal prism results in movement of the horizontally displaced image. The central image is unaffected by the movement of either prism. Apertures A and B act as a Scheiner disc and double the central image of the mire when the system is out of focus. This 'Scheiner disc' therefore acts as a focusing device for one-position instruments.

185

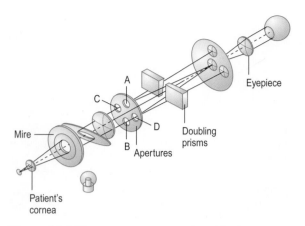

Figure 18.4 The aperture stop employed in the doubing system of the Bausch & Lomb one-position keratometer.

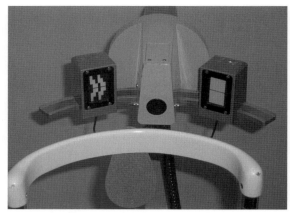

Figure 18.6 The mires of a Javal–Schiötz keratometer.

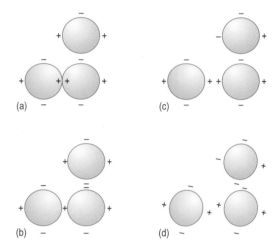

Figure 18.5 Four examples of mire alignment when using the Bausch & Lomb one-position keratometer. (See text for explanation.)

The Bausch & Lomb one-position keratometer has to be rotated about its anterior–posterior axis to find one of the principal meridians of the toroidal surface, but once this has been found no further rotation of the instrument is necessary to obtain a radius measurement along the second principal meridian. **Figure 18.5** shows:

- correct vertical doubling but insufficient horizontal doubling (Figure 18.5a)
- too much vertical doubling but correct horizontal doubling (Figure 18.5b)

- correct vertical and horizontal doubling (Figure 18.5c)
- an astigmatic cornea showing the principal meridians incorrectly set (Figure 18.5d).

Fixed doubling

With this instrument, the doubling is fixed for a particular separation of the mire images. It is only by altering the separation of the mires themselves that a reading can be taken. These keratometers are called **two-position** instruments and are based on the Javal–Schiötz design (**Figure 18.6**). The mires are in the form of a 'step' and a 'square' (see Figure 18.3) and are attached to the front of two small light boxes which, via a gearing arrangement, are made to move equally in opposite directions along a circular arc, the centre of curvature of which corresponds to the patient's eye. Doubling is achieved with a prism placed behind the objective lens (**Figure 18.7**). The whole instrument can rotate about its anterior–posterior axis to enable the measurement of the corneal radius along any meridian. The stepped mire has a green filter over it whereas the square one is covered with a red filter. These filters help the practitioner to recognise when the mires overlap because any areas of superimposition appear yellow. **Figure 18.8** shows the Javal–Schiötz mires when:

- the mire separation is too large (Figure 18.8a)

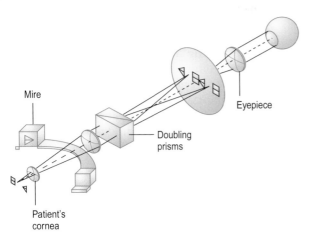

Figure 18.7 The doubling arrangement employed in the Javal–Shiötz two-position keratometer.

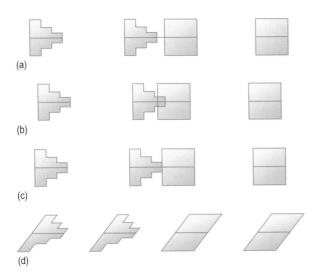

(a)

(b)

(c)

(d)

Figure 18.8 Four examples of mire alignment when using the Javal–Shiötz two-position keratometer. (See text for explanation).

- the mire separation is too small (Figure 18.8b)
- the mire separation is correct (Figure 18.8c)
- an astigmatic cornea showing the principal meridians incorrectly set (Figure 18.8d).

The telecentric keratometer

The telecentric keratometer is a third type of keratometer that is seldom encountered in general practice. However, the design of this

instrument has features that make it superior to both the Bausch & Lomb and the Javal–Schiötz instruments. First, the corneal image is formed at the focal plane of the cornea (F) rather than near it. This means that d and x are now equal, which eliminates the approximation in the keratometer equation that assumes $d = x$. As previously mentioned, this assumption can lead to errors of up to 5% in the corneal radius. Another interesting difference with this instrument is that it uses a pair of lenses to produce a variable prism power by decentration.

When using a 'conventional' keratometer the practitioner must ensure that the eyepiece image is in the plane of the eyepiece crosswire. This is achieved by making sure that the observed mire image and the eyepiece cross-wire image are both clear at the same time. If the mire image is not formed in the plane of the eyepiece cross-wire, then the radius reading obtained using a conventional keratometer is inaccurate. A major advantage of the telecentric keratometer is that it does not require the observed mire image to be formed exactly in the plane of the eyepiece cross-wire and therefore does not require an adjustable eyepiece or eyepiece cross-wire. Of the three keratometer types available, the telecentric keratometer is probably the easiest to use. For further information about this keratometer, the reader is referred to Douthwaite (2006).

Practical use of the keratometer

Method

Focusing the eyepiece
Most keratometers have a graticule incorporated in the eyepiece. This should be focused at a distant object before taking a reading, to prevent observer accommodation giving inaccurate results.

Positioning the patient
The patient should be comfortably seated, with the forehead positioned firmly against the headrest. Fixation needs to be accurate and stable. To assist this, the eye not being assessed is occluded to prevent it from assuming fixation and there-

fore causing undesired peripheral readings from the eye being measured.

Line up the optical system

To help line up the optical system and locate the patient's cornea use:

- the sights attached to the instrument
- the light from a pen torch directed through the eyepiece, looking for the corneal reflection.

Using the keratometer

The instrument body is positioned initially at a greater distance from the cornea than necessary and slowly moved forward until the mire images come into clearly into view and are centrally located.

- When using the Bausch & Lomb one-position instrument (see Figure 18.2), the operator must sharply focus and superimpose the doubled circular image before carefully aligning the plus and minus targets adjacent to it. When the principal meridians are *not* at 90 and 180, the body of the instrument requires rotation to align these supplementary targets correctly. After this has been achieved the main circular target and the supplementary targets need to be re-focused before a final reading of the radii and axis can be achieved.
- When using the Javal–Schiötz two-position instrument (see Figure 18.6), the instrument body is initially set up in the horizontal plane. Looking through the eyepiece, the operator sees four images: two of each mire either side of the centre. The colour coding helps to ensure that the images are in the correct order and in final alignment. The central pair needs to be brought together until just touching. If the astigmatism is along 180, the thin black line running through the centre of the mires is perfectly aligned; if not, the instrument body requires rotation until alignment is achieved. The axis can then be read off using the external protractor scale. Before doing this the actual mire focusing and alignment should once more be

checked. The instrument body then requires rotation through 90° to take the second reading in exactly the same way.

- *Remember*, whichever instrument is employed and however good fixation the patient maintains, the images constantly need fine refocusing to obtain the best results. Ideally three readings in each meridian should be taken and these averaged.

Instrument calibration

As with all measuring devices it is important that the instrument be regularly calibrated. This can easily be achieved using steel ball bearings, which have an accuracy of ±0.001 mm. Five readings of each ball should be taken and a minimum of three balls used, so that a calibration line can be plotted and accuracy checked.

Extending the range

Radii that are out of range of the instrument, e.g. steep in keratoconus and flat in post-refractive surgery, can be obtained by placing either a +1.25 DS or a −1.00 DS trial lens in front of the keratometer objective. Prior re-calibration is necessary using steel balls of known and validated radius. The Javal–Shiötz instrument is particularly useful, because the range at the steep end extends as far as 5.50 mm.

Tips

- Keratometry readings are expressed as *along* a particular meridian.
- Remember that not all eyes have their principal meridians at 90 and 180.
- The principal meridians are not at 90 degrees to each other in an irregular cornea.
- Instruments usually show the corneal radius in both millimetres and dioptres, the former being the one commonly used in the UK.

Example 18.1

In a certain keratometer, the power of the doubling prism is 2.50 Δ and the distance between the mires is 17 cm. Find the corneal radius if the distance between the mires and the corneal image is 16 cm.

To find the corneal radius we need to make use of the keratometer equation:

$$r = 2\frac{h'}{h}d$$

We are given h (the distance between the mires) and d (the distance between the mires and the corneal image). We therefore need to find h'. We can do this by using the imaging doubling equation:

$$h' = \frac{dP}{100}$$

Substituting into the imaging doubling equation gives:

$$h' = \frac{16 \times 2.5}{100} = 0.40\,cm$$

Substituting into the keratometer equation gives:

$$r = 2\left(\frac{0.40}{17}\right)16 = 0.75294\,cm$$

The corneal radius is therefore 7.5294 mm.

Autokeratometers

Autokeratometers determine the radii and principal meridians along the visual axis. They can also measure peripheral radii at predetermined positions away from the corneal apex. Some instruments use computerised image processing to determine the flattest and steepest corneal radii, with corresponding meridians and powers.

Autokeratometers achieve the measurement by calculating the distance between the reflected images from light-emitting diodes (LEDs). The mires are usually circular and placido in nature, designed to reveal any corneal distortion. They also include distance indicators that enable the reading to take place. Despite the speed of measurement, steady fixation by the patient is essential. This is often assisted by an LED, but the practitioner should carefully observe the eye during measurement, as well as ensuring that the patient maintains a wide palpebral aperture.

Keratoscopy

As already mentioned, the keratometer provides only an estimation of corneal curvature based on an approximate 3–4 mm chord for its anterior surface. Keratometry also assumes that the cornea is spherical, which of course it is not. The corneal shape is actually likened to an oblate ellipse, which flattens gradually towards its periphery. As a result of this flattening, and as different keratometers reflect their mires from different regions of the cornea, two readings of the same cornea with two different keratometers may not give the same radius. This is because light reflected from the cornea during keratometry does not come from its centre, but from two small areas on either side of the instrument's axis. The size of this area is dependent on the effective aperture of the keratometer objective.

The technique of keratoscopy allows the corneal contour to be assessed more comprehensively than was done by traditional keratometry.

The first keratoscope on which modern photokeratoscopy is based was the placido disc, which is a series of concentric rings that are usually illuminated. The rings are projected on to the cornea and the observer looks at their reflection. The assessment of corneal topography is made by judgement of the regularity of the image. Although this method is a simple way to make a gross assessment of any corneal irregularities, it cannot provide a detailed quantifiable assessment of the contour. One of the earliest attempts at quantifying the corneal contour was the Wesley–Jessen photoelectronic keratoscope (PEK). A Polaroid photograph was taken of a series of concentric rings and the diameter of each ring was measured. From these, the shape factor was calculated, but the lack of immediacy of the measurement and problems in the reproducibility of lenses ordered using the system restricted its commercial success.

Corneal topographers

Videokeratoscopy gives a more detailed assessment of overall corneal topography by means of modern computer analysis. These instruments are now in more general use and are mainly placido disc-based, computerised, corneal topography systems. The tear film reflects the image of the rings that is captured by a digital video camera. An algorithm detects and identifies the position of the rings. The detection of the border of the keratographic axis, which differs depending on the system, is applied to the digital image, which then reconstructs the corneal curvature.

The presentation of data collected from the range of instruments now on the market is increasingly sophisticated. The results are presented as a topographic contour map of the cornea with colour used to indicate the curvature and power distribution for the area under analysis. Curvature map options vary but the most common for contact lens practice are the tangential radius of curvature and the axial representation because these tend to differ more in the mid to outer corneal periphery, revealing clinically significant differences. The colour scale normally runs from red at the steep end to blue at the flat end. The options are an 'absolute' scale, which is consistent and allows direct comparison between each eye and on different occasions, or the 'normalised' scale, which auto-

matically adjusts the map into smaller colour intervals for greater detail, but can vary depending on the particular cornea. Elevation maps depict relative height differences that can be useful for gas-permeable lens fluorescein patterns. High red areas indicate where fluorescein is displaced and the low blue areas where fluorescein pools. Difference maps are specifically helpful for monitoring corneal change or distortion because they represent a mathematical subtraction of two selected maps. However, it can be all too easy to allow the impressive appearance of the output to mask the actual clinical value of the data collected. The useful information is the corneal contour maps themselves, which simply show the contour changes across the corneal surface and can help detect and differentiate regular and irregular corneal surfaces. **Figure 18.9** (a) shows a spherical cornea, (b) high regular astigmatism (bowtie) and (c) keratoconus.

Topographers also provide a value for the apical radius of the cornea, r_o. It is important to note that this differs from any reading for the same eye obtained by keratometry, because these instruments measure the cornea over an area about 3 mm from the apex.

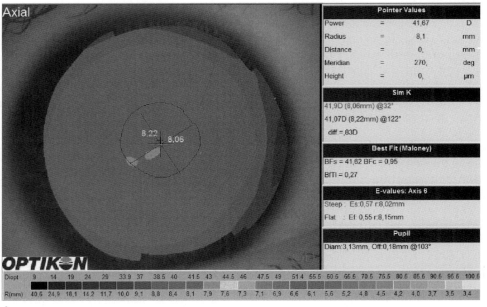

A

Figure 18.9 (a) A topographical contour map for a spherical cornea. (See also colour plates.)

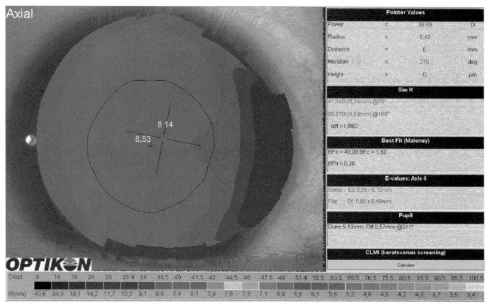

B

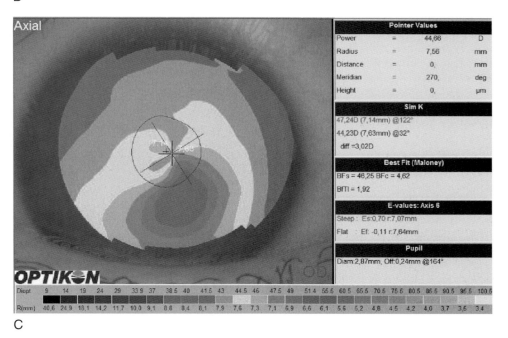

C

Figure 18.9 Topographical contour maps for (b) a cornea with high astigmatism and (c) a cornea with keratoconus. (See also colour plates.)

In contact lens practice, videokeratoscopy is particularly useful for assessing corneal eccentricity (*e*-value), surface irregularity and irregular astigmatism. Subtractive plots are also used to monitor corneal changes with orthokeratology or when ceasing wear of polymethyl methacrylate (PMMA) or low DK lenses. Misalignment of the visual axis during measurement can give misleading results.

Extreme asymmetry or distortion is frequently seen with the following:

- Irregular astigmatism, distinguished from regular astigmatism in the corneal map by the lack of a symmetrical 'bowtie' pattern (Figure 18.9(b))
- Keratoconus
- Penetrating keratoplasty
- Contact lens-induced corneal warpage
- Refractive surgery
- Trauma

A more recent version of a corneal topographer is based on slitlamp technology, using diffusely reflected light from the corneal surfaces, iris and lens, e.g. the Orbscan. The images are acquired by video camera. The slit beam scans with over 7000 independently and directly measured data points. Ray-trace triangulation determines elevations of the anterior and posterior corneal surface, giving limbus-to-limbus data. The corneal mapping produced gives a full analysis of the cornea under view, including a variety of refractive power maps; central (optical axis identification) and peripheral measurement data for contact lens fitting, full corneal pachometry and three-dimensional virtual reality models.

Another recent device, the Oculus Pentacam, has also moved away from placido technology and uses a rotating Scheimpflug camera. It provides data from the anterior and posterior corneal shape, corneal pachymetry data and anterior chamber depth and angle. There is also the option of anterior lens surface imaging and, if the pupil is large enough, posterior lens surface imaging.

Almost all corneal topography devices incorporate contact lens-fitting software, which ranges from pre-loaded designs of current lens manufacturers' stock lenses to those topographical units that allow practitioners to create their own contact lens designs.

Concluding points for Chapter 18

Indications for keratometry

- During initial contact lens assessments to provide baseline data and determine the location of the astigmatic surface.
- During aftercare examination to compare the previously recorded base line data with current data, to allow any changes or distortions to be revealed.
- In RGP lens fitting, to provide data to assist empirical fitting or initial lens selection.
- Measurement of on-eye contact lens flexure.
- Measurement of non-invasive break-up time (NIBUT).

Limitations of keratometry

- Only the central region of the cornea is assessed; peripheral information is difficult to obtain by standard methods.
- Changes in keratometer readings of 0.05 mm are not clinically significant as a result of instrument inaccuracy.
- A patient's keratometer readings may depend on the instrument used, as a result of calibration differences.
- In contact lens fitting, the power of the lens and the action of the eyelids are influential factors so lens fit and position can be assessed only during an on-the-eye assessment.

Reference

Douthwaite W A (2006) *Contact Lens Optics and Lens Design*. Elsevier, Oxford

Further recommended reading

Efron N (2002) *Contact Lens Practice*. Butterworth-Heinemann, Oxford

Elliott D B (2003) *Clinical Procedures in Primary Eyecare*. Butterworth-Heinemann, Oxford

Mountford J, Ruston D, Dave T (2004) Corneal topography and its measurement. *Orthokeratology Principles and Practice*. Butterworth-Heinemann, Oxford

Rabbetts R B (1998) *Bennett & Rabbetts' Clinical Visual Optics*. Butterworth-Heinemann, Oxford

Tunnacliffe A H (1993) *Introduction to Visual Optics*. Association of the British Dispensing Opticians, London

Ray-tracing Through RGP Contact Lenses and Contact Lens–Tear Lens Systems

Andrew Keirl

CHAPTER

19

Introduction

Although Chapter 17 was designed to form an introduction to the optics of contact lenses, Chapter 19 is the first to be devoted to the subject of contact lens optics. Some readers will find Chapter 19 the most difficult chapter in this text so far. However, after close inspection of worked examples the reader will discover that the 'tools' required to master this material have all been covered in previous chapters, e.g. effectivity, equivalent air distance, surface power calculation, step-along ray-tracing and step-back ray tracing. It is also important to be able to draw clear diagrams showing all the required vergences and to organise the computation in a careful and logical manner. The reader must also realise that all of the skills mastered in this chapter are used many times again in this book.

Chapter content

- Ray-tracing through RGP contact lenses
- Ray-tracing through a contact lens–tear lens system

Ray-tracing through RGP contact lenses

Although a rigid gas permeable (RGP) contact lens has a centre thickness in the region of only 0.50 mm, its steep surface curves mean that it must be considered to be a thick lens. Thin lens equations cannot be applied to a contact lens. When considering a contact lens in isolation, i.e. not actually on the eye, the standard ray-tracing techniques outlined in Chapter 1 can be applied to a contact lens in air. It is necessary at this point to introduce and use relevant contact lens terminology. For the radius of curvature of the front surface of the contact lens (r_1) the term **front optic zone radius** (FOZR) is used and for the radius of curvature of the back surface of the contact lens (r_2) the term **back optic zone radius** (BOZR) is used.

Example 19.1

Calculate the back vertex power (BVP) of the following RGP contact lens in air:

FOZR (r_1) = 7.30 mm
BOZR (r_2) = 7.80 mm
t = 0.50 mm
n = 1.50

As the contact lens is a thick lens, the equivalent air distance (EAD) will be used when ray-tracing. The BVP can be calculated using a standard step-along ray-trace.

The first task is to calculate the two surface powers. The radius of curvature r is substituted in metres and **standard form** used as appropriate, e.g. 6.50 mm is written as 6.5^{-3} m and not 0.0065 m. At least four decimal places are used for all stages of the calculation and rounding up or down does not take place during the calculation itself. Accuracy is very important when dealing with steep surface powers. To find surface powers we must use:

$$F_1 = \frac{n'-n}{r_1}$$

193

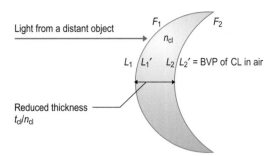

Light from a distant object

Reduced thickness
t_{cl}/n_{cl}

Figure 19.1 Diagram for Example 19.1. The RGP contact lens is treated as a thick lens in air.

and

$$F_2 = \frac{n - n'}{r_2}$$

Sign convention tells us that both r_1 and r_2 are positive:

$$F_1 = \frac{1.50 - 1.00}{+7.3^{-3}} = +68.49315\,D$$

$$F_2 = \frac{1.00 - 1.50}{+7.8^{-3}} = -64.1026\,D$$

The equivalent air distance (with t_{cl} substituted in metres) is given by:

$$EAD = \frac{t_{cl}}{n_{cl}} = \frac{5^{-4}}{1.5} = 3.333^{-4}\,m$$

Now the EAD has been determined, the refractive index between the two surfaces can be assumed to be **equal to that of air** ($n = 1$). This value is used in the following ray-trace and a step-along ray-trace is used to find the BVP. As always, the computation takes place in two columns, one for vergences and the other for distances. The vergences for this problem are illustrated in **Figure 19.1**.

Vergences (D)   Distances (m)

$L_1 = 0.0000$

$F_1 = +68.49315\,D$

$L_1' = L_1 + F_1 = 68.49315\,D \quad \rightarrow \quad l_1' = \dfrac{n}{L_1'}$

$$l_1' = \frac{1}{+68.4931} = +0.0146\,m$$

$$l_2 = l_1' - \left(\frac{t_{cl}}{n_{cl}} \right)$$

$$L_2 = \frac{n}{l_2} \qquad \leftarrow \qquad l_2 = +0.0147 - 3.333^{-4}$$
$$= +0.014267\,m$$

$$L_2 = \frac{1}{+0.014267} = +70.09346\,D$$

$$L_2' = L_2 + F_2$$

$$L_2' = +70.09346 + (-64.1026) = +5.9909\,D$$

As $L_1 = 0.0000$, $L_2' = BVPF_v' = +5.9909\,D$

The BVP of this RGP contact lens in air is therefore +5.9909 D.

Example 19.2

A rigid gas permeable contact lens has a BVP of −8.00 D, a centre thickness of 0.45 mm and a refractive index of 1.51. If the BOZR is 8.00 mm find the FOZR.

In this example the unknown value is the front surface radius. We therefore have to use a step-back ray-trace to find F_1 and hence the FOZR (r_1). We are given the BOZR (r_2), the centre thickness (t_{cl}), the refractive index of the lens material (n_{cl}) and the BVP (L_2'). Using what we know, we can calculate the back surface power (F_2) and the reduced thickness (t_{cl}/n_{cl}). The back surface power is given by:

$$F_2 = \frac{n - n'}{r_2}$$

Sign convention tells us that r_2 is positive:

$$F_2 = \frac{1.00 - 1.51}{+8.00^{-3}} = -63.7500\,D$$

The EAD (with t_{cl} substituted in metres) is given by:

$$EAD = \frac{t_{cl}}{n_{cl}} = \frac{4.5^{-4}}{1.51} = 2.9801^{-4}\,m$$

Now that the EAD has been determined, the refractive index between the two surfaces can be assumed to be **equivalent to that of air** ($n = 1$). This value is used in the following ray-trace. A step-back ray-trace is used to find the front surface power. As always, the computation takes place in two columns, one for vergences and the other for distances.

Vergences (D)   Distances (m)

$L_2' = -8.0000\,D$

$L_2 = L_2' - F_2$

$L_2 = -8.0000 - (-63.7500)$

$L_2 = +55.7500\,D \;\rightarrow\; l_2 = \dfrac{1}{+55.7500} = +0.01794\,m$

$$l_1' = l_2 + \left(\dfrac{t_{cl}}{n_{cl}}\right) \text{ (step-back)}$$

$$\leftarrow \quad l_1' = +0.01794 + 2.9801^{-4}$$
$$= +0.018235\,m$$

$L_1' = \dfrac{1}{+0.018235} = +54.8389\,D$

As $L_1 = 0.0000$, $L_1 = F_1$

$F_1 = +54.8389\,D$

The FOZR can be found using:

$r_1 = \dfrac{n' - n}{F_1}$

$r_1 = \dfrac{1.51 - 1}{+54.8389} = +9.2999^{-3}\,m$

The FOZR of the contact lens is therefore +9.2999 mm.

The above examples illustrate that a contact lens in air is treated exactly the same way as a spectacle lens in air. However, when dealing with a contact lens, expect to find strong surface powers.

Ray-tracing through a contact lens–tear lens system

As described in Chapter 17, when a RGP contact lens is fitted to an eye, a contact lens–tear lens system is formed. This system consists of two elements: the **contact lens** and the **liquid** or **tear** lens. The form, and therefore the power, of the tear lens depends on the radius of curvature of the back surface of the contact lens (BOZR), the radius of curvature of the anterior corneal surface (the keratometer reading) and the refractive index of the tears. To make calculations easier, the contact lens, liquid lens and cornea are imagined to be separated by two infinitely thin 'air films' or 'air gaps' (see Figure 17.1).

As a minimum, a contact lens–tear lens system consisting of an RGP contact lens and a tear lens are made up of the following.

Four surface radii

- The front surface radius of the contact lens (FOZR or r_1)
- The back surface radius of the contact lens (BOZR or r_2)
- The front surface radius of the tear lens (r_3)
- The back surface radius of the lens (r_4)

These radii are shown in **Figure 19.2**.

As the form (shape) of the tear lens depends on the BOZR of the contact lens and the radius of curvature of the anterior corneal surface:

$r_3 = r_2$ and

$r_4 = $ keratometer reading

Sign convention tells us that all the above radii are positive.

Four surface powers

As long as there is a different refractive index either side, each surface of course generates a power. As a minimum, the four surface powers of a contact lens–liquid lens system are:

- $F_1 = $ the power of the front surface of the contact lens

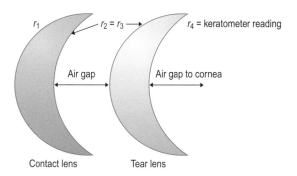

Figure 19.2 The surface radii of the contact lens–tear lens system.

- F_2 = the power of the back surface of the contact lens
- F_3 = the power of the front surface of the tear lens
- F_4 = the power of the back surface of the tear lens

These surface powers are shown in **Figure 19.3**.

As each surface is imagined to be separated by an infinitely thin air gap:

$$F_1 = \frac{n_{cl} - 1}{r_1}$$

$$F_2 = \frac{1 - n_{cl}}{r_2}$$

$$F_3 = \frac{n_{tears} - 1}{r_3}$$

$$F_4 = \frac{1 - n_{tears}}{r_4}.$$

Remember:

- The refractive index of air is usually taken to be 1.00
- The refractive index of tears is usually taken to be 1.336
- An infinitely thin air gap has no optical effect
- $r_2 = r_3$
- r_4 = the keratometer reading

The contact lens–tear lens vergences

Step-along and step-back ray-tracing techniques can be applied to the contact lens–tear lens system to find the vergences either side of each surface. As a minimum, there are eight vergences within the contact lens–tear lens system.

The contact lens vergences are shown in Figure 19.4:

- L_1 is the vergence arriving at the front surface of the contact lens
- L_1' is the vergence leaving the front surface of the contact lens
- L_2 is the vergence arriving at the back surface of the contact lens
- L_2' is the vergence leaving the back surface of the contact lens.

The vergence leaving the back surface of the contact lens (in isolation) is equal to the back vertex power of the contact lens in air:

- L_2' is equal to the BVP of the RGP contact lens in air.

The tear lens vergences are shown in **Figure 19.5**. Because the contact lens and the tear lens

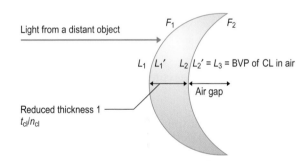

Figure 19.4 The contact lens vergences.

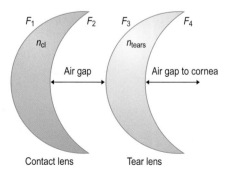

Figure 19.3 The surface powers of the contact lens–tear lens system.

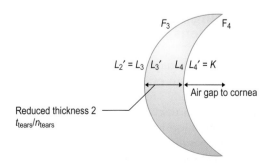

Figure 19.5 The tear lens vergences.

are separated by an infinitely thin air gap, which has no optical effect, $L_2' = L_3$.

- L_3 is the vergence arriving at the front surface of the tear lens
- L_3' is the vergence leaving the front surface of the tears lens
- L_4 is the vergence arriving at the back surface of the tear lens
- L_4' is the vergence leaving the back surface of the tears lens

The vergence leaving the back surface of the contact lens–tear lens system is equal to the BVP of the contact lens–tear lens system:

- $L_4' = BVP$ (F_v') of the contact lens–tear lens system.

As the tear lens and the cornea are separated by an infinitely thin air gap that, once again has no optical effect, the vergence leaving the back surface of the contact lens–tear lens system is equivalent to the patient's ocular refraction (assuming a distant object and $L_1 = 0$):

- $L_4' = K$.

So:

- L_2' is equal to the BVP of the RGP contact lens in isolation in air
- L_4' is equal to the BVP of the contact lens–tear lens system, which is also equal to the patient's ocular refraction

The above text makes several references to a minimum of four surfaces. This is the case if an RGP contact lens with *spherical* surfaces is fitted to a cornea with a *spherical* front surface. If one or more of the surfaces in the system is *toroidal*, the number of radii, surface powers and vergences increase. Astigmatic systems are dealt with in Chapter 23.

Example 19.3
Find the BVP of the following RGP contact lens–tear lens system:

FOZR (r_1) of the contact lens = 8.90 mm
BOZR (r_2) of the contact lens = 8.10 mm
The corneal radius = 7.80 mm
The centre thickness of the contact lens
 = 0.50 mm

The centre thickness of the tear lens = 0.30 mm
The refractive index of contact lens material
 = 1.49
The refractive index of tears = 1.336

As the examples are now becoming more complicated, it is important to organise the information given in the question in order to decide what is known, what can be initially calculated and what type of ray-trace is required to solve the problem. In this example, we are given r_1, r_2 and the corneal radius. Remember that the shape of the tear lens formed depends on the BOZR of the contact lens and the corneal radius. The front surface radius of the tear lens (r_3) is the same as the BOZR (r_2) and the back surface radius of the tear lens (r_4) is the same as the anterior corneal radius. Sign convention dictates that all these radii are positive. So:

$r_1 = +8.90$ mm
$r_2 = +8.10$ mm
$r_3 = +8.10$ mm (same as r_2)
$r_4 = +7.80$ mm (same as the anterior corneal radius).

The refractive index of contact lens material is 1.490 and the refractive index of the tears is 1.336. The first task is to calculate the four surface powers. Remember that each surface is separated by an infinitely thin and optically insignificant air gap (see Figure 19.3). The four surface powers can be found using the following equations:

$$F_1 = \frac{n_{cl} - 1}{r_1} = \frac{1.49 - 1.00}{+8.90^{-3}} = +55.0562\,D$$

$$F_2 = \frac{1 - n_{cl}}{r_2} = \frac{1.00 - 1.49}{+8.10^{-3}} = -60.4938\,D$$

$$F_3 = \frac{n_{tears} - 1}{r_3} = \frac{1.336 - 1.00}{+8.10^{-3}} = +41.4815\,D$$

$$F_4 = \frac{1 - n_{tears}}{r_4} = \frac{1.00 - 1.336}{+7.80^{-3}} = -43.0769\,D$$

The next task is to determine an EAD for *both* the contact lens and the tear lens. The EAD for the *contact lens* (with t_{cl} substituted in metres) is given by:

$$EAD = \frac{t_{cl}}{n_{cl}} = \frac{5^{-4}}{1.49} = 3.3557^{-4}\,m$$

197

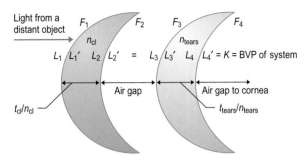

Figure 19.6 Diagram for Example 19.3.

The equivalent air distance for the *tear lens* (with t_{tears} substituted in metres) is given by:

$$\text{EAD} = \frac{t_{tears}}{n_{tears}} = \frac{3^{-4}}{1.336} = 2.2455^{-4}\,\text{m}$$

Now that the EAD has been determined, the refractive index between the two surfaces of the contact lens and the two surfaces of the tear lens can be assumed to be **equal to that of air** ($n = 1$). This value is used in the following ray trace. A step-along ray-trace is needed find the BVP. As always, the computation takes place in two columns, one for vergences and the other for distances. The vergences for this problem are illustrated in **Figure 19.6**. The ray-trace needs to be extended because we are dealing with four refracting surfaces. A step-along ray-trace is performed, first, through the contact lens and, second, through the tear lens.

Vergences (D)   Distances (m)

$L_1 = 0.0000$

$F_1 = +55.0562\,\text{D}$

$L_1' = L_1 + F_1 = 55.0562\,\text{D} \quad \rightarrow \quad l_1' = \dfrac{n}{L_1'}$

$$l_1' = \frac{1}{+55.0562} = +0.01816\,\text{m}$$

$$l_2 = l_1' - \left(\frac{t_{cl}}{n_{cl}}\right)$$

$L_2 = \dfrac{n}{l_2} \qquad \leftarrow \qquad$ $l_2 = +0.01816 - 3.3557^{-4}$
$= +0.017828\,\text{m}$

$L_2 = \dfrac{1}{+0.017828} = +56.0925\,\text{D}$

$L_2' = L_2 + F_2$

$L_2' = +56.0925 + (-60.4938) = -4.4013\,\text{D}$

L_2' is equivalent to the BVP of the lens in isolation if measured in air using a focimeter. Important! As a result of the thin air gap, $L_2' = L_3$. We must now perform a second step-along ray-trace through the tear lens. The reader is advised to follow all of the vergences carefully on Figure 19.6.

$L_3 = -4.4013$

$F_3 = +41.4815\,\text{D}$

$L_3' = L_3 + F_3 = +37.0801\,\text{D} \quad \rightarrow \quad l_3' = \dfrac{n}{L_3'}$

$$l_3' = \frac{1}{+37.0801} = +0.02697\,\text{m}$$

$$l_4 = l_3' - \left(\frac{t_{tears}}{n_{tears}}\right)$$

$L_4 = \dfrac{n}{l_4} \qquad \leftarrow \qquad$ $l_4 = +0.02697 - 2.2455^{-4}$
$= +0.02674\,\text{m}$

$L_4 = \dfrac{1}{+0.02674} = +37.3915\,\text{D}$

$L_4' = L_4 + F_4$

$L_4' = +37.3915 + (-43.0769) = -5.6854\,\text{D}$

As $L_1 = 0.0000$, $L_4' = \text{BVP}$ and $F_v' = -5.6854\,\text{D}$

The BVP of this RGP contact lens–tear lens system is therefore −5.6854 D.

For this contact lens–tear lens system to correct this particular eye, the ocular refraction must also equal −5.6854 D. This first example of a step-along ray-trace through a RGP contact lens–tear lens system demonstrates the following:

- A methodical and careful approach is required.
- All surface powers should be calculated before ray-tracing.
- Both equivalent air distances should be calculated before ray-tracing.
- The step-along ray-tracing technique is simply extended to accommodate the second element (the tear lens).
- A high degree of accuracy is required.

At least four decimal places are used through-out; rounding should not take place and the memory functions of the calculator should be exploited.

Example 19.4

Find the FOZR of the following RGP contact lens:

The subject's spectacle correction (F_{sp}) is
   −10.00 D at 14 mm
The BOZR (r_2) of the contact lens is 8.00 mm
The corneal radius is 7.40 mm
The centre thickness of the contact lens is
   0.40 mm
The centre thickness of the tear lens is 0.10 mm
The refractive index of the contact lens is 1.49
The refractive index of the tears is 1.336

In this example, the FOZR (r_1) of the contact lens is required. To find r_1 we need to calculate F_1 and, if a distance object is assumed, $L_1 = 0$ and $F_1 = L_1'$. To find L_1' a step-back ray-trace is required, starting from L_4'. If the RGP contact lens–tear lens system is to correct this patient's refractive error, the vergence *leaving the system* is equal to the patient's ocular refraction. So K and L_4' are therefore the same. The question provides us with the patient's spectacle refraction F_{sp}, which can be converted into an ocular refraction. This vergence is the starting point for the step-back ray-trace because $L_4' = K$. The first task is therefore to find K:

$$K = \frac{F_{sp}}{1-(dF_{sp})} = \frac{-10.00}{1-(0.0140\times-10.00)}$$
$$= -8.7719 D$$

The surface powers can now be calculated. We have been given r_2 (the BOZR) and the corneal radius. The shape of the tear lens formed depends on the BOZR of the contact lens and the corneal radius, so it follows that the front surface radius of the tear lens (r_3) is the same as the BOZR and the back surface radius of the tear lens (r_4) is the same as the anterior corneal radius. Sign convention dictates that all these radii are positive. So:

$r_2 = +8.00$ mm
$r_3 = +8.00$ mm (same as r_2)

$r_4 = +7.40$ mm (same as the anterior corneal radius)

The unknown radius is r_1 (the FOZR).

Remember that each surface is separated by an infinitely thin and optically insignificant air gap (see Figure 19.3). The three surface powers can be found using the following equations:

$$F_2 = \frac{1-n_{cl}}{r_2} = \frac{1.00-1.49}{+8.00^{-3}} = -61.2500 D$$

$$F_3 = \frac{n_{tears}-1}{r_3} = \frac{1.336-1.00}{+8.00^{-3}} = +42.0000 D$$

$$F_4 = \frac{1-n_{tears}}{r_4} = \frac{1.00-1.336}{+7.40^{-3}} = -45.4054 D$$

The EADs for both the contact lens and the tear lens can now be found. The EAD for the *contact lens* (with t_{cl} substituted in metres) is given by:

$$EAD = \frac{t_{cl}}{n_{cl}} = \frac{4^{-4}}{1.490} = 2.6846^{-4} m$$

The EAD for the *tear lens* (with t_{tears} substituted in metres) is given by:

$$EAD = \frac{t_{tears}}{n_{tears}} = \frac{1^{-4}}{1.336} = 7.4850^{-5} m$$

Now that the EADs have been determined, the refractive index between the two surfaces of the contact lens and the two surfaces of the tear lens can be assumed to be **equal to that of air** ($n = 1$). This value is used in the following ray-trace. A step-back ray-trace is needed to find the front surface power. As always, the computation takes place in two columns, one for vergences and the other for distances. The vergences for this problem are again illustrated in Figure 19.6 and the ray-trace once again needs to be extended to accommodate the four refracting surfaces. This time, a step-back ray trace is performed first through the tear lens and then through the contact lens.

Vergences (D)   Distances (m)

$L_4' = -8.7719 D$

$L_4 = L_4' - F_4$

$L_4 = -8.7719 - (-45.4054) = +36.6335 D$

$L_4 = +36.6335\,\mathrm{D} \quad \rightarrow \quad l_4 = \dfrac{1}{+36.6335} = +0.02730\,\mathrm{m}$

$$l'_3 = l_4 + \left(\dfrac{t_{\mathrm{tears}}}{n_{\mathrm{tears}}}\right) \quad \text{(step-back)}$$

$L'_3 = \dfrac{1}{+0.02737} \qquad \leftarrow \quad l'_3 = +0.02730 + 7.4850^{-5}$
$ = +36.5333\,\mathrm{D} \qquad\qquad\quad = +0.02737\,\mathrm{m}$

$L_3 = L'_3 - F_3$

$L_3 = +36.5333 - (+42.0000) = -5.4667\,\mathrm{D}$

As a result of the thin air gap, $L_3 = L'_2$. We must now perform the second step-back ray-trace through the contact lens. The reader is again advised to follow all of the vergences carefully on Figure 19.6.

$L_3 = L'_2 = -5.4667\,\mathrm{D}$

$L'_2 = -5.4667\,\mathrm{D}$

$L_2 = L'_2 - F_2$

$L_2 = -5.4667 - (-61.25) = +55.7833\,\mathrm{D}$

$L_2 = +55.7833\,\mathrm{D} \quad \rightarrow \quad l_2 = \dfrac{1}{+55.7833} = +0.01793\,\mathrm{m}$

$$l'_1 = l_2 + \left(\dfrac{t_{\mathrm{cl}}}{n_{\mathrm{cl}}}\right) \quad \text{(step-back)}$$

$L'_1 = \dfrac{1}{+0.01819} \qquad \leftarrow \quad l'_1 = +0.01793 + 2.6846^{-4}$
$ = +54.9602\,\mathrm{D} \qquad\qquad\quad = 0.01819\,\mathrm{m}$

As $L_1 = 0.0000$, $L_1 = F_1$

$F_1 = +54.9602\,\mathrm{D}$

The FOZR can be found using:

$$r_1 = \dfrac{n' - n}{F_1}$$

$$r_1 = \dfrac{1.49 - 1.00}{+54.9602} = +8.9155^{-3}\,\mathrm{m}$$

The FOZR of the contact lens is therefore +8.9155 mm

This example of a step-back ray-trace through an RGP contact lens–tear lens system demonstrates the following:

- A methodical and careful approach is required.

- The importance of placing particular values in the correct position in the ray-tracing sequence, i.e. $K = L'_4$.
- The ocular refraction should be calculated before ray-tracing.
- All surface powers should be calculated before ray-tracing.
- Both equivalent air distances should be calculated before ray-tracing.
- The step-back ray-tracing technique is simply extended to accommodate the second element (the contact lens).
- A high degree of accuracy is required.
- At least four decimal places are used throughout, rounding should not take place and the memory functions of the calculator should be exploited.

Example 19.5
An RGP contact lens with the following specification is fitted to an eye:

BOZR = 8.20 mm
Centre thickness = 0.55 mm
BVP = +4.00 D
Refractive index = 1.51

The lens fits 0.20 mm steeper than the cornea with a clearance of 0.10 mm. Find the ocular refraction and the FOZR of the lens. The refractive index of the tears is 1.336.

This example requires some considerable thought and is an example of how both step-back and step-along ray-tracing methods can be incorporated into one question. Here, we are given the BVP of the contact lens, which is always measured in air and is equivalent to L'_2 in our ray-tracing sequence. Thanks to the thin air gap, L'_2 is also equal to L_3. We can therefore easily **step along** from L_3 to find L'_4, which of course is equal to the ocular refraction K. To find the FOZR all we have to do is **step back** from L'_2. Some of the data in this example is also provided in a less than straightforward way. 'The lens fits 0.20 mm steeper than the cornea' means that if the BOZR of the RGP contact lens is 8.20 mm the anterior corneal radius must be 0.20 mm *flatter*. The anterior corneal radius is therefore 8.40 mm. In addition, 'with a clearance of 0.10 mm' means that the centre thickness of the tear lens is 0.10 mm.

So in summary, the data provided by the question (in one way or another) is as follows:

- BOZR of contact lens is 8.20 mm
- Centre thickness of contact lens is 0.55 mm
- BVP of the contact lens is +4.00 D
- Refractive index of the contact lens material is 1.51
- The anterior corneal radius is 8.40 mm
- The centre thickness of the tear lens is 0.10 mm
- The refractive index of the tears is 1.336

We have been given r_2 (the BOZR) and the corneal radius. The shape of the tear lens formed, as always, depends on the BOZR of the contact lens and the anterior corneal radius, so it follows that the front surface radius of the tear lens (r_3) is the same as the BOZR (r_2) and the back surface radius of the tear lens (r_4) is the same as the anterior corneal radius. Sign convention dictates that all these radii are positive. So:

$r_2 = +8.20$ mm
$r_3 = +8.20$ mm (same as r_2)
$r_4 = +8.40$ mm (same as the anterior corneal radius)

The unknown values are r_1 (the FOZR) and the ocular refraction K.

Remember that each surface is separated by an infinitely thin and optically insignificant air gap (see Figure 19.3). The three surface powers can be found using the known radii and the following equations:

$$F_2 = \frac{1 - n_{cl}}{r_2} = \frac{1.00 - 1.51}{+8.20^{-3}} = -62.1951\,D$$

$$F_3 = \frac{n_{tears} - 1}{r_3} = \frac{1.336 - 1.00}{+8.20^{-3}} = +40.9756\,D$$

$$F_4 = \frac{1 - n_{tears}}{r_4} = \frac{1.00 - 1.336}{+8.40^{-3}} = -40.0000\,D$$

The EAD for both the contact lens and the tear lens can now be found. The EAD for the *contact lens* (with t_{cl} substituted in metres) is given by:

$$EAD = \frac{t_{cl}}{n_{cl}} = \frac{5.5^{-4}}{1.51} = 3.6424^{-4}\,m$$

The EAD for the *tear lens* (with t_{tears} substituted in metres) is given by:

$$EAD = \frac{t_{tears}}{n_{tears}} = \frac{1^{-4}}{1.336} = 7.4850^{-5}\,m$$

Now that the EADs have been determined, the refractive index between the two surfaces of the contact lens and the two surfaces of the tear lens can be assumed to be **equal to that of air** ($n = 1$). This value is used in the following ray traces. A step-back ray-trace is performed first to find the front surface power and FOZR. As always, the computation takes place in two columns, one for vergences and the other for distances, and a distant object ($L_1 = 0$) is assumed. The vergences for this problem are again illustrated in Figure 19.6. The step-back ray-trace is performed through the contact lens to find L_1' and therefore F_1. The starting vergence for the computation is L_2' (the BVP given in the question).

Vergences (D)   Distances (m)

$L_2' = +4.0000\,D$

$L_2 = L_2' - F_2$

$L_2 = +4.0000 - (-62.1951) = +66.1951\,D$

$L_2 = +66.1951\,D \rightarrow l_2 = \frac{1}{+66.1951} = +0.01511\,m$

$$l_1' = l_2 + \left(\frac{t_{cl}}{n_{cl}}\right) \quad \text{(step-back)}$$

$L_1' = \frac{1}{+0.01547} \quad \leftarrow \quad l_1' = +0.01511 + 3.6424^{-4}$
$= +64.6367\,D \qquad\qquad = 0.01547\,m$

As $L_1 = 0.00$, $L_1 = F_1$

$F_1 = +64.6367\,D$

The FOZR can be found using:

$$r_1 = \frac{n' - n}{F_1}$$

$$r_1 = \frac{1.51 - 1.00}{+64.6367} = +7.89026^{-3}\,m$$

The FOZR of the contact lens is therefore +7.89026 mm.

The second part of the question to find the ocular refraction can be solved using a step-along ray-trace. The vergences for this problem are again illustrated in Figure 19.6. The step-back ray trace is performed through the tear lens to find L_4' and therefore K. The starting vergence for the computation is L_3, which thanks to the thin air gap is the same as L_2' (the BVP given in the question):

$$L_3 = +4.0000\,D$$

$$F_3 = ++40.97561\,D$$

$$L_3' = L_3 + F_3 = +44.97561\,D \quad \rightarrow \quad l_3' = \frac{n}{L_3'}$$

$$l_3' = \frac{1}{+44.97561} = +0.02223\,m$$

$$l_4 = l_3' - \left(\frac{t_{tears}}{n_{tears}}\right)$$

$$L_4 = \frac{n}{l_4} \qquad \leftarrow \qquad l_4 = +0.022223 - 7.4850^{-4}$$
$$= +0.02216\,m$$

$$L_4 = \frac{1}{+0.02216} = +45.1275\,D$$

$$L_4' = L_4 + F_4$$

$$L_4' = +45.1275 + (-40.0000) = +5.1275\,D$$

As $L_1 = 0.0000$, $L_4' = $ BVP and $F_v' = +5.1275\,D$

The BVP of this RGP contact lens–tear lens system is therefore +5.1275. If this contact lens–tear lens system corrects this particular eye, the ocular refraction must also equal +5.1275 D.

Concluding points for Chapter 19

Chapter 17 was in part designed to form an introduction to this chapter as the concept of a RGP contact lens–tear lens system was introduced. This concept has been developed in Chapter 19, which has so far probably been the most difficult chapter in this book. However, all the skills mastered in this chapter are used many times in this book.

Further recommended reading

Douthwaite W A (2006) *Contact Lens Optics and Lens Design*. Elsevier, Oxford

Efron N (2002) *Contact Lens Practice*. Butterworth-Heinemann, Oxford

Spectacle Magnification and Relative Spectacle Magnification

Andrew Keirl

Introduction

Chapter 6 introduced the concept of spectacle magnification with respect to both thin and thick spectacle lenses. After a brief revision of these concepts, Chapter 20 discusses the spectacle magnification of a contact lens system (rigid gas permeable contact lens and tear lens) and worked examples are presented. It is vital that the material introduced in Chapter 19 has been fully understood before attempting the problems in this chapter because all the tools introduced in the former are employed in the latter.

Chapter content

- Revision of the basic definitions and principles of spectacle magnification for thin and thick spectacle lenses
- Spectacle magnification produced by a contact lens–liquid lens system
- Relative spectacle magnification

Revision of the basic definitions and principles of spectacle magnification for thin and thick spectacle lenses

Spectacle magnification (described in Chapter 6) is a term that is used to compare the size of the retinal image formed in the uncorrected eye with the size of the retinal image formed in the corrected eye. The general definition of spectacle magnification (SM) is:

$$SM = \frac{h'_c}{h'_u}$$

where h'_c is the size of the retinal image formed in the corrected ametropic eye and h'_u the size of the retinal image formed in same eye when uncorrected. The uncorrected retinal image size is also known as the **basic** retinal image size.

Spectacle magnification produced by a 'thin' lens is a result of its power and position only because a thin lens has *no* form or thickness. The magnification produced by a thin spectacle lens is also known as the **power factor** (PF). For thin spectacle lenses, SM (or power factor) can be given by:

$$\frac{\text{Ocular refraction}}{\text{Spectacle refraction}} \quad \text{or} \quad SM\,(PF) = \frac{K}{F_{sp}}$$

This definition and equation apply to hypermetropia and myopia, both axial and refractive. An alternative expression to find the spectacle magnification or power factor of a thin lens is:

$$SM\,(PF) = \frac{1}{1 - (dF_{sp})}$$

where d is the vertex distance in metres and F_{sp} is the power of the thin lens in dioptres.

If a patient is corrected using a contact lens (assuming a thin contact lens), the contact lens magnification can be given by:

$$\frac{\text{Ocular refraction}}{\text{Contact lens correction}} \quad \text{or} \quad SM_{cl} = \frac{K}{F_{cl}}$$

With the positions of the correcting contact lens and principal point of a reduced eye almost coinciding, we can state that the SM in contact lens wear approximates to unity (SM = 1). Thus, contact lenses do not alter the retinal image size significantly from that of the uncorrected eye. However, this statement is an approximation that arises from the use of the standard reduced eye. Unity is actually achieved only when the correcting lens is positioned *in the plane of the entrance pupil of the eye*. A contact lens is positioned some 3 mm in front of the entrance pupil of the eye. In addition, a contact lens is not thin and must be considered as a thick and steeply curved lens. Remember also that the form of a lens affects the positions of the principal planes and the steep curves of a contact lens lead to displacement of the principal planes.

Spectacle magnification produced by a thick or real spectacle lens is a result of the power and position of the lens and *also* the **form** and **thickness** of the lens. The magnification resulting from the form and thickness of a lens is known as the **shape factor** (SF), given by:

$$SF = \frac{1}{1 - (t/n)F_1}$$

where t is the centre thickness of the spectacle lens in metres, n the refractive index of the lens material and F_1 the power of the front surface of the lens in dioptres. An alternative expression to find the shape factor of a thick lens is:

$$SF = \frac{F_v'}{F_E}$$

where F_v' is the back vertex power of the thick lens and F_E the power of the equivalent thin lens. This expression is useful in contact lens problems.

So, unlike a thin lens, the magnification produced by a *thick* spectacle lens has **two** components: the **power factor** and the **shape factor**. The total spectacle magnification for a thick lens is, therefore, the product of the power factor and the shape factor.

Spectacle magnification produced by a contact lens–liquid lens system

In a similar way to a spectacle lens, the SM of both a contact lens and a contact lens–tear lens system is determined by its power and shape factors.

The power factor of a contact lens

Figure 20.1 shows a lens placed in contact with an eye and a distant object subtending an angle ω_o at the lens. If the eye were removed from Figure 20.1, the lens would form an inverted image h' in the plane of the back vertex focal point, F_v' of the lens. The image would be formed at a distance f_v' from the lens. The ray of light forming the image passes undeviated through the optical centre of the lens and continues to form an angle ω_o to the optical axis. If the eye is to be corrected by the lens, F_v' must coincide with the far point of the eye, M_R and h' therefore becomes the object for the eye. The corrected visual angle, ω, is measured from h' to the eye's entrance pupil E. The symbol a gives the position of the entrance pupil of the eye measured from the corneal surface. SM can therefore be defined as:

$$SM = \frac{w}{w_o}$$

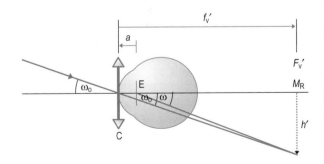

Figure 20.1 Spectacle magnification of a contact lens: The visual angle of the distant object is ω_o. The correcting contact lens produces an image at the far point plane. This becomes the object for the eye and subtends a visual angle of ω at the eye's entrance pupil (E). (After Douthwaite 2006, with permission of Elsevier Ltd.)

From Figure 20.1:

$$\tan w_o = \frac{-h'}{f'_v}$$

and

$$\tan w = \frac{-h'}{f'_v - a}$$

So:

$$SM = \frac{-h'/(f'_v - a)}{-h'/f'_v}$$

which, after some rearrangement, becomes:

$$SM = \frac{1}{1 - (aF'_v)}$$

The above expression can be used to calculate the power factor of a contact lens in isolation or a contact lens–tear lens system. In the above expression, a is the distance (in metres) from the contact lens to the entrance pupil of the eye, and F'_v is the back vertex power of the contact lens or contact lens-tear lens system. The distance a is always positive.

The shape factor of a contact lens

For a thick lens, the positions of the principal planes P and P' can be found using the methods reviewed in Chapter 1. The second principal point P' is of most interest in ophthalmic optics, because, if placed at P', the equivalent thin lens replaces the back vertex power (BVP) of a thick lens and it is the BVP that corrects a patient's refractive error. Thus, a thick contact lens could be replaced by the equivalent thin lens placed at P'. As stated above, the form of a lens affects the positions of the principal planes, and the steep curves of a contact lens lead to displacement of these principal planes away from the convex surface. A strongly curved lens results in P' being further from the eye, which increases the spectacle magnification and therefore the retinal image size. **Figure 20.2** shows a thick contact lens of focal length f'_v forming an image of size h_1. This thick lens can of course be replaced by an equivalent thin lens of focal length f'_E placed at P'. The image formed by the equivalent thin lens is h_2. The ratio of these two

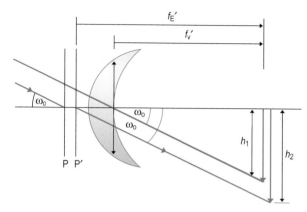

Figure 20.2 The shape factor: compares the image sizes produced by a thin flat lens and a thick curved lens with the same back vertex focal length. The object is standing on the axis of the system at infinity, producing a visual angle ω_o. (After Douthwaite 2006, with permission of Elsevier Ltd.)

image sizes is called the **shape factor** (SF). Thus:

$$SF = \frac{h_2}{h_1}$$

which is equivalent to:

$$\frac{f'_E}{f'_v}$$

If these distances are expressed in metres, then their reciprocals give the lens powers in dioptres. So:

$$SF = \frac{F'_v}{F_E}$$

where F'_v is the BVP of the thick lens or lens system and F_E the power of the equivalent thin lens placed at the second principal point. As outlined in Chapter 1, there are several methods for calculating the power of the equivalent thin lens.

For a system of no more than two thin lenses, use:

$$F_E = F_1 \times \frac{L'_2}{L_2}$$

or

$$F_E = F_1 + F_2 - dF_1F_2$$

For a thick spectacle or contact lens use:

$$F_E = F_1 \times \frac{L_2'}{L_2}$$

or if no more than two surfaces:

$$F_E = F_1 + F_2 - (t/n)F_1 F_2$$

For a multi-lens system:

$$F_E = F_1 \times \frac{L_2'}{L_2} \times \frac{L_3'}{L_3} \times \frac{L_4'}{L_4}$$

The final equation can be extended to include further vergences and it is this last equation that is used to find the equivalent power and therefore the shape factor of a contact lens–tear lens system.

A summary of SM equations

The reader is advised to inspect the following equations carefully in order to appreciate the sometimes subtle differences between them and their use.

Thick spectacle lenses (two surfaces)
Power factor

If the entrance pupil is assumed to coincide with the reduced surface use:

$$PF = \frac{1}{1-(dF_v')}$$

where d (in metres) is the vertex distance and F_v' the back vertex power of the thick spectacle lens.

If the position of the entrance pupil *and* a vertex distance d is stated use:

$$PF = \frac{1}{1-(aF_v')}$$

where a (in metres) is the *sum* of these two distances in metres. The above expression for the power factor can also be applied to a contact lens in isolation where a is the position of the entrance pupil and F_v' is the back vertex power of the contact lens measured in air.

Shape factor

$$SF = \frac{1}{1-\left(\frac{t}{n}\right)F_1}$$

where t/n is the reduced thickness of the spectacle lens in metres. The above expression for shape factor can be applied to a contact lens in isolation but *not* a contact lens–tear lens system.

Alternatively use

$$SF = \frac{F_v'}{F_E}$$

where F_v' is the BVP of the thick spectacle lens and F_E the equivalent thin lens power. This alternate expression will be used to find the shape factor for a contact lens–tear lens system.

SM of a contact lens–tear lens system
Power factor

The power factor of a contact lens–tear lens system is calculated using:

$$PF = \frac{1}{1-(aF_v')}$$

where a (in metres) is the position of the entrance pupil and F_v' the BVP of the contact lens–tear lens system (L_4').

Shape factor

The shape factor of a contact lens-tear lens system is found using:

$$SF = \frac{F_v'}{F_E}$$

where F_v' is the BVP of the contact lens–tear lens system (L_4') and F_E the power of the equivalent thin lens:

$$F_E = F_1 \times \frac{L_2'}{L_2} \times \frac{L_3'}{L_3} \times \frac{L_4'}{L_4}$$

Remember, when calculating the power factor of spectacle lenses, it is generally assumed that the entrance pupil of the eye coincides with the reduced surface of the eye. However, when dealing with contact lenses, the entrance pupil is usually assumed to be 3mm behind the corneal surface. This needs to be considered, particularly in 'compare and contrast'-type questions.

Example 20.1

A patient is corrected with:

1. **A thick spectacle lens $F_v' = +7.50\,D$ @ 14mm, $F_2 = -2.50\,D$, $t = 7.50\,mm$, $n = 1.50$.**
2. **An RGP contact lens, BOZR (back optic zone radius) 7.90mm which is fitted 0.1mm steeper than the corneal radius with an apical clearance of 0.10mm, $t_{cl} = 0.30\,mm$, $n_{cl} = 1.45$, $t_{tears} = 0.10\,mm$ and $n_{tears} = 1.336$.**

If the entrance pupil lies 3mm behind the cornea, calculate the SM in both cases.

Spectacle lens

As the lens is thick we must calculate the shape factor and the power factor. An analysis of the data provided shows that we have not been given the front surface power of the lens F_1. This must be found using a step-back ray trace as shown in Chapter 1. As the lens is thick, the equivalent air distance (EAD) t/n must be used in the computation. As always, the initial vergence in a step-back ray-trace is L_2' which is the same as the BVP (F_v').

Vergences (D)   Distances (m)

$L_2' = +7.5000\,D$

$L_2 = L_2' - F_2$

$L_2 = +7.5000 - (-2.5000) = +10.0000\,D$

$L_2 = +10.0000\,D \quad \rightarrow \quad l_2 = \dfrac{1}{+10.0000} = +0.10\,m$

$\qquad\qquad l_1' = l_2 + (t/n) \text{ (step-back)}$

$L_1' = \dfrac{1}{+0.105} \quad \leftarrow \quad l_1' = +0.10 + 5^{-3} = +0.105\,m$

$\quad = +9.5238\,D$

As $L_1 = 0.00$, $L_1 = F_1$

$F_1 = +9.5238\,D$

The front surface power of the spectacle lens is therefore +9.5238 D.

To calculate the power factor use:

$$PF = \frac{1}{1-(aF_v')}$$

where a is the position of the entrance pupil of the eye (in metres) and F_v' the BVP of the thick spectacle lens. As both the position of the entrance pupil from the corneal surface and the vertex distance have been given:

$$a = 14 + 3 = 17\,mm$$

$$PF = \frac{1}{1-(0.017 \times +7.50)} = 1.1461$$

To calculate the shape factor use:

$$SF = \frac{F_v'}{F_E}$$

where F_v' is the BVP of the thick spectacle lens (L_2') and F_E the power of the equivalent thin lens:

$$F_E = F_1 \times \frac{L_2'}{L_2}$$

$$F_E = +9.5238 \times \frac{+7.50}{+10.00} = +7.14286\,D$$

$$SF = \frac{F_v'}{F_E} = \frac{+7.50000}{+7.14286} = 1.05$$

The total SM is the product of the power factor and the shape factor, so:

$$SM = PF \times SF = 1.1461 \times 1.05 = 1.2034\times$$

This is equivalent to a 20.34% increase in magnification.

Contact lens

An analysis of the data given for the contact lens is as follows:

$BOZR = 7.90\,mm$
$t_{cl} = 0.30\,mm$
$n_{cl} = 1.45$
$t_{tears} = 0.10$
$n_{tears} = 1.336$

Corneal radius = 8.00 mm (the BOZR is 0.1 mm steeper than the corneal radius).

As is always the case with an RGP contact lens, the shape of the tear lens formed depends on the BOZR of the contact lens and the corneal radius. The front surface radius of the tear lens (r_3) is the same as the BOZR (r_2) and the back surface radius of the tear lens (r_4) is the same as the anterior corneal radius. Sign convention dictates that all theses radii are positive.

So:

$r_2 = +7.90\,\text{mm}$

$r_3 = +7.90\,\text{mm}$ (same as r_2)

$r_4 = +8.00\,\text{mm}$ (same as the anterior corneal radius)

We are missing the front optic zone radius (FOZR or r_1) and therefore power of the front surface of the contact lens. F_1 is needed in order to find the power of the equivalent thin lens and hence the shape factor. A step-back ray-trace is therefore needed that starts at L_4' and ends at L_1'. If a distant object is assumed $L_1' = F_1$. The final vergence L_4' is equal to the patient's ocular refraction, K, which we can find because we have been given the spectacle refraction and vertex distance. The spectacle refraction is +7.50 D at 14 mm, so using:

$$K = \frac{F_{sp}}{1 - dF_{sp}}$$

$$K = \frac{+7.50}{1 - (0.014 \times +7.50)} = +8.3799\,\text{D}$$

L_4' is therefore +8.3799 D

The refractive index of contact lens material is 1.450 and that of the tears is 1.336. The first task is to calculate the surface powers for which we have radii. Remember that each surface is separated by an infinitely thin and optically insignificant air gap (see Figure 19.3). The three surface powers can be found using the following equations:

$$F_2 = \frac{1.00 - n_{cl}}{r_2} = \frac{1.00 - 1.45}{+7.90^{-3}} = -56.9620\,\text{D}$$

$$F_3 = \frac{n_{tears} - 1}{r_3} = \frac{1.336 - 1.00}{+7.90^{-3}} = +42.5316\,\text{D}$$

$$F_4 = \frac{1 - n_{tears}}{r_4} = \frac{1.00 - 1.336}{+8.00^{-3}} = -42.0000\,\text{D}$$

The next task is to determine an EAD for both the contact lens and the tear lens. The EAD for the *contact lens* (with t_{cl} substituted in metres) is given by:

$$\text{EAD} = \frac{t_{cl}}{n_{cl}} = \frac{3^{-4}}{1.450} = 2.0690^{-4}\,\text{m}$$

The EAD for the *tear lens* (with t_{tears} substituted in metres) is given by:

$$\text{EAD} = \frac{t_{tears}}{n_{tears}} = \frac{1^{-4}}{1.336} = 7.4850^{-5}\,\text{m}$$

Now that the EADs have been determined, the refractive index between the two surfaces of the contact lens and the two surfaces of the tear lens can be assumed to be **equal to that of air** ($n = 1$). This value is used in the following ray-trace. A step-back ray-trace is needed to find the front surface power. As always, the computation takes place in two columns, one for vergences and the other for distances. The vergences for this problem are again illustrated in Figure 19.6 and the ray-trace once again needs to be extended to accommodate the four refracting surfaces. This time a step-back ray-race is performed first through the tear lens and then through the contact lens starting with the vergence L_4', which is equivalent to the ocular refraction K.

Vergences (D)   Distances (m)

$L_4' = +8.3799\,\text{D}$

$L_4 = L_4' - F_4$

$L_4 = +8.3799 - (-42.0000) = +50.3799\,\text{D}$

$L_4 = +50.3799\,\text{D} \;\rightarrow\; l_4 = \dfrac{1}{+50.3799} = +0.01985\,\text{m}$

$$l_3' = l_4 + \left(\frac{t_{tears}}{n_{tears}} \right) \quad \text{(step-back)}$$

$L_3' = \dfrac{1}{+0.01992}$ $\leftarrow$ $l_3' = +0.01985 + 7.4850^{-5}$

$= +50.1906\,\text{D}$ $= +0.01992\,\text{m}$

$L_3 = L_3' - F_3$

$L_3 = +50.1906 - (+42.53165) = +7.6590\,\text{D}$

As a result of the thin air gap, L_3 is equal to L_2'. We must now perform the second step-back ray-trace through the contact lens. The reader is once again advised to follow all the vergences carefully on **Figure 19.6**.

$L_3 = L_2' = +7.6590\,\text{D}$

$L_2' = +7.6590\,\text{D}$

$L_2 = L_2' - F_2$

$L_2 = +7.6590 - (-56.9620) = +64.6210\,\text{D}$

$$L_2 = +64.6210\,\mathrm{D} \quad \rightarrow \quad l_2 = \frac{1}{+64.6210} = +0.01547\,\mathrm{m}$$

$$l_1' = l_2 + \left(\frac{t_{cl}}{n_{cl}}\right) \quad \text{(step-back)}$$

$$L_1' = \frac{1}{+0.01568} \quad \leftarrow \quad l_1' = +0.01547 + 2.0690^{-4}$$
$$= +64.7684\,\mathrm{D} \qquad\qquad = 0.01568\,\mathrm{m}$$

As $L_1 = 0.00$, $L_1 = F_1$

$$F_1 = +64.7684\,\mathrm{D}$$

Now, we that have found all the vergences within the contact lens–tear lens system (see Figure 19.6), we can calculate the SM of the system. To do this we need to find the power factor and the shape factor. The power factor of a contact lens–tear lens system is calculated using:

$$PF = \frac{1}{1 - (aF_v')}$$

where a (in metres) is the position of the entrance pupil and F_v' the BVP of the contact lens–tear lens system (L_4'):

$$PF = \frac{1}{1 - (0.003 \times +8.3799)} = 1.0258$$

The shape factor of a contact lens–tear lens system is found using:

$$SF = \frac{F_v'}{F_E}$$

where F_v' is the BVP of the contact lens–tear lens system (L_4') and F_E the power of the equivalent thin lens:

$$F_E = F_1 \times \frac{L_2'}{L_2} \times \frac{L_3'}{L_3} \times \frac{L_4'}{L_4}$$

$$F_E =$$
$$+63.7684 \times \frac{+7.6590}{+64.6210} \times \frac{+50.19062}{+7.6590} \times \frac{+8.3799}{+50.3799}$$
$$= +8.2383\,\mathrm{D}$$

$$SF = \frac{F_v'}{F_E} = \frac{+8.3799}{+8.2383} = 1.0172$$

The total SM is the product of the power factor and the shape factor, so:

$$SM = PF \times SF = 1.0258 \times 1.0172 = 1.0434 \times$$

This is equivalent to a **4.34% increase** in magnification, which of course is less than the SM produced when a spectacle lens is worn. This explains why some patients with hypermetropia achieve better visual acuity with spectacles than with contact lenses.

The ratio of the SMs with contact lenses and spectacles is given as:

$$\frac{SM_{specs}}{SM_{CL}} = \frac{1.2034}{1.0434} = 1.1533$$

This means that, for this individual, the retinal image size is **15.33% larger** with spectacles than with contact lenses.

Relative spectacle magnification

Relative spectacle magnification (RSM) is defined as:

$$\frac{\text{The retinal image size in the corrected ametropic eye } (h_c')}{\text{The retinal image size in the standard emmetropic eye } (h_{em}')}$$

The above definition is true for a distant object and is used when a comparison is needed between two different eyes:

$$RSM = \frac{K}{F_{sp}} \times \frac{K_{em}'}{K'}$$

In the standard emmetropic eye, K_{em}' is always equal to +60.00 D, so:

$$RSM = \frac{K}{F_{sp}} \times \frac{+60.00}{K'}$$

where K' is the dioptric length of the ametropic eye ($K' = K + F_e$), K the ocular refraction and F_{sp} the spectacle refraction. The significance of RSM is the comparison of corrected retinal image sizes in myopic and hypermetropic eyes with the retinal image size in a standard emmetropic eye.

The RSM in axial ametropia

The corrected retinal image in an axially hyper-metropic eye is increased by the size of the spectacle correction, but it is still smaller than the image in the emmetropic eye. In axially myopic eyes the correction that makes the image smaller still leaves an image that is larger than that for the standard emmetropic eye.

The RSM in refractive ametropia

The axial length of the refractively ametropic eye is the same as the axial length in the emmetropic eye. Remember that, in refractive ametropia, the problem is with the eye's power and not its length. This means that the image size in the uncorrected refractively ametropic eye is the same as the image size in the standard emmetropic eye. In addition, the relative spectacle magnification is equal to the SM in refractive ametropia. In refractive hypermetropia the RSM is >1 and in refractive myopia <1.

Anisometropia

Anisometropia (meaning a difference in the right and left corrections) gives rise to several optical problems when corrected with spectacle lenses. These include:

- Horizontal and vertical differential prismatic effects which occur when the individual views through points away from the optical centres of the lenses.
- A difference in right and left spectacle magnifications, resulting in aniseikonia (a difference in cortical image sizes). Unequal magnifications also give rise to poor cosmesis when wearing spectacle lenses.

As a contact lens moves with the eye, the problems of prismatic effects that occur when an individual views through points away from the optical centre of the lens are minimised. Induced vertical and horizontal differential prismatic effects that result from contact lens decentration can produce problems in the eyes of anisometropic individuals, but such problems occur only very occasionally. An optical problem that may be an issue with the anisometropic contact lens wearer is that of aniseikonia because the effects of changes in retinal image size may be influenced by the type of ametropia present.

In refractive anisometropia, the right and left uncorrected retinal images sizes are equal, because the right and left axial lengths are the same. A correction with a positive lens of course magnifies the retinal image whereas the use of a negative results in a smaller retinal image being formed. The use of spectacle lenses in refractive ametropia may produce binocular vision difficulties, resulting from a disparity in the sizes of the right and left retinal images. As a pair of contact lenses produces a much smaller disparity in the image sizes than spectacles, the correction of refractive anisometropia with contact lenses is often more satisfactory than correction with spectacles. In axial anisometropia, however, the opposite situation occurs because correction with spectacles gives less disparity in image size than correction with contact lenses. The preceding statements are best illustrated with an example.

Example 20.2
A patient's prescription is as follows:
Right −10.00 DS   Left −5.00 DS
The vertex distance is 15 mm. Compare the relative spectacle magnifications when the patient:

(1) has axial myopia and is fitted with
    (a) spectacles and (b) contact lenses
(2) has refractive myopia and is fitted with
    (a) spectacles and (b) contact lenses.

Assuming the constants of the reduced eye, comment on the aniseikonia found in each situation.

This example is not optically or mathematically difficult, but it is rather long and tedious! A methodical and logical approach is therefore required. Relative spectacle magnification is given by:

$$RSM = \frac{K}{F_{sp}} \times \frac{+60.00}{K'}$$

For each eye, we therefore need to find the ocular refraction, K, and the dioptric length, K'. The expressions required are:

$$K = \frac{F_{sp}}{1 - dF_{sp}} \text{ and}$$

$$K' = K + F_e$$

where F_{sp} is the spectacle refraction, d the vertex distance in metres and F_e the power of the reduced surface. The right and left ocular refractions are calculated as follows:

Right eye

$$K = \frac{-10.00}{1 - (0.015 \times -10.00)} = -8.6956\,D$$

Left eye

$$K = \frac{-5.00}{1 - (0.015 \times -5.00)} = -4.6512\,D$$

Axial myopia corrected with spectacles

Right eye:

$$K = -8.6956\,D$$
$$F_{sp} = -10.00D$$

As error is axial, $F_e = +60.00\,D$
K' is required:

$$K' = K + F_e$$

$$K' = -8.6956 + (+60.00) = +51.3043\,D$$

$$RSM = \frac{K}{F_{sp}} \times \frac{+60.00}{K'}$$

$$RSM = \frac{-8.6956}{-10.00} \times \frac{+60.00}{+51.3043} = 1.01695\times$$

The right image is therefore 1.695% larger than in the standard emmetropic eye.

Left eye:

$$K = -4.6512\,D$$
$$F_{sp} = -5.00D$$

As error is axial, $F_e = +60.00\,D$
K' is required:

$$K' = K + F_e$$

$$K' = -4.6512 + (+60.0000) = +55.3488\,D$$

$$RSM = \frac{K}{F_{sp}} \times \frac{+60.00}{K'}$$

$$RSM = \frac{-4.6512}{-5.00} \times \frac{+60.00}{+55.3488} = 1.00840\times$$

The left image is therefore 0.840% larger than in the standard emmetropic eye.

With spectacles, the ratio of the right and left images is:

$$\frac{R}{L} = \frac{1.01695}{1.00840} = 1.0085 \equiv 0.848\%$$

Correction with spectacles results in almost no difference in image size and therefore good binocular vision.

Axial myopia corrected with contact lenses

As both the spectacle lens and the contact lens are assumed to be thin in this example, K and F_{cl} are considered to be equal.
Right eye:

$$K = -8.6956\,D$$
$$F_{cl} = -8.6956\,D$$

As error is axial, $F_e = +60.00\,D$

$$K' = K + F_e$$

$$K' = -8.6956 + (+60.00) = +51.3043\,D$$

$$RSM = \frac{K}{F_{cl}} \times \frac{+60.00}{K'}$$

$$RSM = \frac{-8.6956}{-8.6956} \times \frac{+60.0000}{+51.3043} = 1.1695\times$$

The right image is therefore 16.95% larger than in the standard emmetropic eye.
Left eye:

$$K = -4.6511\,D$$
$$F_{sp} = -5.00\,D$$

As error is axial, $F_e = +60.00\,D$

$$K' = K + F_e$$

$$K' = -4.6511 + (+60.00) = +55.3488\,D$$

$$RSM = \frac{K}{F_{cl}} \times \frac{+60.00}{K'}$$

$$RSM = \frac{-4.6511}{-4.6511} \times \frac{+60.00}{+55.3488} = 1.0840\times$$

The left image is therefore 8.40% larger than in the standard emmetropic eye.

With contact lenses the ratio of the right and left images is:

$$\frac{R}{L} = \frac{1.1695}{1.0840} = 1.0789 \equiv 7.789\%$$

Correction with contact lenses is likely to produce aniseikonia and therefore disturb binocular vision.

The above calculation show that spectacles are better if the anisometropia is axial because aniseikonia is less and binocular vision more comfortable.

Refractive myopia corrected with spectacles

As the myopia is now refractive in origin, the axial length k' is 22.22 mm and the dioptric length K' +60.00 D.

Right eye:

$$K = -8.6956\,D$$
$$F_{sp} = -10.00\,D$$

As error is refractive $K' = +60.00\,D$

$$RSM = \frac{K}{F_{sp}} \times \frac{+60.00}{K'}$$

$$RSM = \frac{-8.6956}{-10.00} \times \frac{+60.00}{+60.00} = 0.8695\times$$

The right image is therefore 13.05% smaller than in the standard emmetropic eye.

Left eye:

$$K = -4.6512\,D$$
$$F_{sp} = -5.00\,D$$

As error is refractive, $K' = +60.00\,D$.

$$RSM = \frac{K}{F_{sp}} \times \frac{+60.00}{K'}$$

$$RSM = \frac{-4.6512}{-5.00} \times \frac{+60.00}{+60.00} = 0.9302$$

The left image is therefore 6.98% smaller than in the standard emmetropic eye.

With spectacles, the ratio of the right and left images is:

$$\frac{R}{L} = \frac{0.8695}{0.9302} = 0.9347 \equiv 6.525\%$$

Correction with spectacles is therefore likely to produce aniseikonia and to disturb binocular vision.

Refractive myopia corrected with contact lenses

Once again, as both the spectacle lens and the contact lens in this example are assumed to be thin, K and F_{cl} are considered to be equal.

Right eye:

$$K = -8.6956\,D$$
$$F_{cl} = -8.6956\,D$$

As error is refractive, $K' = +60.00\,D$

$$K' = K + F_e$$

$$RSM = \frac{K}{F_{cl}} \times \frac{+60.00}{K'}$$

$$RSM = \frac{-8.6956}{-8.6956} \times \frac{+60.00}{+60.00} = 1.00\times$$

The right image is therefore the same size as the image in the standard emmetropic eye.

Left eye:

$$K = -4.6511\,D$$
$$F_{sp} = -5.00\,D$$

As error is refractive, $K' = +60.00\,D$

$$RSM = \frac{K}{F_{cl}} \times \frac{+60.00}{K'}$$

$$RSM = \frac{-4.6511}{-4.6511} \times \frac{+60.00}{+60.00} = 1.00\times$$

The left image is therefore the same size as the image in the standard emmetropic eye.

With contact lenses, the ratio of the two images is unity. There is no difference in image sizes so binocular vision is good. The above calculation therefore shows that contact lenses are better if the anisometropia is refractive in origin because there is no aniseikonia and binocular vision is good.

The above example shows that use of spectacle lenses in refractive ametropia may result in a disparity in the sizes of the right and left retinal images, which may produce binocular vision difficulties. As a pair of contact lenses produces a much smaller disparity in the image sizes than spectacles, the correction of refractive anisometropia with contact lenses is often more satisfactory than correction with spectacles. In axial anisometropia, however, correction with spectacles gives less disparity in image size than correction with contact lenses.

Anisometropia is more likely to be axial than refractive. So if individuals with anisometropia and comfortable binocular vision when wearing spectacles changes to contact lenses, they can expect to encounter adaptation problems when transferring to contact lenses.

A study by Winn *et al.* (1986) suggested that all subjects with anisometropia are better corrected using contact lenses if the aim of correction is to minimise aniseikonic effects.

Concluding points for Chapter 20

This chapter has introduced little additional material. However, material from earlier chapters has been developed and extended to include the magnification of a contact lens–tear lens system. Both examples are rather long and there is considerable room for error, so a logical approach and a sensible layout are recommended. It is interesting to note that most of the work in Example 20.1 is ray-tracing!

References

Winn B, Ackerly R G, Brown C A, Murray F K, Prais J, St John M F (1986) The superiority of contact lenses in the correction of all anisometropia. *Transactions of the British Contact Lens Association Annual Clinical Conference*, pp 95–100

Further recommended reading

Douthwaite W A (2006) *Contact Lens Optics and Lens Design*. Elsevier, Oxford

Efron N (2002) *Contact Lens Practice*. Butterworth-Heinemann, Oxford

Rabbetts R B (1998) *Bennett & Rabbetts' Clinical Visual Optics*. Butterworth-Heinemann, Oxford

Tunnacliffe A H (1993) *Introduction to Visual Optics*. Association of the British Dispensing Opticians, London

Contact Lens Materials: Physical and Optical Properties

CHAPTER

21

Caroline Christie

Introduction

This chapter discusses the physical and optical properties of materials used for the manufacture of rigid gas permeable (RGP), hydrogel and silicone hydrogel contact lenses. Some practitioners assume that the differences between contact lens materials, particularly hydrogel materials, often seem too subtle to be of clinical significance and are therefore not worthy of study. However this chapter demonstrates that a basic understanding of the chemistry of the materials used by clinicians on a daily basis is critical to understanding many of the clinical observations seen during their use. Properties discussed include transmissibility, wettability and surface deposition.

Chapter content

- Material classification
- Properties of rigid gas permeable materials
- Hydrogel soft lenses
- Silicone hydrogels

Material classification

Contact lens materials can broadly be classified as one of two types: water-containing soft (hydrogel) and non-water-containing, RGP materials. Within the hydrogel classification, there are two subcategories: conventional hydrogel lenses and silicone–hydrogel lenses.

The international standard for the classification of both rigid and soft contact lens materials is BS EN ISO 11539: 1999. This is a published European Standard and so has the status of, and replaces, the former British Standard for contact lenses. Each material is classified by a six-part code:

Prefix: code name for material
Stem: *filcon* for hydrogel lenses (>10% water by mass) and *focon* for rigid lenses

Series suffix: a capital letter to indicate version level of chemical formula
Group suffix: see Table 21.1
Dk range: numerical code that gives a range rather than exact value (Table 21.2)
Surface modification code: a lower case m denotes that the lens surface has been modified from the bulk material.

Example 1: Paragon HDS
Paflufocon B III 3

Paflu = USAN prefix
focon = stem, indicating a hard lens material
B = USAN series suffix, indicating the second formulation of this polymer
III = group suffix, indicating a material containing both silicon and fluorine
3 = *Dk* in the 31–60 range of ISO units.

Example 2: ACUVUE 2
Etafilcon A IV 2

Eta = USAN prefix
filcon = stem, indicating a soft lens material >10% water by mass
A = USAN series suffix, indicating the first formulation of the polymer
IV = water content >50% and ionic
1 = *Dk* in the 16–30 range of ISO units.

Table 21.1 Group suffixes for rigid and soft contact lenses

Group Suffix	Rigid lenses	Soft lenses
I	No silicon or fluorine	<50% water content, non-ionic
II	Silicon but no fluorine	>50% water content, non-ionic
III	Contains both silicon and fluorine	<50% water content, ionic
IV	Fluorine but no silicon	>50% water content, ionic

Table 21.2 The classification of oxygen permeability, Dk

Number	Dk units (ISO)	Example
0	<1	PMMA
1	1–15	Frequency 55
2	16–30	Soflens 66
3	31–60	Paragon HDS
4	61–100	Purevision
5	101–150	Fluroperm 151
6	151–200	Menicon Z
7	For any new category	None at present

Example 3: AirOptix

Lotrafilcon B I 5 m

Lotra = USAN prefix
filcon = stem, indicating a soft lens material >10% water by mass
B = USAN series suffix, indicating the second formulation of the polymer
I = water content <50% and non-ionic
5 = Dk in the 101–150 range of ISO units
m = surface modified (plasma treated)

Rigid gas permeable lenses

Oxygen permeability

To enable practitioners to compare the oxygen permeability of one material with that of another, a form of measurement is required. This is known as the Dk value. The oxygen per-

meability of a rigid lens material is the product of the diffusion coefficient (D) for that gas and the solubility of the gas (k) in that material. Dk (permeability) and Dk/t (transmissibility) are in vitro measurements of oxygen transfer capability for a given material and lens, respectively. Units of measurement for Dk are:

$$cm^2 s^{-1} ml^{-1} O_2/ml \times mmHg$$

In diffusion, a high concentration of randomly moving gas molecules is on one side of the material. These molecules bounce against each other and against the molecules of the material, and eventually some move through to the other side of the material. The rate at which this happens is the diffusion coefficient. The solubility of the gas in a material is the measurement of its ability to dissolve in that material. Diffusion is the oxygen transfer mechanism more often attributed to RGP lenses, solubility more often being attributed to hydrogel lenses, where the fluid is in fact the liquid within the lens. In both cases, the energy and speed of movement of the gas molecules can be increased by raising the temperature. This, in turn, increases the rate of transfer of gas through the material, giving a higher Dk measurement.

The landmark study by Holden and Mertz (1984) confirmed the correlation between oxygen supply and corneal swelling, and established that a minimum Dk/t of 87×10^{-9} was required to limit overnight oedema to 4%. Extrapolating figures for oedema-free daily wear gave the value 24×10^{-9}. More recent work by Harvitt and Bonnano (1999) raised the figures for safe oedema-free wear to 125×10^{-9} for overnight wear and 37×10^{-9} for daily wear.

Although material may be supplied quoting the Dk value for that specific material, if it has a low refractive index it will be thicker compared with a material with a higher refractive index. The Dk/t will therefore be reduced in comparison with the higher refractive index material, even though the Dk values may be the same. A working knowledge of the Dk values and the ability to make an intelligent material selection are at the core of successful fitting.

The *ACLM Year Book* (Kerr and Ruston 2007; website www.aclm.org.uk) lists the Dk values in the new ISO terms and all materials should now **215** be graded according to this list. These new *Dk* values are approximately **75%** of the old Fatt units. It can be very difficult to know which figure is being quoted; if in doubt, the only option is to contact the manufacturer for the information to ensure that the *Dk* figures quoted are ISO values.

Centre and average lens thickness

For physiological reasons, lenses should be as thin as possible, although in practical terms making them too thin is counterproductive because they are very likely to distort, especially in the higher minus powers, and also to become too brittle. In most cases, a realistic minimum centre thickness is 0.14 mm, even for high minus powers.

Refractive index

With regard to lens thickness and refractive index, contact lenses are no different to spectacle lenses because the use of higher refractive index materials results in a thinner finished lens. However, modern RGP materials tend to have a lower refractive index than PMMA (polymethyl methacrylate) and are therefore thicker.

Examples

	Dk (ISO)	Specific gravity	Refractive index
Boston XO	100	1.27	1.415
Boston EO	58	1.23	1.429

Specific gravity

Specific gravity is the ratio of the mass of a solid or liquid to the mass of an equal volume of distilled water at 4°C. Materials with high specific gravity have greater mass than equal volumes of materials with lower specific gravity. The mass of an RGP lens can be altered significantly just by selecting a material with a high or low specific gravity, even when all other parameters remain the same. Fitting hypermetropic patients is a special challenge because plus lenses have an anteriorly located centre of

gravity compared with minus lenses, and these patients often have flat corneas, which create low riding lenses. By combining a minus carrier with a low specific gravity material, the mass of a plus lens can be greatly reduced, which minimises its tendency to ride inferiorly.

Thickness and lenticulation

Thickness depends on back vertex power (BVP), design and material. Centre thickness (t_c) and edge thickness (t_e) are both important.

Suggested minimum thicknesses for different materials (BVP −3.00 DS) are:

Material	t_c (mm)	t_e (mm)
PMMA	0.10	0.12
Silicon acrylate	0.15	0.13
Fluorosilicon acrylate	0.14	0.15

BVPs greater than −6.00 D or +4.00 D should ideally be lenticulated to reduce excess thickness and mass. Lenticulation reduces thickness in the centre for plus and towards the edge for minus lenses, by making the front optic zone diameter (FOZD) smaller.

Tips

- When using modern lens materials, do not order centre thickness <0.14 mm, because of possible lens flexure, particularly with toric corneas and tight lids.
- Edge thickness should be a minimum of 0.12 mm. A 'knife edge' causes discomfort and can be fragile, especially with plus lenses.
- Minus lenses usually give a natural lid attachment.
- A negative carrier helps give lid attachment with a low-riding or plus lens.
- A positive carrier helps reduce a high-riding tendency.

Surface properties

As soon as an RGP lens is placed on the cornea, it is covered by the tear film (the pre-lens tear layer). This covering is thinner and less stable than the tear film that covers the cornea in the absence of a contact lens, and the very presence of the contact lens destabilises the tear film. As a result of the relative thickness of the contact lens (150 μm at the centre, 100 μm at the edge, with a 80–100 μm clearance) and the tear film (about 7 μm) the contact lens appears as a very irregular, thick, foreign body, over which it is difficult to maintain a continuous tear film. It is not surprising, therefore, that good surface properties, which would help to establish a stable tear film, are considered vital to the well-being of a successful RGP contact lens wearer.

Wettability

If an accurate comparison of the effect of one variable, such as the influence of the surface material on the wetting angle, is to be made, all other factors must remain constant. Unfortunately, this generally has not been true from study to study, and contact angles reported by various contact lens manufacturers are very difficult to compare. Although still frequently quoted, wetting angles do not give the practitioner a true indication of just how a patient's tear film wets the lens.

To observe in vivo wettability, it becomes necessary to examine the pre-lens tear film (PLTF). PLTF observations are made using a slitlamp and Tearscope, the latter being a cold diffuse light source that is reflected by the tear film, rendering it visible without the installation of a staining agent such as fluorescein; this allows the practitioner to make a non-invasive observation of the tear film in its normal state.

Deposit resistance

Resistance to surface deposits is a vital property of RGP materials, without which the lens surface would soon become increasingly coated with a protein plaque and not wet properly. As a result, wearing times would be reduced. As material manufacturers tried to increase oxygen permeability by adding more silicone, the level of methacrylic acid increased, creating a stronger positive charge. It reached a point where the benefits of higher oxygen permeability were lost as a result of the build-up of protein on the

lens surface. This point was known as the **silicone acrylate barrier**. The materials industry then had to find a monomer that would increase oxygen permeability, without adversely affecting the lens surface; this monomer was to be fluorine. The fluorosilicon acrylates offered higher oxygen permeability without the compromise of protein build-up and were offered to the profession as having 'deposit resistance'. Although protein films can still observed on lenses after a prolonged period of time, these can removed by the use of efficient abrasive cleaners and avoided by planned replacement schemes. However, protein is not the only contaminant that we find on the surface of RGP lenses. Lipids from the meibomian glands can be expressed on to the surface of the lens, causing a greasy and smeary surface; these can be removed with the use of alcohol-based cleaners.

Silicon acrylates (siloxanes)

Silicon acrylates are co-polymers with varying proportions of PMMA, which provides lens rigidity, and silicon, which controls the degree of oxygen permeability. Also included are cross-linking agents to improve the strength of the material and wetting agents such as methacrylic acid to improve the naturally hydrophobic properties of silicon permeability. They give superior oxygen permeability and physiological performance when compared with PMMA, and most have stood the test of time in terms of dimensional stability, optical and mechanical results. They are routinely fitted for daily wear and to a limited degree were once used for extended wear.

Examples	Dk (ISO)
Polycon II	14
Boston IV	19
Paraperm EW	54

Advantages of silicon acrylates

- Low to medium *Dk* materials available
- Good dimensional stability
- Limited lens flexure
- Good scratch resistance.

Disadvantages of silicon acrylates

- They attract protein from the tears
- Some materials are brittle and have a tendency to break
- Increased incidence of 3 and 9 o'clock staining
- Some incidence of contact lens-associated papillary conjunctivitis (CLAPC)

Fluorosilicon acrylates

Fluorosilicon acrylates are composed of fluoro-monomers and siloxyacrylate monomers. The addition of fluorine atoms to replace some of the hydrogen present in methacrylate monomers improves surface wettability, tear film stability and deposit resistance, as well as increasing oxygen permeability. The solubility of oxygen in fluoro materials is enhanced and so higher Dk values can be achieved. Alternatively, moderate Dk values can be achieved with a lower siloxyacrylate content, and hence provide improved wettability. The silicon content ranges from 5–7% with Boston EO and ES to 16–18% with Fluoroperm 90.

Examples	Dk (ISO)
Boston ES	18
Quantum	55
Boston EO	58
Paragon HDS	58
Fluoroperm 151	99
Boston XO	100
Quantum 2	130
Menicon Z	163

Advantages of fluorosilicon acrylates

- Extremely high Dk values possible
- Improved wettability
- Less deposit problems
- Lower incidence of CLAPC
- Suitable for flexible or extended wear.

Disadvantages of fluorosilicon acrylates

- Require careful manufacture
- Dimensional stability depends on material and manufacture
- Corneal adhesion in some cases (especially in overnight wear).

Surface modifications have been put forward as a way of improving the surface properties of an RGP lens. This can be done in two ways: either a surface coating or surface treatment. It is the latter that has become the more common, assisting with initial comfort, surface wettability and deposit resistance.

The Novalens has a Dk of 55×10^{-11} and a 'soft' coating of hydroxyl groups that gives the surface characteristics of a hydrophilic lens. The lens itself does not absorb water but has good wettability and improved comfort. It may be cleaned and soaked in almost all conventional RGP products, but use of solutions with high levels of alcohol or microscopic bead cleaners should be avoided.

A different approach is adopted by Hybrid FS and FS Plus from Contamac, where the latter's 'fluid surface technology' combines a fluorosilicon acrylate material with a hydrophilic component. Before hydration, the hydrophilic molecules are distributed throughout the polymer matrix, so that after lens manufacture the surface is guaranteed to contain a number of hydrophilic sites. After hydration the hydrophilic components attract water molecules, which it is claimed improve both comfort and oxygen permeability. No additional treatment is necessary and lenses are compatible with all standard solutions. The Hybrid FS has a refractive index of 1.4465 in both the dehydrated and the hydrated states, a Dk of 23 and a water uptake of less than 0.94%. The Hybrid FS Plus has the same refractive index but a Dk of 60 and a water uptake of less than 0.85%.

Hydrophilic soft lenses

Soft lens materials are termed 'hydrogels' and they transmit oxygen primarily through the water molecules within the lens structure. The first material used in the manufacture of soft lenses was the polymer poly(hydroxyethyl methacrylate), abbreviated to p-HEMA (water content 38%). As our understanding of the cornea's oxygen requirements improved, and knowing that the ability of a soft lens material to transmit oxygen is related to its water content and thickness, polymer chemists looked carefully at ways of increasing the water content of soft lenses from the original 38% material. New monomers were introduced to combine with p-HEMA, or form co-polymers without HEMA, resulting in lens materials with water contents of between 55% and 80%. These monomers included methacrylic acid (MA), N-vinylpyrrolidone (NVP), polyvinyl pyrrolidone (PVP) and polyvinyl alcohol (PVA).

The water content of a hydrogel material refers to the percentage of water uptake of a fully hydrated hydrophilic lens. This percentage varies according to the variations in chemical structure and formulation of the lens material. Materials absorbing less than 4% of water by weight are known as hydrophobic materials and those absorbing more than 4% are called hydrophilic materials. If you increase the water content of a hydrophilic polymer, you increase the oxygen permeability of the material. However, this can make the material more fragile and prone to deposit formation and dehydration. Consequently, contact lenses that have higher water content are usually thicker in design. Materials with higher water content are often more prone to deposit formation and require more frequent replacement. The water content of soft lens materials available today varies from 38% to 80%. In Europe these lenses are classified as 'low', 'mid' and 'high' water content lenses.

Low water lenses (38–50% water) provide the cornea with sufficient oxygen for limited daily wear. Mid-water content lenses (50–60% water), when in ultrathin design, can have better transmissibility than thicker high-water-content designs. High water lenses (60% or more), although providing good oxygen transmissibility, are prone to discoloration, deposition and dimensional changes. Ideally, patients should be prescribed lenses that allow the highest levels of oxygen to the cornea, on a planned replacement or disposable basis, thus reducing physiological complications such as oedema, conjunctival and limbal hyperaemia, upper tarsal plate changes and, indeed, inflammation and infection.

In the USA, the water content of hydrophilic soft contact lenses is classified as low or high, low being less than 50% and high being 50% or greater.

Ionic charge

A soft lens material may carry an electric charge, or be neutral. Ionic charge results in increased deposit formation, because most deposits from the patient's tears are positively charged, especially protein, attracting them to the negative ionic charge of the lens material. This difference in ionic charge can result in a significant clinical difference between ionic and non-ionic lens materials because increased deposition can result in reduced comfort and vision for the patient. With disposable and planned replacement lenses, deposits do not usually build up to a significant level before the lens is replaced with a fresh, sterile, new lens. Ionic materials are also more 'reactive' with certain types of lens care solutions. High-water-content ionic materials are prone to parameter changes if allowed to soak in hydrogen peroxide overnight, and consequently require longer periods of neutralisation. Most of today's lens care systems do not use hydrogen peroxide and those that do involve shorter periods of exposure to it.

Polymers are categorised into four groups, linking water content to ionic properties. The classification relates to materials rather than clinical considerations and, in this context, high water content is defined as greater than 50%. Ionic polymers contain more than 0.2% MA and these high-water-content lenses may therefore contain the negatively charged carboxylic acid. The polymers are more sensitive to both temperature and the composition of care products, so lens parameters may show variability with environmental factors. Examples from these four groups include:

- Group I: low water, non-ionic: Bausch & Lomb Optima 38
- Group II: high water, non-ionic: Bausch & Lomb SofLens 66
- Group III: low water, ionic: Ciba Vision Durasoft 2, Bausch & Lomb Purevision
- Group IV: high water, ionic: Johnson & Johnson ACUVUE 2, Coopervision Frequency 55.

UV-blocking materials

Before manufacture an additional monomer can be added to the lens material that significantly increases the material's ability to block the transmission of ultraviolet radiation (UVR). This advantage is of increasing clinical significance because research continues to identify the detrimental effect of acute and chronic exposure of ocular tissues to UVR. Tissue changes and conditions include photokeratitis, cataract, pterygium, pingueculae and macular degeneration. UVR has a wavelength between 100 and 380 nm and can be classified as:

- UVA: 315–380 nm (4% is absorbed by the ozone layer)
- UVB: 280–315 nm (70–90% is absorbed by the ozone layer)
- UVC: 100–280 nm (absorbed by the earth's atmosphere).

A class II UV-blocking device must block at least 95% of UVB and 70% UVA to meet the American National Standard Institution (ANSI) standards. A greater percentage of UVB block is required (95% as opposed to 70%) because research has confirmed that UVB radiation has the most detrimental effect on ocular tissue and is responsible for most of the UV-induced ocular tissue changes that do occur.

Although there are clear benefits for UV-blocking contact lenses, in high UVR environments extra protection is required to protect the eyelids and surrounding tissue, as well as reduce the glare resulting from high levels of visible light. Activities such as skiing, sailing and using a sun bed are included, and extra protection can be achieved only by wearing good quality sunglasses/goggles.

Oxygen permeability

The oxygen performance of a soft lens is considered, by many, to be the most important property when considering the physiological response of a soft lens on the eye. This is because the cornea needs a constant supply of oxygen, which it receives primarily from atmospheric air. A contact lens acts as a barrier between the cornea and its oxygen supply. In a contact lens-wearing patient, oxygen is able to reach the cornea in two different ways depending on the type of contact lens worn:

1. By oxygen dissolving in the tears, which are then pumped behind the lens during the lens movement that results from blinking. This is relevant only for patients wearing RGP lenses, because there is minimal tear exchange behind a soft lens. This results from the greater total diameter and less movement of soft lenses compared with an RGP lens.
2. By diffusing directly through the lens material and into the tear layer that is formed between the lens and the cornea. Oxygen in the tear layer can then diffuse into the layers of the cornea via the corneal epithelium.

Dk measurement

The most widespread technique used for the measurement of oxygen permeability and oxygen transmissibility of a contact lens is termed the 'polarographic technique', which was pioneered by Fatt in the early 1970s. In this technique, a soft lens is draped over the surface of a polarographic oxygen sensor. The apparatus allows the measurement of the transmissibility of the lens sample at 35°C and, knowing the thickness of the lens sample, allows the Dk/t of the material to be calculated.

As the understanding of the polarographic technique improved, it became apparent that there were two important potential errors in the system: the first resulted from a layer of water located between the sample lens and the sensor, which acted as a barrier to oxygen transmission. This error is called the 'boundary effect', which effectively results in a lower Dk measurement than the true Dk of the lens material.

The second error results from the lateral flow of oxygen through the lens material to the sensor, and enhances oxygen transmission. It is called the 'edge effect', which effectively results in a higher Dk measurement than the true Dk of the lens material. The 'edge effect' error is overcome by reducing the original boundary effect measurement by about 24%. Although manufacturers most frequently quote Dk values that are boundary effect corrected for their lenses, few allow for this 'edge effect'. Unless it is known what correction factors have been applied to the quoted Dk for a particular lens material, it is impossible to compare the oxygen performance of different lens types.

Thickness

The thickness of a contact lens (t) is measured in millimetres and can be taken at the centre of the lens (t_c) or the edge (t_e). The average thickness across the lens from the centre to any point on the lens can also be calculated (t_{avg}). The thinner a given lens material, the greater the amount of oxygen that passes through it. The centre thickness for currently available $-3.00\,D$ soft spherical contact lenses varies from $0.035\,mm$ to $0.17\,mm$.

Transmissibility

Knowing both the Dk value of a contact lens material and its centre thickness allows the calculation of the central transmissibility of any soft contact lens. This unit is very important, because it is the transmissibility of a lens that is more clinically relevant to the practitioner and patient than the Dk value alone. Knowing the transmissibility allows comparison of the oxygen performance of different soft lenses. Although higher-water-content soft lenses have a higher Dk value, they can be significantly thicker than lower-water-content lenses. Consequently, higher-water-content lenses can have lower transmissibility values and therefore reduced oxygen transmissibility compared with thinner mid-water-content (50–60%) lens designs.

Transparency

Transparency refers to the clearness or clarity of a material. No material can be completely transparent, but the degree of transparency is usually measured as a plus percentage. This percentage is the amount of incident light of a certain wavelength that passes through the material. Most clear (non-tinted) contact lens materials have a visible light transmission of 92–98%. Many manufacturers use a light tint within the lens material, which is termed a 'handling tint', and this has the advantage of making the lens more visible to the patient when handling; however, it does result in a very slight reduction in transparency.

Modulus

The modulus of elasticity is a measure of how flexible the material is. The lower the modulus, the more easily the lens wraps to follow the eye's natural contours and gives a closer physical fit and potentially fewer mechanical complications such as contact lens associated papillary conjunctivitis (CLAPC) and superior epithelial accurate lesions (SEALs).

Biocompatible lenses

Biocompatibility has been defined as 'the ability of a material to interface with a natural substance without provoking a biological response'. The advantages of biocompatible materials are their good physiological response with reduced evaporation of tears, corneal desiccation and deposits. The Proclear material includes phosphorylcholine (PC), a naturally occurring component of the cell membrane of red blood cells, which has a high affinity for water and is resistant to protein adsorption. Similar technology is used for intraocular lenses as well as cardiac, orthopaedic and other health-care products. Coopervision Proclear, Omafilcon A (62% water content), contains PC synthetic polymers with a high affinity for water. These PC polymers create a permanent water layer on the surface.

Advantages of biocompatible materials include the following:

- Good dehydration resistance
- Sustained oxygen permeability
- Good deposit resistance
- All-day comfort, giving rise to longer wearing times.

Disadvantages of biocompatible materials include:

- Slightly less durable (however, many lenses now available in disposable modalities).

Silicone elastomer lenses

Silicone elastomer lenses differ from other lenses in several ways. They can be flexed, stretched and turned inside out. They have excellent elastic properties, partly conform to the shape of the cornea in wear, and have extremely high Dk values in the region of 200–300. They are unlike hydrophilic lenses because their natural state is dry and they are extremely tough. As they do not absorb water to any significant extent, fluorescein can be used in their fitting. As a result of the amorphous nature of the raw silicone rubber material, lenses are produced by a moulding and vulcanisation technique, which also assists in maintaining good reproducibility. The main difficulty with silicone is that its natural surface is extremely hydrophobic, and it has been necessary to devise methods of rendering the surface permanently hydrophilic without interfering with any of its optical or physical properties. The final stage of manufacture is therefore surface treatment by ionic bombardment. As a result of the following advantages and disadvantages, silicone elastomers have remained very much a minority lens with limited therapeutic and paediatrie applications.

Advantages

- Extremely high Dk (highest of all lens types)
- Low dehydration so little variation in comfort or fitting with environmental factors

Disadvantages

- Complex to fit, requiring as much precision as RGP lenses
- Negative pressure effect producing lens adhesion, if not correctly fitted
- Breakdown in surface coating and difficulties with wetting
- Rapid build-up of deposits
- Reduction in comfort and vision over time
- Very limited parameter availability
- Expensive

Silicone hydrogels

Silicone hydrogel materials were introduced in the UK during 1999 and represented a major advance in materials technology by combining silicone rubber with hydrogel monomers. These materials also use the classification suffix *filcon*. The silicone constituent permits very high oxygen permeability, whereas the hydrogel component ensures that the lenses are soft and comfortable; in addition, it provides fluid transport through the lens. Silicone-based materials are inherently hydrophobic and the first generation of materials required surface treatment to render the finished lenses hydrophilic and comfortable. The treatment must first have no effect on the oxygen transmission and second become an integral part of the lens, so that it cannot be removed with wear, handling or via interaction with disinfecting solutions.

The surfaces of both the Ciba Vision Focus Night & Day and AirOptix materials are permanently modified in a gas plasma reactive chamber to create a permanent, ultrathin (25 nm), high refractive index, continuous hydrophilic surface.

Bausch & Lomb Purevision lenses have their surfaces treated in a plasma chamber, which transforms the silicone components on the surface of the lens into hydrophilic silicate compounds. Glassy silicate 'islands' result and the hydrophilicity of these areas 'bridges' over the underlying hydrophobic balafilcon A material.

The flow of oxygen through the lenses is not impeded by these surface modifications. Both surface treatments become an integral part of the lens and are not a surface coating that can be easily 'stripped' away from the base material during daily handling and cleaning.

The ACUVUE Advance material was the first non-surface-treated silicone hydrogel to become commercially available, closely followed by ACUVUE Oasys. Both Advance and Oasys use an internal wetting agent (HydraClear) based on PVP, which is designed to provide a hydrophilic layer at the surface of the material to 'shield' the silicone at the material interface, thereby reducing the degree of hydrophobicity

Table 21.3 Silicone hydrogel comparison

Lens name	ACUVUE Advance with Hydraclear	ACUVUE Oasys with Hydraclear Plus	AirOptix	Purevision	Night and Day
Manufacturer	J & J	J & J	CibaVision	Bausch & Lomb	CibaVision
Lens material	Galyfilcon	Senofilcon A	Lotrafilcon B	Balafilcon A	Lotrafilcon A
FDA group	I	I	I	III	I
Modulus (MPa)	0.43	0.72	1.00	1.50	1.52
Power range (increments)	+4.00 to −6.00 (0.25 D) −6.50 to −12.00 (0.50 D) +4.50 to +8.00 (0.50 D)	+4.00 to −6.00 (0.25 D) −6.50 to −12.00 (0.50 D) +4.50 to +8.00 (0.50 D)	+6.00 to−6.00 (0.25 D) −6.50 to 10.00 (0.50 D)	+6.00 to −6.00 (0.25 D) −6.50 to −12.00 (0.50 D)	+6.00 to −8.00 (0.25 D) −8.50 to −10.00 (0.50 D)
Diameter (mm)	14.0	14.0	14.2	14.0	13.8
Base curve (mm)	8.30, 8.70	8.40	8.60	8.60	8.40, 8.60
Centre thickness (−3.00) (mm)	0.07	0.07	0.08	0.09	0.08
Wetting agent	Yes	Yes	No	No	No
Surface treatment	No	No	Yes	Yes	Yes
Water content (%)	47	38	33	36	24
O_2 available at central cornea,[a] open eye (%)	97	98	98	98	99
O_2 available at central cornea,[a] closed eye (%)	Not applicable	96	96	94	97
Dk/t at centre[b]	87×10^{-9}	147×10^{-9}	138×10^{-9}	110×10^{-9}	175×10^{-9}
Visibility tint	Yes	Yes	Yes	Yes	No
UV blocking	Class I: 93% UVA, 99% UVB	Class I: 96% UVA, 100% UVB	None	None	None
Recombination wearing schedule	2 week DW	2 week DW 1 week EW	DW + FW 6N EW 1/12 replacement	30N CW 6N EW FW and DW 1/12 replacement	30N CW 6N EW FW and DW 1/12 replacement

[a]Compared with 100% available with no lens wear for a −3.00D lens (data on file).
[b]Manufacturers quoted values, Fatt units at 35°C; boundary and edge corrected for a −3.00 D lens.
DW Daily Wear, FW Flexi-wear (occasional overnight use), EW Extended Wear (six nights), CW Continous Wear (thirty nights)

typically seen at the surface of silicone hydrogels.

By overcoming the need for surface treatment, the cost savings are considerable and it is therefore likely that future generation silicone hydrogel lenses will avoid the need for surface treatment. Table 21.3 summarises the key properties of currently available silicone hydrogel materials.

Advantages

- High *Dk* values available
- Ideal for patients with conjunctival injection and neovascularisation with their current lenses
- Suitable for extended (6 nights) and continuous (30 nights) wear
- Therapeutic indications
- Good dehydration characteristics
- Easier handling because of lens rigidity
- Good tensile strength with low breakage rate.

Disadvantages

- Increased lipid deposition (this is, however, patient and material dependent)
- Limited availability in complex designs, but improving all the time
- No daily disposable modality yet
- First-generation materials were stiffer and therefore more prone to cause SEALs and CLAPC

Concluding points for Chapter 21

The differences between contact lens materials, particularly hydrogel materials, often seem too subtle to be of clinical significance or too complex to be worthy of study! However, a basic understanding of the chemistry of the materials used by clinicians on a daily basis is critical to understand many of the clinical observations seen during their use. Subtle differences in the chemical composition of materials can impact on a wide variety of important clinical factors, including oxygen transmissibility, dehydration, parameter stability and deposition.

References

Harvitt D M, Bonnano J A (1999) Re-evaluation of the oxygen diffusion model for predicting minimum contact lens *Dk/t* values needed to avoid corneal anoxia. *Optometry and Vision Science* **76**:712–19

Holden B A, Mertz G W (1984) Critical oxygen levels to avoid corneal oedema for daily and extended wear contact lenses. *Investigative Ophthalmology and Vision Science* **25**:1161–7

Kerr C, Ruston D (2007) *The ACLM Contact Lens Year Book 2006*. Association of Contact Lens Manufacturers, London

Further recommended reading

Efron N (2002) *Contact Lens Practice*. Butterworth-Heinemann, Oxford

Sweeney D F, ed. (2000) *Silicone Hydrogels: The rebirth of continuous wear contact lenses*. Butterworth-Heinemann, Oxford

The Correction of Astigmatism with Rigid Gas Permeable Contact Lenses

Caroline Christie and Andrew Keirl

Introduction

The fitting of rigid gas permeable (RGP) toric lenses is often considered a mysterious and complex art. Practitioners are often confused by the mathematical approach with columns of calculations and formulae. The object of this chapter is to demystify the process and provide a practical approach that can be applied both in examinations and on a day-to-day basis in contact lens practice.

Although a very high percentage of patients can be fitted with spherical or aspheric RGP lens forms, there remains a significant percentage who can attain optimal vision and/or comfort only with a toric RGP lens.

Most human eyes exhibit at least a small amount of astigmatism. As discussed in Chapter 7, astigmatism can arise from either the cornea or the crystalline lens. Corneal astigmatism occurs when the corneal surfaces are toroidal in shape. The keratometer (see Chapter 18) can be used to measure the anterior corneal radii and hence the corneal astigmatism. It is not easy, and certainly impossible with a keratometer, to measure astigmatism arising from the posterior corneal surface. However, there is evidence to suggest that in cases of corneal astigmatism, the anterior and posterior corneal surfaces would have approximately parallel principal meridians. This is especially true in high degrees of corneal astigmatism, and means that about 10% of any anterior corneal astigmatism is neutralised by astigmatism arising from the posterior corneal surface. This fact is allowed for in the calibration the keratometer, which measures the *anterior* corneal radius and the *total* corneal power. Astigmatism resulting from the posterior corneal surface rarely exceeds 1.00 D and can be caused by one or both of the crystalline lens surfaces having toroidal geometry or by a tilted or decentred crystalline lens. Astigmatism caused by the crystalline lens is often termed **lenticular astigmatism**, and can account for any difference between the corneal astigmatism measured by a keratometer and the total spectacle (or ocular) astigmatism found during a subjective refraction.

Chapter content

- Corneal astigmatism
- Residual astigmatism
- The correction of astigmatism with toric RGP contact lenses
- Induced astigmatism
- Ray-tracing through spherical and toric RGP contact lenses
- Toric periphery lenses

Revision of terminology

Before proceeding further, it is necessary to review some relevant terminology.

Ocular refraction

This is the dioptric power required to correct an individual's ametropia measured in the plane of the reduced (corneal) surface. The usual symbol for ocular refraction is K.

Spectacle refraction

This is the dioptric power required to correct an individual's ametropia measured in the plane of the spectacle lens. The usual symbols for spectacle refraction are F_{sp} (thin lens) and F'_v (thick lens).

Ocular astigmatism

An individual with astigmatic ametropia has two ocular refractions. These are measured in the plane of the reduced (corneal) surface, correspond to the principal meridians of the reduced (corneal) surface and are at right angles to each other. The difference between the two ocular refractions is known as the ocular astigmatism. When measured in the corneal plane, the vertex distance is assumed to be zero.

Spectacle astigmatism

If astigmatic ametropia is to be corrected a by a spectacle lens at a given vertex distance, a spectacle refraction for each principal meridian is needed. The difference between the two spectacle refractions is known as the spectacle astigmatism.

Corneal astigmatism

Astigmatism that arises from the corneal surface is known as corneal astigmatism. This is a result of the corneal surfaces being toroidal in form. When measured using a keratometer, it can be estimated by using the clinical rule: 0.1 mm difference in radii = 0.50 DC.

Lenticular astigmatism

Astigmatism that arises from the crystalline lens is known as lenticular astigmatism. This can be caused by the crystalline lens surfaces

being toroidal in form. It can also occur if the lens is tilted or decentred, or if isolated changes in refractive index develop as a result of cortical lens opacities.

Total astigmatism

This is the amount of astigmatic ametropia that can be measured subjectively during an eye examination and is the sum of any corneal and/or lenticular astigmatism present. A routine subjective refraction does not differentiate between corneal and lenticular astigmatism.

Two more terms explained later on in this chapter are residual astigmatism and induced astigmatism.

Corneal astigmatism

To discuss the origin of corneal astigmatism, we will carry out several simple calculations using data from Gullstrand's exact schematic eye. Using this model we will assume the anterior corneal radius in the horizontal meridian to be 7.70 mm. To make this eye astigmatic we will make the vertical meridian of the anterior surface 10% shorter (6.93 mm), which gives a steeper and more powerful vertical meridian to the anterior surface, resulting in **with-the-rule astigmatism**. Using 1.376 for the refractive index of the cornea the principle powers of the anterior corneal surface in air are:

$$F_{1_{180}} = \frac{n_{cornea} - 1}{r_{180}} = \frac{1.376 - 1.000}{+7.70^{-3}} = +48.8312\,D$$

$$F_{1_{90}} = \frac{n_{cornea} - 1}{r_{90}} = \frac{1.376 - 1.000}{+6.93^{-3}} = +54.2568\,D$$

This gives 5.4256 D ($F_{1_{90}} - F_{1_{180}}$) of anterior surface astigmatism (with the rule).

We must also consider astigmatism generated by the posterior corneal surface. Again, using Gullstrand's exact schematic eye, the radius of curvature of the posterior corneal surface is 6.80 mm. If we also make the vertical meridian of the posterior corneal surface 10% shorter (6.12 mm), this gives a steeper and more powerful vertical meridian to the posterior corneal surface. Using 1.376 for the refractive index of the cornea and 1.336 for the refractive index of the aqueous, the principal powers of the posterior corneal surface are:

$$F_{2_{180}} = \frac{n_{aqueous} - n_{cornea}}{r_{2_{180}}} = \frac{1.336 - 1.376}{+6.80^{-3}}$$
$$= -5.8823\,D$$

$$F_{2_{90}} = \frac{n_{aqueous} - n_{cornea}}{r_{2_{90}}} = \frac{1.336 - 1.376}{+6.12^{-3}}$$
$$= -6.5359\,D$$

This gives $-0.6536\,D$ $(F_{2_{90}} - F_{2_{180}})$ of posterior surface astigmatism.

Assuming that the cornea is thin, the total uncorrected corneal astigmatism is (5.4256 + [−0.6536]) = 4.7720 DC (with the rule). The total uncorrected corneal astigmatism is lower than the astigmatism generated by the anterior surface in isolation, and is approximately 88% of the anterior corneal astigmatism. The posterior surface astigmatism is therefore neutralising around 12% of the anterior surface astigmatism.

If a spherical RGP contact lens is placed on the eye in this illustration, a tear lens is formed as usual. However, because of the geometry of the anterior corneal surface, a *toric* tear lens is formed. The tear lens in this case has a spherical front surface (mimicking the spherical back surface of the RGP contact lens) and a toroidal back surface (mimicking the toroidal front surface of the cornea). Any astigmatic neutralisation is therefore by means of the *back* surface of the tear lens. Using the *anterior* corneal radii given above, $n = 1.336$ for the refractive index of tears and assuming an infinitely thin air gap between the anterior corneal surface and the posterior tear lens, the power of the back surface of the tear lens is:

$$F_{2_{180}} = \frac{n_{air} - n_{tears}}{r_{2_{180}}} = \frac{1.000 - 1.336}{+7.70^{-3}} = -43.6364\,D$$

$$F_{2_{90}} = \frac{n_{air} - n_{tears}}{r_{2_{90}}} = \frac{1.000 - 1.336}{+6.93^{-3}} = -48.4848\,D$$

The cylinder generated by the back surface of the tear lens is

$$F_{2_{90}} - F_{2_{180}} = (-48.4848 - [-43.6364])$$
$$= -4.8484\,DC$$

If this value is compared with the anterior surface astigmatism of the cornea in this illustration, 0.5772 DC of anterior surface astigmatism (5.4256 + [−4.8484]) is left uncorrected. This means that 89% ([4.8484/5.4256] × 100) of the anterior corneal astigmatism is neutralised by the back surface of the tear lens. The remaining 11% of uncorrected anterior corneal astigmatism is neutralised by the posterior corneal astigmatism. Thus the corneal astigmatism has been almost completely neutralised. This neutralisation does not in any way depend on the refractive index of the contact lens itself. It should be noted that any astigmatism arising from the crystalline lens remains uncorrected. It should also be noted that in theory, any degree of corneal astigmatism can be neutralised by the fitting of a spherical RGP lens. However, in practice, when corneal astigmatism reaches 2.50 D, a back surface toric RGP lens is required to produce a comfortable physical fit.

The amount of astigmatism neutralised by an RGP contact lens can also be determined by using the ratio of the posterior tear lens surface power to the anterior corneal surface power. Using the usual values for the refractive index of the tears and the cornea, this is given by:

$$\frac{336}{376} = 0.89$$

So 89% of the anterior corneal astigmatism is neutralised by the back surface of the tear lens.

In contact lens practice, the keratometer is often used to measure the radius of curvature of the anterior corneal surface. However, the measuring scale on a keratometer is usually calibrated to give two values: the radius of curvature of the *anterior* corneal surface and the *total* corneal power. To measure the latter, the keratometer is calibrated on the assumption that the cornea has a refractive in index of 1.3375 and not 1.376. This lower refractive index gives a power reading that is closer to the total corneal power rather than the anterior corneal power.

When a patient's spectacle astigmatism is similar to the measured corneal astigmatism, the eye can often be corrected using a spherical

RGP lens and its accompanying tear lens. If corrected in this way, the three main aims of contact lens fitting are usually achieved:

1. Good comfort
2. Good vision
3. Acceptable physiological response

When an astigmatic eye is fitted with a spherical RGP contact lens, up to 2.50 D of corneal astigmatism can be considered to be neutralised. The wearer therefore obtains a good visual acuity with a spherical lens. As the axial edge lift of an aspheric lens is less than that of an equivalent multicurve, it is sometimes possible to fit higher degrees of astigmatism with an aspheric design because of the reduced edge stand-off in the steeper meridian.

There are three basic types of toric RGP lenses used in contact lens practice today:

1. Front surface toric (a RGP contact lens used to correct residual or lenticular astigmatism)
2. Back surface toric (a RGP contact lens with a toroidal back surface, used to improve the fit of the lens on a highly toric cornea)
3. Bitoric (a RGP lens with front and back toroidal surfaces used to correct the astigmatism induced by a rigid back surface toric RGP lens)

Residual astigmatism

In some cases, the over-refraction of a spherical RGP contact lens reveals the presence of astigmatism. This astigmatism must be the result of ocular components other than the cornea because the contact lens–tear lens system neutralises any corneal astigmatism present. The source of this manifest astigmatism is most probably the crystalline lens. This astigmatism, still present during the contact lens over-refraction using a spherical RGP lens, is called **residual astigmatism**, and can be defined as 'the amount of astigmatism remaining uncorrected when a rigid spherical contact lens is placed on the eye'. Residual astigmatism is also known as **lenticular astigmatism** and can be thought of as the total ocular astigmatism minus the corneal astigmatism. A front surface toric

RGP lens is required for the correction of spectacle (ocular) astigmatism in the absence of significant corneal astigmatism.

As an example, consider a patient with the spectacle prescription:

−2.00 DS/−2.00 DC × 80 and keratometry readings of 8.10 mm along 80 and 8.00 mm along 170.

The RGP trial lens 8.10/9.20/−3.00 D is placed on the eye and an over-refraction is performed. Using the above keratometry readings the corneal astigmatism can be estimated as about 0.50 DC. The expected over-refraction would therefore be +1.00 DS/−1.50 DC × 80. In this example, the tear lens has corrected the anterior corneal astigmatism and the over-refraction has revealed the presence of residual (lenticular) astigmatism, which of course reduces the patient's visual acuity. Residual astigmatism can be corrected using an RGP lens with a front toroidal surface.

This patient would require a toric lens, not to improve the physical fit but to achieve adequate visual correction. The back surface of this lens is spherical, the toricity being on the front surface of the lens. If the lens rotates, the astigmatic correction rotates with the lens, leading to unstable vision. To avoid this, lenses generally have at least 1.5 prism base-down added to stabilise the lens during blink and eye excursions, although the amount of prism may need to be increased in higher minus powers. Although the lens may orient satisfactorily in the primary gaze position and after blink, it is possible that it may roll along the lower lid as the patient looks from side to side. A truncation of 0.3 mm can prevent this from happening. In such cases, it is important to ensure that the laboratory places the bevel posteriorly, so that the full thickness of the lens edge does not lie on the lower lid margin.

The final order for this patient would be:

8.10:9.20 −2.00 DS/−1.50 DC × 80
1.5 Δ ballast base 270.

A material with a high Dk/t (oxygen transmissibility) should be used because the incorporation of prism increases the lens thickness and reduces oxygen transmission.

Another reason for fitting a front surface toric RGP lens is when a patient presents with no spectacle (ocular) astigmatism but has corneal astigmatism. In this case there must also be lenticular astigmatism that is equal but opposite to the corneal cylinder. One neutralises the other and the net result is no spectacle (ocular) astigmatism.

The correction of astigmatism with toric RGP contact lenses

Spherical and aspheric RGP contact lenses are frequently used to correct ocular astigmatism that is close or equal to any corneal astigmatism. As corneal astigmatism increases, it becomes more and more difficult to fit a well-centred stable lens that is comfortable to wear and does not insult corneal integrity. Lenses with very small total diameters (<9.00 mm) can be successful, as they fit only the central area of the cornea, which may be more spherical than the periphery. Reducing the overall lens size helps reduce the distance from the cornea to the lens in the steepest meridian. However, small lenses often give rise to flare and may still be prone to flexing. Lenses with narrow axial edge lift (typically 0.10 mm) can be used to avoid excessive edge clearance or stand-off in the steeper meridian.

Most aspheric designs also have a narrow edge lift and give reduced clearance along the steeper meridian. In some cases they can mask up to about 3.50 D of astigmatism. They are usually fitted in alignment or flat to avoid flexure. The differential bearing effect of a spherical lens on a significantly astigmatic cornea can lead a reduction in the magnitude of the cylinder by as much as 50%.

With modern lens design and manufacturing techniques, corneal astigmatism of greater than 2.50 D should fall into the criteria for fitting a fully toric back surface lens. However, it should be noted that spherical lenses fitted to corneas that are steeper in the horizontal meridian (against the rule), tend to be come unstable and decentre at even lower levels of corneal toricity, e.g. 1.50 DC.

Potential problems of fitting a spherical/aspheric RGP lens to a toric cornea include:

- Unstable lens with poor centration
- Lens flexure if the lens is not fitted in alignment with the flattest corneal meridian:
  - unstable vision
  - partial correction of the corneal astigmatism
- Excessive edge stand-off on the steep meridian
  - 3 and 9 o'clock staining
- Corneal moulding and spectacle blur
- Reduced comfort

Goals of toric gas permeable lens fitting include:

- Distributing the mass of the lens over as wide an area as possible to avoid distorting the cornea to an unacceptable degree
- Minimising interference with the tear film particularly in the horizontal meridian thus reducing the risk of 3 & 9 o'clock staining
- Avoiding areas of 'hard' touch/bearing which can give rise to local disruption such as staining and VLK (vascularised limbal keratitis)
- Optimising vision
- Providing good comfort and improved wearing times

As an example, consider the following:

Ocular refraction −2.00 DS/−4.50 DC × 180
Keratometry readings of 8.10 mm along 180/7.30 mm along 90.

Assuming that the lens is fitted in alignment, an initial order may be:

8.10 × 7.30 : 9.80 −2.00 D along flat, −6.50 D along steep.

No axis or prism ballast is necessary because the toroidal corneal surface stabilises the lens. It should be noted that if the lens were fitted in alignment to the flat and the steep meridians, a very thin tear lens with poor tear exchange would result. When fitting back surface toric RGP lenses, the usual approach is to align the lens with the flatter corneal meridian and fit slightly flat on the steeper meridian. The flatter fit along the steeper meridian increases the edge clearance and promotes tear exchange. The ideal fluorescein fit will show a faint band of

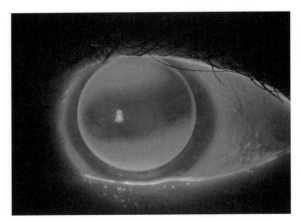

Figure 22.1 A spherical RGP contact lens fitted to a toroidal cornea. (See also colour plates.)

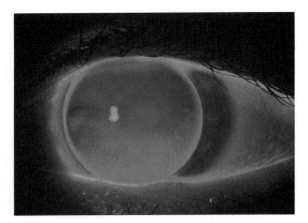

Figure 22.2 The same eye as shown in Figure 22.1 but fitted with a back surface toric RGP lens. (See also colour plates.)

fluorescein (alignment) in the central part of the lens along the horizontal meridian. In the mid periphery the lens will be supported on the cornea so there will be no visible fluorescein under the lens (appearing dark on the slit lamp). In the vertical meridian the lens will be fractionally flatter than the cornea (0.10–0.15 mm) to allow for vertical movement and aid lid attachment. Therefore in the vertical meridian the fluorescein will gradually increase towards the periphery. Figure 22.1 shows a spherical RGP lens fitted to a toroidal cornea whereas Figure 22.2 shows the same eye fitted with a back surface toric RGP lens.

Induced astigmatism

Whenever a back surface toric RGP lens is fitted to an eye, a tear lens is formed with both a posterior and an anterior toroidal surface. This bitoric tear lens mimics the toroidal posterior surface of the contact lens and the toroidal anterior corneal surface. The formation of this bitoric tear lens results in an over-correction of the corneal astigmatism. This astigmatic over-correction is called **induced astigmatism** and is present at the contact lens/tear lens interface. It is totally independent of the corneal astigmatism, which is partially neutralised by the posterior surface of the tear lens. If the tear lens is assumed to be thin, induced astigmatism is given by:

$$\frac{n_{\text{tears}} - n_{\text{lens}}}{\text{Steep radius}} - \frac{n_{\text{tears}} - n_{\text{lens}}}{\text{Flat radius}}$$

Assuming a typical RGP refractive index of 1.470 this gives:

$$\frac{1.3363 - 1.470}{\text{Steep radius}} - \frac{1.336 - 1.470}{\text{Flat radius}}$$

Induced astigmatism can be neutralised by making the front surface of the RGP contact lens toroidal. Such a lens is known as a **compensated parallel bitoric lens**. The purpose of the front toroidal surface is to correct the induced astigmatism. It has nothing to do with the corneal astigmatism which is neutralised by the posterior surface of the tear lens. A bitoric RGP lens used to correct induced astigmatism does not need to be rotationally stable to maintain good vision because both the cause (the back surface of the lens) and the correction (worked on the front surface) maintain alignment (are parallel to each other) if the lens rotates on the eye. In practice, the lens manufacturer calculates the amount of front surface astigmatism required to correct any induced astigmatism.

Ray-tracing through astigmatic systems

Example 22.1

A patient with a spectacle prescription of +6.00 DS/−1.00 DC × 180 at 14 mm is fitted with a

spherical RGP contact lens. The specification of the contact lens is as follows:

BOZR 8.00 mm
Centre thickness 0.50 mm
Refractive index 1.49

The lens is fitted with a corneal clearance of 0.10 mm and the keratometry readings are 8.00 mm along 180 and 7.80 mm along 90. Find the FOZR of the lens.

Most questions involving astigmatism usually result in things being done twice. This question is no exception because a ray-trace has to be performed through each principal meridian of the system. As the question is asking for the FOZR (r_1) the required ray-trace is a step-back ray-trace. Step-back ray-traces start from L_4' which is of course equal to the ocular refraction K. This can be found using the spectacle prescription and the vertex distance. As the cornea is toroidal there are two values for r_4 (along 90 and along 180) and therefore two values for F_4 (along 90 and along 180). The values for r_4 are the same as the keratometer readings. The front surface radius of the tear lens (r_3) is the same as the BOZR of the contact lens (r_2). The refractive index of tears is assumed to be 1.336. There is a considerable amount of preparation involved in this example before the ray-tracing can start. The required values are:

- K along 180 and K along 90
- F_4 along 180 and F_4 along 90
- The equivalent air distance (EAD) for the tear lens
- F_3
- The EAD for the contact lens
- F_2

To find the ocular refractions use:

$$K = \frac{F_{sp}}{1-(dF_{sp})}$$

Along 180

$$K = \frac{+6.00}{1-(0.014\times+6.00)} = +6.5502\,D$$

Along 90

$$K = \frac{+5.00}{1-(0.014\times+5.00)} = +5.3763\,D$$

When calculating surface powers, remember that an infinitely thin air gap separates all the surfaces. To find the back surface powers of the tear lens (F_4) use:

$$F_4 = \frac{1-n_{tears}}{r_4}$$

Along 180

$$F_{4_{180}} = \frac{1-n_{tears}}{r_{4_{180}}} = \frac{1.000-1.336}{+8.00^{-3}} = -42.0000\,D$$

Along 90

$$F_{4_{90}} = \frac{1-n_{tears}}{r_{4_{90}}} = \frac{1.000-1.336}{+7.80^{-3}} = -43.0769\,D$$

To find the front surface power of the tear lens (F_3) use:

$$F_3 = \frac{n_{tears}-1}{r_3}$$

The BOZR (r_2) = r_3 = 8.00 mm:

$$F_3 = \frac{1.336-1.000}{+8.00^{-3}} = +42.0000\,D$$

To find the back surface power of the RGP contact lens (F_2) use:

$$F_2 = \frac{1-n_{cl}}{r_2}$$

$$F_2 = \frac{1.000-1.490}{+8.00^{-3}} = -61.2500\,D$$

The EAD for both the contact lens and the tear lens can now be found. The EAD for the *contact lens* (with t_{cl} substituted in metres) is given by:

$$EAD = \frac{t_{cl}}{n_{cl}} = \frac{5^{-4}}{1.490} = 3.3557^{-4}\,m$$

The EAD for the *tear lens* (with t_{tears} substituted in metres) is given by:

$$EAD = \frac{t_{tears}}{n_{tears}} = \frac{1^{-4}}{1.336} = 7.4850^{-5}\,m$$

Now that the EADs have been determined, the refractive index between the two surfaces of the contact lens and the two surfaces of the tear lens can be assumed to be **equal to that of air** ($n = 1$). This value is used in the following

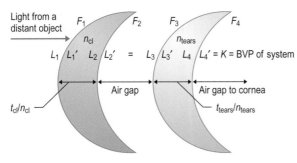

Figure 22.3 Diagram for Example 22.3.

ray-trace. A step-back ray-trace is needed to find the front surface power and hence the FOZR. As always, the computation takes place in two columns, one for vergences and the other for distances. The vergences for this problem are illustrated in **Figure 22.3** and the ray-trace once again needs to be extended to accommodate the four refracting surfaces. The starting point for any step-back ray-trace is the vergence closest to the eye, which in this example is L_4'. Two identical step-back ray-traces must be performed, one for each principal meridian of the system. It is important to take care when selecting powers, distances etc. Although some values are common to both, do not mix values from the two meridians. The first ray-trace is along the 180 meridian.

Along 180

Vergences (D)   Distances (m)

$L_4' = +6.5502\,D$

$L_4 = L_4' - F_4$

$L_4 = +6.5502 - (-42.0000) = +48.5502\,D$

$L_4 = +48.5502\,D \quad \rightarrow \quad l_4 = \dfrac{1}{+48.5502} = +0.02060\,m$

$$l_3' = l_4 + \left(\dfrac{t_{\text{tears}}}{n_{\text{tears}}}\right) \text{ (Step-back)}$$

$L_3' = \dfrac{1}{+0.02067}$ $\leftarrow$ $l_3' = +0.02060 + 7.4850^{-5}$

$\quad = +48.3744\,D$ $= +0.02067\,m$

$L_3 = L_3' - F_3$

$L_3 = +48.3744 - (+42.0000) = +6.3744\,D$

As a result of the thin air gap, $L_3 = L_2'$. We must now perform the second step-back ray-trace through the contact lens. The reader is advised to follow all the vergences carefully on Figure 22.3.

$L_3 = L_2' = +6.3744\,D$

$L_2' = +6.3744\,D$

$L_2 = L_2' - F_2$

$L_2 = +6.3744 - (-61.25) = +67.6244\,D$

$L_2 = +6.3744\,D \quad \rightarrow \quad l_2 = \dfrac{1}{+67.6244} = +0.01479\,m$

$$l_1' = l_2 + \left(\dfrac{t_{\text{cl}}}{n_{\text{cl}}}\right) \text{ (step-back)}$$

$L_1' = \dfrac{1}{+0.01512}$ $\leftarrow$ $l_1' = +0.01479 + 3.3557^{-4}$

$\quad = +66.1239\,D$ $= 0.01512\,m$

As $L_1 = 0.0000$, $L_1 = F_1$

$F_1 = +66.1239\,D$

The FOZR can be found using:

$$r_1 = \dfrac{n' - n}{F_1}$$

$$r_1 = \dfrac{1.490 - 1.000}{+66.1239} = +7.4103^{-3}\,m$$

The FOZR of the contact lens along the 180 meridian is therefore +7.4103 mm.

We now have to do the same again for the 90 meridian!

Along 90

Vergences (D)   Distances (m)

$L_4' = +5.3763\,D$

$L_4 = L_4' - F_4$

$L_4 = +5.3763 - (-43.0769) = +48.4533\,D$

$L_4 = +48.4533\,D \quad \rightarrow \quad l_4 = \dfrac{1}{+48.4533} = +0.02064\,m$

$$l_3' = l_4 + \left(\dfrac{t_{\text{tears}}}{n_{\text{tears}}}\right) \text{ (step-back)}$$

$$L_3' = \frac{1}{+0.02071} \quad \leftarrow \quad l_3' = +0.02064 + 7.4850^{-5}$$
$$= +0.02071\,\text{m}$$
$$= +48.2782\,\text{D}$$

$$L_3 = L_3' - F_3$$

$$L_3 = +48.2782 - (+42.0000) = +6.2782\,\text{D}$$

as a result of the thin air gap, $L_3 = L_2'$. We must now perform the second step-back ray-trace through the contact lens. The reader is again advised to follow all the vergences carefully on Figure 22.3.

$$L_3 = L_2' = +6.2782\,\text{D}$$

$$L_2' = +6.2782\,\text{D}$$

$$L_2 = L_2' - F_2$$

$$L_2 = +6.2782 - (-61.25) = +67.5282\,\text{D}$$

$$L_2 = +67.5282\,\text{D} \quad \rightarrow \quad l_2 = \frac{1}{+67.5282} = +0.01481\,\text{m}$$

$$l_1' = l_2 + \left(\frac{t_{cl}}{n_{cl}}\right) \quad \text{(step-back)}$$

$$L_1' = \frac{1}{+0.01514} \quad \leftarrow \quad l_1' = +0.01481 + 3.3557^{-4}$$
$$= 0.01514\,\text{m}$$
$$= +66.0319\,\text{D}$$

As $L_1 = 0.0000$, $L_1 = F_1$

$$\mathbf{F_1 = +66.0319\,D}$$

The FOZR can be found using:

$$r_1 = \frac{n' - n}{F_1}$$

$$r_1 = \frac{1.490 - 1.000}{+66.0319} = +7.4207^{-3}\,\text{m}$$

The FOZR of the contact lens along the 90 meridian is therefore +7.4207 mm.

The FOZR of the contact lens is +7.4103 mm along 180 and +7.4207 mm along 90. The lens effectively has a *spherical* front surface. This example reinforces the fact that a spherical RGP contact lens corrects corneal astigmatism.

Relatively speaking this is a long question. Nothing new, but long! As mentioned at the conclusion of previous examples, it is important to organise the initial data and take a logical approach with regard to the ray-tracing.

Example 22.2

A subject has a spectacle refraction F_{sp} of +5.00 DS/–2.00 DC × 180 at 12 mm. The corneal radius is 7.70 mm along 90 and 7.90 mm along 180. A spherical RGP contact lens is fitted as follows:

BOZR 7.85, n = 1.52, BVP = +5.50 DS

Calculate the residual astigmatism if the tear lens is 0.15 mm thick and the refractive index of the tears is 1.336.

To calculate the residual astigmatism we must compare the patient's ocular astigmatism with the astigmatic correction produced by the contact lens–tear lens system. We therefore need to calculate the patient's ocular refraction and perform two step-along ray-traces to find the BVP (L_4') for each principal meridian of the contact lens–tear lens system. The difference in the two values of L_4' gives the astigmatic correction produced by the contact lens–tear lens system. As the BVP of the contact lens has been given, the step-along ray trace can start at L_3. Remember that $L_2' = L_3 = $ contact lens BVP in air. If you are not sure about this refer to Figure 22.3.

As the cornea is toroidal there are two values for r_4 (along 90 and along 180) and therefore two values for F_4 (along 90 and along 180). The values for r_4 are the same as the keratometer readings. The front surface radius of the tear lens (r_3) is the same as the BOZR of the contact lens (r_2). The refractive index of tears is assumed to be 1.336. There is some preparation involved in this example before the ray-tracing can start. The required values are:

- K along 180 and K along 90
- F_4 along 180 and F_4 along 90
- The EAD for the tear lens
- F_3

To find the ocular refractions use:

$$K = \frac{F_{sp}}{1 - (dF_{sp})}$$

Along 180

$$K = \frac{+5.00}{1-(0.012 \times +5.00)} = +5.3191\,\text{D}$$

Along 90

$$K = \frac{+3.00}{1-(0.012 \times +3.00)} = +3.1120\,\text{D}$$

When calculating surface powers, remember that an infinitely thin air gap separates all the surfaces. To find the back surface powers of the tear lens (F_4) use:

$$F_4 = \frac{1-n_{\text{tears}}}{r_4}$$

Along 180

$$F_{4_{180}} = \frac{1-n_{\text{tears}}}{r_{4_{180}}} = \frac{1.000-1.336}{+7.90^{-3}} = -42.5316\,\text{D}$$

Along 90

$$F_{4_{90}} = \frac{1-n_{\text{tears}}}{r_{4_{90}}} = \frac{1.000-1.336}{+7.70^{-3}} = -43.6364\,\text{D}$$

To find the front surface power of the tear lens (F_3) use:

$$F_3 = \frac{n_{\text{tears}}-1}{r_3}$$

The BOZR (r_2) = r_3 = 7.85 mm:

$$F_3 = \frac{1.336-1.000}{+7.85^{-3}} = +42.8025\,\text{D}$$

As the step-along ray-trace starts from L_3 only the EAD for the tear lens is required. The EAD for the *tear lens* (with t_{tears} substituted in metres) is given by:

$$\text{EAD} = \frac{t_{\text{tears}}}{n_{\text{tears}}} = \frac{1.5^{-4}}{1.336} = 1.1227^{-4}\,\text{m}$$

Now that the EAD has been determined, the refractive index between the two surfaces of the tear lens can be assumed to be **equal to that of air** ($n = 1$). This value is used in the following ray-trace. A step-along ray-trace is needed find the back vertex power (L_4') of the system. As always, the computation takes place in two columns, one for vergences and the other for distances. The vergences for this problem are

illustrated in Figure 22.3 although the ray-trace starts at L_3. This is because the BVP of the RGP contact lens in air has been given in the question and the contact lens BVP in air = $L_2' = L_3$. Two identical step-along ray-traces must be performed, one for each principal meridian of the system. It is important to take care when selecting powers, distances, etc. Although some values are common to both, do not mix values from the two meridians. The first ray-trace is along the 180 meridian.

Along 180

Vergences (D)   Distances (m)

The BVP of the RGP contact lens in air = +5.50 D and BVP = $L_2' = L_3$

$$L_3 = +5.5000\,\text{D}$$

$$F_3 = +42.8025\,\text{D}$$

$$L_3' = L_3 + F_3 = +48.3025\,\text{D} \quad \rightarrow \quad l_3' = \frac{n}{L_3'}$$

$$l_3' = \frac{1}{+48.3025} = +0.0207\,\text{m}$$

$$l_4 = l_3' - \left(\frac{t_{\text{tears}}}{n_{\text{tears}}}\right)$$

$$L_4 = \frac{n}{l_4} \quad \leftarrow \quad \begin{aligned} l_4 &= +0.0207 - 1.1227^{-4} \\ &= +0.0206\,\text{m} \end{aligned}$$

$$L_4 = \frac{1}{+0.0206} = +48.5659\,\text{D}$$

$$L_{4_{180}}' = L_4 + F_{4_{180}}$$

$$L_{4_{180}}' = +48.5659 + (-42.5316) = +6.0343\,\text{D}$$

As $L_1 = 0.0000$

$$L_4' = \text{BVP}$$

The BVP of this RGP contact lens–tear lens system along the 180 meridian is therefore +6.0343 D.

Along 90

Careful inspection of the information given in this example, the above computation and Figure 22.3 shows that the ray-trace along 90 is exactly the same as that along 180, up to and including the vergence L_4. The only difference in the two ray-races is the value of F_4 which is different along 180 and 90. This difference is a result of the corneal surface being toroidal. There is no other toroidal surface in this contact lens–tear lens system. We can therefore start the calculation at L_4.

Vergences (D)   Distances (m)

$$L_4 = \frac{1}{+0.0206} = +48.5659\,\text{D}$$

$$L'_{4_{90}} = L_4 + F_{4_{90}}$$

$$L'_{4_{90}} = +48.5659 + (-43.6364) = +4.9296\,\text{D}$$

As $L_1 = 0.0000$

$$L'_4 = \text{BVP}$$

The BVP of this RGP contact lens–tear lens system along the 90 meridian is therefore +4.9296 D

The sph–cyl form, the BVP of this contact lens-tear lens system is:

+6.0343 DS/–1.1047 DC × 180

The ocular refraction for this subject in sph–cyl form is:

+5.3191 DS/–2.2071 DC × 180

The residual astigmatism is the *difference* between the above cylinder values, which in this example is –1.1024 DC × 180.

Example 22.3

A certain cornea has the following radii of curvature:

Anterior corneal radii: 7.95 along 180/7.25 along 90

Posterior corneal radii: 7.00 along 180/6.50 along 90

The corneal thickness is 0.50 mm; $n_{\text{cornea}} = 1.376$ and $n_{\text{aqueous}} = 1.336$

Find the BVP of the cornea and hence the total corneal astigmatism. What would be the astigmatism recorded by a keratometer calibrated for a refractive index of 1.3375?

As is usually the case in astigmatism, two ray-traces need to be performed, one along each principal meridian of the cornea. As the BVP is required, two step-along ray-traces are necessary. The first task is to calculate the four surface powers. For the anterior corneal surface use:

$$F_1 = \frac{n_{\text{cornea}} - 1}{r_{180}}$$

For the posterior corneal surface use:

$$F_2 = \frac{n_{\text{aqueous}} - n_{\text{cornea}}}{r_{2_{180}}}$$

All surface radii are entered in metres. The anterior corneal surface powers

$$F_{1_{180}} = \frac{n_{\text{cornea}} - 1}{r_{180}} = \frac{1.376 - 1.000}{+7.95^{-3}} = +47.2956\,\text{D}$$

$$F_{1_{90}} = \frac{n_{\text{cornea}} - 1}{r_{90}} = \frac{1.376 - 1.000}{+7.25^{-3}} = +51.8621\,\text{D}$$

The posterior corneal surface powers

$$F_{2_{180}} = \frac{n_{\text{aqueous}} - n_{\text{cornea}}}{r_{2_{180}}} = \frac{1.336 - 1.376}{+7.00^{-3}}$$
$$= -5.7143\,\text{D}$$

$$F_{2_{90}} = \frac{n_{\text{aqueous}} - n_{\text{cornea}}}{r_{2_{90}}} = \frac{1.336 - 1.376}{6.50^{-3}}$$
$$= -6.1538\,\text{D}$$

The next task is to determine the EAD of the cornea. The EAD of the cornea (with the centre thickness substituted in metres) is given by:

$$\text{EAD} = \frac{t_{\text{cornea}}}{n_{\text{cornea}}} = \frac{5^{-4}}{1.376} = 3.6337^{-4}\,\text{m}$$

Now that the EAD has been determined, the refractive index between the two surfaces of the cornea can be assumed to be **equal to that of air** ($n = 1$). This value is used in the following ray-trace. To find the total corneal astigmatism, the cornea must be treated as a thick lens. A step-along ray-trace is used to find the BVP for each principal meridian of the cornea. As always, the computation takes place in two columns, one for vergences and the other for distances.

Along 180

Vergences (D)   Distances (m)

$$L_1 = 0.00$$

$$F_{1_{180}} = +47.2956\,\text{D}$$

$$L_1' = L_1 + F_1 = 47.2956\,D \quad \rightarrow \quad l_1' = \frac{n}{L_1'}$$

$$l_1' = \frac{1}{+47.2946} = +0.02114\,m$$

$$l_2 = l_1' - \left(\frac{t_{cl}}{n_{cl}} \right)$$

$$L_2 = \frac{n}{l_2} \qquad \leftarrow \qquad l_2 = +0.02114 - 3.6337^{-4}$$
$$= +0.02078\,m$$

$$L_2 = \frac{1}{+0.02078} = +48.1226\,D$$

$$L_{2_{180}}' = L_{2_{180}} + F_{2_{180}}$$

$$F_{2_{180}} = -5.7143\,D$$

$$L_2' = +48.1226 + (-5.7143) = +42.4083\,D$$

As $L_1 = 0.00$, $L_2' = $ BVP and $F_v' = +42.4083\,D$

The BVP of this cornea along the 180 meridian is therefore +42.4083 D.

Along 90

Vergences (D)	Distances (m)

$$L_1 = 0.00$$

$$F_{1_{90}} = +51.8621\,D$$

$$L_1' = L_1 + F_1 = +51.8621\,D \quad \rightarrow \quad l_1' = \frac{n}{L_1'}$$

$$l_1' = \frac{1}{+51.8621} = +0.01928\,m$$

$$l_2 = l_1' - \left(\frac{t_{cl}}{n_{cl}} \right)$$

$$L_2 = \frac{n}{l_2} \qquad \leftarrow \qquad l_2 = +0.01928 - 3.6337^{-4}$$
$$= +0.01892\,m$$

$$L_2 = \frac{1}{+0.01892} = +52.8582\,D$$

$$L_{2_{90}}' = L_{2_{90}} + F_{2_{90}}$$

$$F_{2_{90}} = -6.1538\,D$$

$$L_2' = +52.8582 + (-6.1538) = +46.7043\,D$$

As $L_1 = 0.00$, $L_2' = $ BVP and $F_v' = +46.7042\,D$

The BVP of this cornea along the 90 meridian is therefore +46.7042 D.

The BVPs of the cornea are therefore **+42.4083 D along 180** and **+46.7042 D along 90**. The total corneal astigmatism is the difference between the two BVPs, which is **4.2959 DC**. The *front* surface astigmatism is the difference in the two anterior corneal powers, which is **4.5665 DC**. The total corneal astigmatism is therefore 94% of the front surface corneal astigmatism.

To find the astigmatism recorded by a keratometer calibrated for a refractive index of 1.3375, we need to calculate the anterior corneal powers using this refractive index:

$$F_{1_{180}} = \frac{1.3375 - 1.000}{+7.95^{-3}} = +42.4528\,D$$

$$F_{1_{90}} = \frac{1.3375 - 1.000}{+7.25^{-3}} = +46.5517\,D$$

The astigmatism recorded by a keratometer is therefore the difference between 46.5517 and 42.4528. This is **4.0989 DC**. A refractive index of 1.3375 allows the keratometer to give values that are closer to the actual total corneal powers than if a refractive index of 1.376 were used.

Example 22.4

A patient has a spectacle prescription of −6.00 DS at a vertex distance of 13 mm. His keratometry readings are 8.45 mm along 180 and 7.95 mm along 90. He is to be fitted with a bitoric RGP contact lens with a BOZR of 8.40 mm along 180 and 8.00 mm along 90. The refractive index of the contact lens material is 1.49. The contact lens has a centre thickness of 0.40 mm and the corneal clearance is 0.10 mm. Calculate the front surface radii of the contact lens.

This is an interesting example because the patient in question has a reasonable amount of corneal astigmatism but a spherical spectacle refraction. This can occur only if there is lenticular astigmatism that is equal but opposite to the corneal astigmatism. The corneal astigmatism is therefore neutralised by the lenticular astigmatism, leaving a spherical spectacle refraction. When an RGP contact lens is fitted, the vast majority of the corneal astigmatism is corrected by the tear lens, leaving the lenticular astigmatism manifest. This lenticular astigmatism reduces visual acuity if not corrected, so a bitoric RGP lens is required. The toroidal back surface provides a stable and comfortable fit whereas the toroidal front surface gives good vision by correcting the lenticular astigmatism and also neutralising any induced astigmatism produced by the formation of a bitoric tear lens.

As previously mentioned, most questions involving astigmatism usually result in things being done twice. In this example, a ray-trace is performed through each principal meridian of the system. As the question is asking for the FOZR (r_1), the required ray-trace is of the step-back variety. Step-back ray-traces start from L_4' which, is equal to the ocular refraction K. This can be found using the spectacle prescription and the vertex distance. As the cornea is toroidal there are two values for r_4 (along 90 and along 180) and therefore two values for F_4 (the posterior surface of the tear lens along 90 and along 180). The values for r_4 are the same as the keratometer readings. As the RGP contact lens in this example is a back surface toric, the front surface radius of the tear lens is also toroidal in form because it mimics the back surface of the RGP contact lens. There are therefore two values for r_3 and two for F_3, one along 180 and the other along 90. The values for r_3 are of course the same as the radii of the toric BOZR of the contact lens (r_2). So r_2 along 180 = r_3 along 180 and r_2 along 90 = r_3 along 90. The tear lens in this example is therefore bitoric. The refractive index of tears is assumed to be 1.336. There is a considerable amount of preparation involved in this example before the ray-tracing can start. The required values are:

- The ocular refraction K
- F_4 along 180 and F_4 along 90
- The EAD for the tear lens
- F_3 along 180 and F_3 along 90
- The EAD for the contact lens
- F_2 along 180 and F_2 along 90

To find the ocular refractions use:

$$K = \frac{F_{sp}}{1 - (dF_{sp})}$$

$$K = \frac{-6.00}{1 - (0.013 \times -6.00)} = -5.5659\,D$$

When calculating surface powers of the tear lens and contact lens, remember that an infinitely thin air gap separates all the surfaces. To find the back surface powers of the tear lens (F_4) use:

$$F_4 = \frac{1 - n_{tears}}{r_4}$$

Along 180

$$F_{4_{180}} = \frac{1 - n_{tears}}{r_{4_{180}}} = \frac{1.000 - 1.336}{+8.45^{-3}} = -39.7633\,D$$

Along 90

$$F_{4_{90}} = \frac{1 - n_{tears}}{r_{4_{90}}} = \frac{1.000 - 1.336}{+7.95^{-3}} = -42.2641\,D$$

To find the front surface power of the tear lens (F_3) use:

$$F_3 = \frac{n_{tears} - 1}{r_3}$$

Along 180

$$F_{3_{180}} = \frac{n_{tears} - 1}{r_{3_{180}}} = \frac{1.336 - 1.000}{+8.40^{-3}} = +40.0000\,D$$

Along 90

$$F_{3_{90}} = \frac{n_{tears} - 1}{r_{3_{90}}} = \frac{1.336 - 1.000}{+8.00^{-3}} = +42.0000\,D$$

To find the back surface power of the RGP contact lens (F_2) use:

$$F_2 = \frac{1 - n_{cl}}{r_2}$$

Along 180

$$F_{2_{180}} = \frac{1 - n_{cl}}{r_{2_{180}}} = \frac{1.000 - 1.490}{+8.40^{-3}} = -58.3333\,D$$

Along 90

$$F_{2_{90}} = \frac{1 - n_{cl}}{r_{2_{90}}} = \frac{1.000 - 1.490}{+8.00^{-3}} = -61.2500\,D$$

The EAD for both the contact lens and the tear lens can now be found. The EAD for the *contact lens* (with t_{cl} substituted in metres) is given by:

$$\text{EAD} = \frac{t_{cl}}{n_{cl}} = \frac{4^{-4}}{1.490} = 2.6846^{-4}\,\text{m}$$

The EAD for the *tear lens* (with t_{tears} substituted in metres) is given by

$$\text{EAD} = \frac{t_{tears}}{n_{tears}} = \frac{1^{-4}}{1.336} = 7.4850^{-5}\,\text{m}$$

Now that the EADs have been determined, the refractive index between the two surfaces of the contact lens and the two surfaces of the tear lens can be assumed to be **equal to that of air** ($n = 1$). This value is used in the following ray trace. A step-back ray-trace is needed to find the front surface power and hence the FOZR. As always, the computation takes place in two columns, one for vergences and the other for distances. The vergences for this problem are illustrated in Figure 22.3 and the ray-trace once again needs to be extended to accommodate the four refracting surfaces. The starting point for any step-back ray-trace is the vergence closest to the eye, which in this example is L_4'. Two identical step-back ray-traces must be performed, one for each principal meridian of the system. It is important to take care when selecting powers, distances, etc. Although some values are common to both, do not mix values from the two meridians. The first ray-trace is along the 180 meridian.

Along 180

Vergences (D)	Distances (m)

$L_4' = -5.5659\,\text{D}$

$L_4 = L_4' - F_4$

$L_4 = -5.5659 - (-39.7633) = +34.1974\,\text{D}$

$L_4 = +34.1974\,\text{D} \quad \rightarrow \quad l_4 = \dfrac{1}{+39.1974} = +0.029242\,\text{m}$

$$l_3' = l_4 + \left(\frac{t_{tears}}{n_{tears}}\right) \quad \text{(step-back)}$$

$L_3' = \dfrac{1}{+0.02932} \quad \leftarrow \quad l_3' = +0.029242 + 7.4850^{-5}$
$= +34.1101\,\text{D} \qquad\qquad\quad = +0.02932\,\text{m}$

$L_3 = L_3' - F_3$

$L_3 = +34.1101 - (+40.0000) = -5.8899\,\text{D}$

As a result of the thin air gap, $L_3 = L_2'$. We must now perform the second step-back ray-trace through the contact lens. The reader is again advised to follow all of the vergences carefully on Figure 22.3.

$L_3 = L_2' = -5.8899\,\text{D}$

$L_2' = -5.8899\,\text{D}$

$L_2 = L_2' - F_2$

$L_2 = -5.8899 - (-58.3333) = +52.4435\,\text{D}$

$L_2 = +52.435\,\text{D} \quad \rightarrow \quad l_2 = \dfrac{1}{+52.4435} = +0.01907\,\text{m}$

$$l_1' = l_2 + \left(\frac{t_{cl}}{n_{cl}}\right) \quad \text{(step-back)}$$

$L_1' = \dfrac{1}{+0.01934} \quad \leftarrow \quad l_1' = +0.01907 + 2.6846^{-4}$
$= +51.7154\,\text{D} \qquad\qquad\quad = 0.01934\,\text{m}$

As $L_1 = 0.0000$, $L_1 = F_1$

$F_1 = +51.7154\,\text{D}$

The FOZR can be found using:

$$r_1 = \frac{n' - n}{F_1}$$

$$r_1 = \frac{1.490 - 1.000}{+51.7154} = +9.4749^{-3}\,\text{m}$$

The FOZR of the contact lens along the 180 meridian is therefore +9.4749 mm.

We now have to do the same again for the 90 meridian!

Along 90

Vergences (D)	Distances (m)

$L_4' = -5.5659\,\text{D}$

$L_4 = L_4' - F_4$

$L_4 = -5.5659 - (-42.2641) = +36.6983\,\text{D}$

$L_4 = +36.6983\,\text{D} \quad \rightarrow \quad l_4 = \dfrac{1}{+36.6983} = +0.02725\,\text{m}$

$$l_3' = l_4 + \left(\frac{t_{tears}}{n_{tears}}\right) \quad \text{(Step-back)}$$

$$L_3' = \frac{1}{+0.02732} \quad \leftarrow \quad l_3' = +0.02725 + 7.4850^{-5}$$
$$= +0.02732\,\text{m}$$
$$= +36.5978\,\text{D}$$
$$L_3 = L_3' - F_3$$
$$L_3 = +36.5978 - (+42.0000) = -5.4022\,\text{D}$$

As a result of the thin air gap, $L_3 = L_2'$. We must now perform the second step-back ray-trace through the contact lens. The reader is again advised to follow all of the vergences carefully on Figure 22.3.

$$L_3 = L_2' = -5.4022\,\text{D}$$
$$L_2' = -5.4022\,\text{D}$$
$$L_2 = L_2' - F_2$$
$$L_2 = -5.4022 - (-61.2500) = +55.8478\,\text{D}$$
$$L_2 = +55.8478\,\text{D} \quad \rightarrow \quad l_2 = \frac{1}{+55.8478} = +0.01791\,\text{m}$$
$$l_1' = l_2 + \left(\frac{t_{cl}}{n_{cl}}\right) \quad \text{(Step-back)}$$
$$L_1' = \frac{1}{+0.01817} \quad \leftarrow \quad l_1' = +0.01791 + 2.6846^{-4}$$
$$= +0.01817\,\text{m}$$
$$= +55.0228\,\text{D}$$

As $L_1 = 0.0000$, $L_1 = F_1$

$$F_1 = +55.0228\,\text{D}$$

The FOZR can be found using:

$$r_1 = \frac{n' - n}{F_1}$$
$$r_1 = \frac{1.490 - 1.000}{+55.0228} = +8.9054^{-3}\,\text{m}$$

The FOZR of the contact lens along the 90 meridian is therefore +8.9054 mm

The FOZR of the contact lens is **+9.4749 mm along 180** and **+8.9054 mm along 90**. The RGP contact lens has a toroidal front surface and the lens itself is a fully compensated parallel bitoric lens. The purpose of the toroidal front surface is to improve vision by the neutralisation of both residual (lenticular) and induced astigmatism. Remember that whenever a back surface

toric RGP lens is fitted to an eye, a tear lens is formed with both a posterior and an anterior toroidal surface. This bitoric tear lens mimics the toroidal posterior surface of the contact lens and the toroidal anterior corneal surface. The formation of this bitoric tear lens results in an over-correction on the corneal astigmatism, which is called **induced astigmatism** and is present at the contact lens–tear lens interface. Induced astigmatism is neutralised by the use of a toroidal front surface. It is totally independent of the corneal astigmatism.

Relatively speaking this is a very long question. As mentioned at the end of the previous examples, it is important to organise the initial data and take a logical approach with regard to the ray-tracing.

Toric periphery lenses

Toric periphery lenses are manufactured with a spherical back optic zone to which a toroidal periphery is added. The lenses are easy to identify by their oval back optic zone diameters. The concept is that the steeper periphery aligns the steeper corneal meridian and the peripheral fit of the lens is improved. It is doubtful whether there is sufficient peripheral area to maintain correct alignment on the cornea and if the lens rotates by 90 degrees, the edge clearance on the flattest meridian is seriously compromised. The difference in the peripheral curves ought to be at least 0.6 mm to help stabilise the lens. Doubt has also been expressed about the consistency of manufacture of these lenses and whether it is actually possible to manufacture them as specified. With modern lathes and the ease of designing a full back surface toric lens, which is likely to give superior performance on the eye, the days of toric periphery lenses must be numbered.

The advantages of toric periphery lenses are:

- There is no optical complication arising from the use of a toric optic zone
- There may be a cost saving over full toric lenses

The disadvantages of toric periphery lenses are:

239

- They are difficult to manufacture and the ordered curves may have to be altered by the laboratory
- They are difficult to check
- The lens may not 'lock on' and stabilise easily
- They are useful only for modest degrees of corneal astigmatism.

Concluding points for Chapter 22

- Spherical and aspheric RGP lenses cannot always provide both optimal fitting and total visual correction.
- A toric back surface lens is often required to improve the physical fitting.
- A toric front surface may be required to improve the visual acuity where residual astigmatism is present.
- Toric RGP lenses are thicker than standard lenses and therefore a higher Dk material should be used.
- Toric RGP lenses are less amenable to modification after manufacture.

As far as the calculations are concerned, questions involving astigmatism present few new challenges. However, things usually have to be done twice, e.g. ray-tracing. Astigmatism questions are also relatively long. It is therefore important to organise the initial data, take a logical approach with regard to the ray-tracing and in examinations, manage the time available.

The reader should be aware of the existence of the numerous and well-designed spreadsheets and other programmes which remove the tedium of performing manual calculations in practice. Such programmes are both accurate and design specific and can be downloaded from (among others) the British Contact Lens Association website (www.bcla.org.uk).

Further recommended reading

Douthwaite W A (2006) *Contact Lens Optics and Lens Design*. Elsevier, Oxford

Edwards K (1999) Toric rigid contact lens problem solving. *Optician* **217**:5704

Efron N (2002) *Contact Lens Practice*. Butterworth-Heinemann, Oxford

Meyler J, Morgan P (1993) Toric rigid lens fitting. *Optician* **213**:5604

Meyler J, Ruston D (1995) Toric RGP contact lenses made easy. *Optician* **209**:5504

Phillips A J, Speedwell L (2006) *Contact Lenses*. Elsevier, Oxford

Rabbetts R B (1998) *Bennett & Rabbetts' Clinical Visual Optics*. Butterworth-Heinemann, Oxford

The Correction of Astigmatism with Soft and Silicone Hydrogel Contact Lenses

Caroline Christie and Andrew Keirl

CHAPTER

23

Introduction

When a soft or silicone hydrogel lens is fitted to an eye, the flexible nature of the lens material means that the geometry of the back surface of the lens conforms to that of the anterior corneal surface, i.e. the lens 'drapes' to fit the cornea and any corneal astigmatism is transmitted to the lens. Small amounts of astigmatism are often ignored when fitting a patient with such lenses. This is particularly true if an aspheric lens is used, with some manufacturers claiming that this type of lens can mask up to 1.50 D of astigmatism. However, if an uncorrected cylinder significantly affects a patient's visual acuity when wearing a soft or silicone hydrogel contact lens, a toric design should be used. In recent years toric contact lenses have become an increasingly viable and popular fitting option. This trend is mainly a result of advances in manufacturing technology,

the advent of CNC (computer numerically controlled) lathes and moulded toric lenses. Improvements in lens design, and also the increased availability of frequent replacement and disposable lenses, also make these lenses a more attractive option. Toric lenses are now available in thinner designs, higher-water-content materials, daily disposable modality and most recently, silicone hydrogel materials, all leading to an improved on-the-eye physiological performance. Some practitioners however, still view the fitting of a toric contact lens as a complex, time-consuming and expensive procedure. Recent studies show that only 40% of practitioners prescribe toric soft lenses if the astigmatism is less than 1.00 D preferring to prescribe spherical lenses. One must ask would the same practitioners prescribe spectacle lenses in this rather reckless way.

Chapter content

- Toric soft and silicone hydrogel lenses
- Soft and silicone hydrogel toric lens fitting
- Compensation for lens rotation
- Calculations involving soft toric contact lenses

Toric soft and silicone hydrogel lenses

These lenses are similar to astigmatic spectacle lenses in that the practitioner specifies a sphere, cylinder and axis. The form of a toric contact lens is also similar to that of a toric spectacle

lens because it has one toroidal surface and, to some degree, one spherical surface. However, as a contact lens rotates when placed on an eye, a toric soft lens must be stabilised in some way to provide the correct cylinder axis. A number of methods have been used historically to provide stability of rotation and these can broadly be divided into two camps:

(1) prism ballast
(2) dynamic stabilisation

Prism ballast

Prism ballast (**Figures 23.1 and 23.2**) was the first and although considerably adapted over the years, is still the most common means of

Figure 23.1 Cross-section through a prism-ballast lens showing thickness differential.

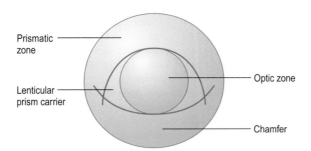

Figure 23.2 A typical modern prism-ballast soft lens. Both the prism and the comfort chamfer appear outside the central optic zone.

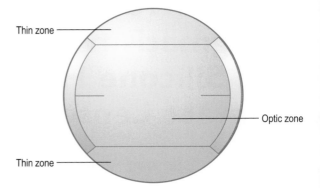

Figure 23.3 Example of a dynamically stabilised toric soft lens.

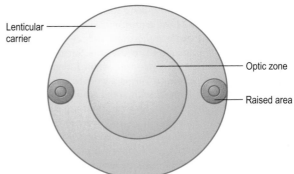

Figure 23.4 Orientation cams.

stabilising a lens on the eye. In principle the lens is produced with an increased thicker profile towards its base; the thinner portion of the lens is located under the upper lid, which then squeezes the thicker portion of the lens towards the lower lid (the so-called watermelon seed principle). Gravity has been shown not to play a part in the axis location. The increased thickness of the lens means that less oxygen is transmitted through the material and can also lead to decreased comfort. To overcome this, manufacturers now remove as much prism as possible from the lens through the use of comfort chamfers. Truncating (the technique of slicing off the bottom of the lens to form a 'shelf' on which it rests, and therefore aligns with the lower lid) is generally no longer carried out with the exception of custom-designed lenses for high levels of oblique astigmatism. Examples of lenses that incorporate prism ballast are the Coopervision Proclear Toric and the Bausch & Lomb Pure Vision Toric.

Dynamic stabilisation

Dynamic stabilisation also relies on the interaction between the lids and the lens. This is achieved by either designing a thin zone that is superior and inferior to the optic zone (**Figure 23.3**) or by placing two raised areas with orientation cams at 3 and 9 o'clock (**Figure 23.4**). In theory the lids again squeeze the thickness differential across the lens, thus maintaining its stability. The advantage of this type of design is that the overall thickness profile can be kept to a minimum, optimising physiological response and patient comfort. Although both types of lens design can be produced with toroidal back surfaces, clinically it is the lens profile that maintains the stability and not the toroidal surface. As mentioned above, truncation is rarely used because it does not lend itself to the mass production moulding methods required for disposable lenses. At the same time, other methods of stabilisation have improved. Although both types of lens design can be produced with toroidal back surfaces, clinically it is the lens profile that maintains the stability and not the toric surface. An example of a lens that uses dual thin zones to achieve stabilisation is the Ciba Torisoft. An example of a lens that utilises orientation cams is the Coopervision Lunelle Toric.

Designs available

Although examples of lenses currently available in the UK stabilised by the main methods illustrated above have been mentioned, it is virtually impossible in an ever-changing market-place, with new products coming on line and older ones being withdrawn, to name every product using a particular stabilisation technique. The authors would advise that for a comprehensive listing of all soft and silicone hydrogel toric lenses, including their design characteristics, material properties, replacement schedule and parameter availability, the reader refer to the most recent version of the *ACLM Contact Lens Year Book* (Kerr and Ruston 2007) or the ACLM website (www.aclm.org.uk).

Soft and silicone hydrogel toric lens fitting

There are three basic ways of fitting a toric lens.

Empirical fitting

Lenses are ordered directly from the manufacturer/laboratory and are based on accurate refraction, including vertex distance, keratometer readings and horizontal visible iris diameter (HVID). No contact lens is inserted on the eye during the initial visit.

Trial lens fitting using a diagnostic lens bank

The most appropriate trial toric lens (only a limited number of lenses are available) is inserted. After settling, the lens is assessed for overall fit, stability and rotational position. An over-refraction (sphere only) is performed. The order for lenses is based on this trial lens assessment. When the lenses arrive from the manufacturer a further reassessment of fit, stability and vision is performed.

Fitting from inventory bank

Clinical data is collected (refraction details including vertex distance, keratometer readings and HVID) and the most appropriate trial lens inserted. Often only a single base curve and diameter are available. After settling, the lens is assessed for overall fit, vision, stability and rotational position. A spherical over-refraction is performed. If all is satisfactory, lenses are issued direct from the inventory bank and, after successful insertion, removal and aftercare instruction, the patient leaves the practice with the toric lenses.

Ten simple fitting steps

1. An *accurate* and *recent* spectacle refraction is required.
2. The spectacle refraction should be adjusted to allow for the vertex distance in *all* meridians.
3. Insert a diagnostic lens from inventory bank that is nearest to refraction.
4. When ordering single diagnostic toric lenses, if the spectacle prescription falls between cylinder powers and/or cylinder axes order:
   (a) lower cylinder power
   (b) if only 10 degrees steps available, adopt the generally held assumption that the lens rotates 5 degrees nasally or order both available axis.
5. Allow about 5–10 minutes settling (some modern lenses may stabilise even faster).
6. Assess for lens fit and orientation:
   (a) full corneal coverage, adequate movement in primary position and on push-up
   (b) note orientation of lens on eye with respect to markings.
7. Where the lens both fits and orients to the desired position, check the visual acuity and issue the lenses for trial.

243

8. Where the lens fits well but orients off-axis in a stable position, apply the left add, right subtract (LARS) or clockwise add, anti-clockwise subtract (CAAS) rules, taking the misorientation into account before inserting the re-calculated lens from the inventory bank or ordering an alternative lens for trial (see 'Mislocation rules' below).
9. Do not over-refract at this stage!
10. If this lens does not work for whatever reason, consider using an alternative toric design. The probability of success falls significantly after the third attempt, so it is important to know when to give up!

Mislocation rules

Once the fit is considered acceptable, lens orientation should be assessed using the markings shown on the lens. There are various markings (Table 23.1) to enable the practitioner to measure the orientation of the lens after settling on the eye. All soft and silicone hydrogel toric contact lenses are marked at their thickest points. This is the base for prism-ballasted types and the horizontal peripheries for dynamically stabilised lenses.

There are several ways to determine the orientation of the lens:

- By rotation of the slitlamp beam, to align with the lens markings and the lens centre.

The angle of mislocation is then read off the slitlamp scale.
- By estimation of the rotation of the lens against the appropriately oriented slitlamp beam, which is surprisingly accurate. It is necessary to know the angular subtense of the various lens markings for this type of estimation procedure.

When physically mislocated, a well-fitting lens exhibits a quick return to axis, a tight-fitting lens shows apparent stability, but a slow return to axis, and a flat or loose-fitting lens show both instability and poor orientation. If the lens is more than 20–30° off the expected axis, this suggests that there is inadequate stabilisation for the lid and corneal features of the patient, and an alternative design should be considered. Provided that the lens is stable and re-orients to the expected position when moved off-axis, the lens can be ordered, making any allowances for misorientation when calculating the cylinder axis. The LARS or CAAS rules should be applied as follows:

- If the lens orients with the markings showing *no rotation*, the cylinder axis is ordered as per the spectacle prescription.
- If the lens rotates left (clockwise), the amount of rotation is *added* to the spectacle cylinder axis.

Table 23.1 Orientation markings for disposable soft and silicone hydrogel toric contact lenses

Name	Manufacturer	Axis markings	separation (° apart)
SofLens Toric and Purevision Toric	Bausch & Lomb	3 markings at 5, 6 and 7 o'clock	30
AirOptix Toric	Ciba Vision	3, 6 and 9 o'clock	
Focus Dailies Toric	Ciba Vision	None	
Frequency 55 Toric	CooperVision	6 o'clock	
Frequency XCEL and XR Toric	CooperVision	6 o'clock	
Proclear Toric	CooperVision	5, 6, 7 o'clock	15
ACUVUE Advance for Astigmatism	J & J Vision Care	Single marking at 6 and 12 o'clock	
ACUVUE 1-day for Astigmatism	J & J Vision Care	Single marking at 6 and 12 o'clock	

- If the lens rotates right (anti-clockwise), the amount of rotation is *subtracted* from the spectacle cylinder axis.
- Up to about 20° of rotation is acceptable, provided that it is stable.

Examples – see Figure 23.5

Diagnostic lens −2.00/−1.25 × 180 which rotates 15 degrees anti-clockwise:

Anti-clockwise rotation, so subtract from cylinder axis
Order (or re-select) −2.00/−1.25 × 165

Diagnostic lens −2.00/−1.25 × 180 which rotates 10 degrees clockwise

Clockwise rotation so add to cylinder axis
Order (or re-select) −2.00/−1.25 × 10

Calculations involving soft toric contact lenses

As previously mentioned, when fitting a soft toric lens, it is generally advisable to avoid an over-refraction. If an over-refraction is required, then spheres only should be used whenever possible. Occasionally, a sph–cyl over-refraction is unavoidable and is therefore necessary. In this situation it is possible that the axis of the over-refraction cylinder does not correspond with the axis of the soft toric contact lens on the eye at the time of the over-refraction. The axes of the over-refraction and the trial contact lens are therefore **oblique** and Stoke's construction (or appropriate formulae) must be used to add these two oblique axes together, so that an alternative soft toric contact lens can be ordered. It is probably sensible to review the procedure for the summation of two obliquely crossed cylinders using Stoke's construction.

All introductory courses in ophthalmic lenses involve the study of lens forms and transposition. A common exercise within this topic is the addition of thin lenses in contact, e.g.

The following thin lenses are placed in close contact:

$$+2.00\,DS/+1.50\,DC \times 45$$

$$-1.00\,DS/+0.50\,DC \times 45$$

What single thin lens replaces the above combination?

The answer is **+1.00 DS/+2.00 DC × 45**. As the cylinder axes (and therefore the principal meridians) are the same, the values for the spheres and cylinders can simply be added together. We now have to look at the situation where we are required to combine cylinders, the axes (and therefore the principal meridians) of which are not the same. As an example of this problem, we combine the thin lenses:

$$+4.00\,DS/+2.00\,DC \times 40$$

with:

$$+2.00\,DS/+3.00\,DC \times 70$$

As we are assuming that the lenses are thin, it is a straightforward process to combine the spheres from the above lenses:

$$+4.00\,DS + (+2.00\,DC) = +6.00\,DS$$

We are therefore left with the problem of adding together the two cylinders:

$$+2.00\,DS \times 40 + (+3.00\,DC \times 70)$$

As the angle between these cylinders is not a right angle (90 degrees), the above cylinders are

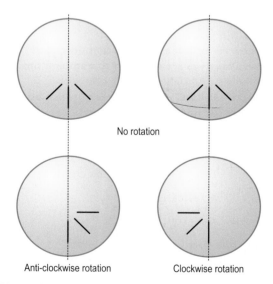

No rotation

Anti-clockwise rotation   Clockwise rotation

Figure 23.5

referred to as **obliquely crossed cylinders**. There are several graphic and mathematical solutions to the problem of obliquely crossed cylinders, but the best known and easiest to use is often referred to as Stoke's construction, which is similar to the parallelogram of forces, except that the angles are doubled.

Stoke's construction

As with *all* graphic constructions or scale drawings, it is advisable to observe the following guidelines:

- Always use graph paper
- Use a convenient but suitably large scale
- Always use the appropriate drawing instruments (protractor, set square, etc.)
- Use clear labels to identify all components of the construction
- Label the construction with the scale used

The construction is simplified if both cylinders are positive. All angles are measured anticlockwise and are taken to be positive.

The symbols used are as follows:

- F_1 and F_2 are the two cylinder powers
- α is the angle between the two cylinder axes
- C is the resultant cylinder
- θ is half the angle between the axis of F_1 and the resultant cylinder
- S is the resultant sphere

So what does Stoke's construction actually do? Very simply, it is a method of adding the powers of two plano-cylinders graphically. The end-result of Stoke's construction is that the plano-cylinders are expressed as a sph–cyl equivalent, i.e. the outcome is a **sphere**, a **cylinder** and an **axis**. This outcome can be demonstrated by placing two plano-cylindrical trial case lenses with their axes crossed at an oblique angle in a focimeter. The user can record two principal powers at 90 degrees to each other. These of course can be written in sph–cyl form. The procedure for Stoke's construction is as follows:

1. If necessary, transpose the given prescriptions (or lens/surface powers) so that they are both in positive cylinder form. At this point, we can ignore any

spheres arising as a consequence of this because they are not considered until the final step.
2. Select the cylinder with its axis numerically nearer to 0 as F_1. If one of the cylinder's axes happens to be **180**, this *must* be regarded as **0**.
3. Using a convenient scale, construct F_1 along its prescribed axis direction.
4. Construct F_2 to scale at the angle 2α from F_1 i.e. twice the difference between the two axes:

$$2\alpha = 2(\text{axis } F_2 - \text{axis } F_1)$$

5. Complete the triangle by joining F_2 to the origin. This gives the magnitude of the new cylinder power C. C is positive because F_1 and F_2 are positive.
6. Measure the angle between F_1 and C. This is 2θ. To express the resultant cylinder's axis in standard notation, θ must be *added* to the axis of F_1. Alternatively, bisect and measure the angle between F_1 and C. The angle between the bisector and the horizontal axis of the origin is the resultant cylinder's axis expressed in standard notation.
7. Find the sphere from:

$$S = \frac{(F_1 + F_2 - C)}{2}$$

8. Add to S any sphere resulting from Step 1.

Example 23.1
We now re-examine the situation described above where we are required to combine the thin lenses +4.00 DS/+2.00 DC × 40 with +2.00 DS/+3.00 DC × 70, the axes of which are not the same (or not 90 degrees apart).

Using Stoke's construction described above, we obtain the following:

1. State both lenses in the plus cylinder form:

   +4.00 DS/+2.00 DC × 40

   +2.00 DS/+3.00 DC × 70

   At this point ignore the spheres.
2. Select the cylinder of the axis numerically nearer to 0 as F_1. In this case, 40 is nearer to 0 than 70 so:

$F_1 = +2.00\,DC \times 40$

Scale: 1 cm = 1 D

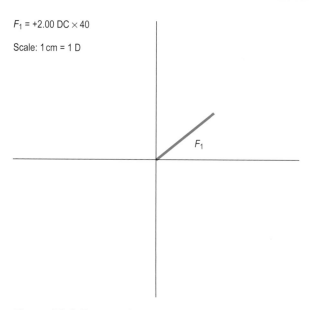

Figure 23.6 Diagram for Example 23.1, Step 3.

$F_1 = +2.00\,DC \times 40$
$F_2 = +3.00\,DC$
$2\alpha = 60$

By measurement:
$C = +4.35\,DC$

Scale: 1 cm = 1 D

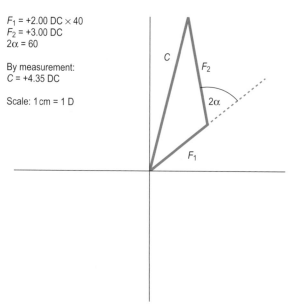

Figure 23.7 Diagram for Example 23.1, Steps 4 and 5.

$F_1 = +2.00\,DC \times 40$

$F_2 = +3.00\,DC \times 70$

3. Construct F_1 to scale along its given axis direction (refer to **Figure 23.6**). A sensible scale should be used, e.g. $1\,cm \equiv 1\,D$.
4. Calculate the angle 2α using:

$$2\alpha = 2(\text{axis } F_2 - \text{axis } F_1)$$

$$2\alpha = 2(70 - 40) = 60$$

Construct F_2 to scale along 60 (using standard notation and measuring anticlockwise) from the end of F_1 (refer to **Figure 23.7**).
5. Complete the triangle by joining F_2 to the origin. This represents the magnitude of C, the resultant cylinder, and is obtained by measurement. It is necessary to allow for any scale used (refer to **Figure 23.7**):

By measurement, $C = +4.35\,D$

6. Measure the angle between F_1 and C. This is 2θ. To express the resultant cylinder's axis in standard notation, θ must be added to the axis of F_1. By measurement, $2\theta = 37$ degrees and $\theta = 18.5$ degrees. The axis of C

is therefore $\theta + 40 = 58.5$ degrees. Alternatively, bisect the angle between F_1 and C. The angle between the bisector and the horizontal axis of the origin is the resultant cylinder's axis expressed in standard notation (refer to **Figure 23.8**).

By measurement, the axis of C = **58.5** degrees

7. Find the sphere from:

$$S = \frac{(F_1 + F_2 - C)}{2}$$

$$S = \frac{+2.00 + (+3.00) - (+4.35)}{2} = +0.325\,D$$

8. Refer to Step 1 and add **S** above to any sphere from step 1:

total sphere =

$$+0.325 + (+4.00) + (+2.00) = +6.325\,DS$$

the resultant sph–cyl is:

$$+6.325\,DS/+4.35\,DC \times 58.5$$

Bitoric lenses

Bitoric lenses are lenses that incorporate *two* toroidal surfaces. Both the convex *and* concave sides of the lens are toroidal in form. The **247**

The angle between the horizontal and the bisector gives the axis of the resultant cylinder

By measurement:
cylinder axis = 58.5

Scale: 1 cm = 1 D

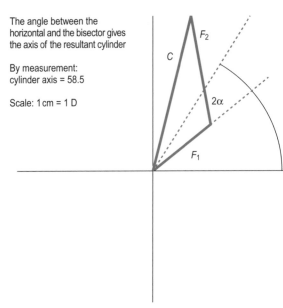

Figure 23.8 Diagram for Example 23.1, Step 6.

principal meridians of these two surfaces may or may not be the same however, the net effect of this lens form is exactly the same as any other astigmatic lens – to produce two principal powers at 90 degrees to each other. We use Example 23.2 to show how Stoke's construction can be applied to this situation.

Example 23.2
A thin bitoric lens of power −4.00 DS/ +2.00 DC × 150 has its convex surface worked with the powers +4.00 DC × 90/+6.00 DC × 180. Find the powers that must be worked on the concave surface to produce the given prescription.

On reading the above question, it is evident that we have been given F (the thin lens power) and F_1 (the front surface power). What we have to find is the back surface power F_2. When using Stoke's construction, we *always* work in sph–cyl form. In sph–cyl form F_1 becomes +4.00 DS/+2.00 DC × 180. As the lens is thin, we can use the expression $F = F_1 + F_2$ which rearranges to become $F_2 = F - F_1$. Substituting values for F and F_1 gives:

$$F_2 = -4.00\,DS/+2.00\,DC \times 150 + (+4.00\,DS/+2.00\,DC \times 180)$$

Now, as Stoke's construction *adds* planocylinders together, we need to apply some simple algebra to the above. If we change the *minus* sign between the two values to a *plus*, we are then *adding* the powers together. We want to do this because Stoke's construction *adds* cylinders. If we do this, the signs of *all* the terms on the right-hand side have to be changed as well, and we obtain:

$$F_2 = -4.00\,DS/+2.00\,DC \times 150 + (-4.00\,DS/-2.00\,DC \times 180)$$

Note the axis has *not* changed. What we have done is simple algebra *not* transposition!

We can now proceed with a standard Stoke's construction in order to add together:

$$-4.00\,DS/+2.00\,DC \times 150$$

and

$$-4.00\,DS/-2.00\,DC \times 180$$

If you work through this problem you should obtain the following results:

1. State both powers in the plus cylinder form:

$$-4.00\,DS/+2.00\,DC \times 150$$

$$-6.00\,DS/+2.00\,DC \times 90$$

Ignore the spheres.

2. Select the cylinder of the axis numerically nearer to zero as F_1:

$$F_1 = +2.00\,DC \times 90$$

$$F_2 = +2.00\,DC \times 150$$

3. Construct F_1 to scale along its given axis direction:

4. Calculate the angle 2α using:

$$2\alpha = 2(\text{axis}\,F_2 - \text{axis}\,F_1)$$

$$2\alpha = 2(150 - 90) = 120$$

Construct F_2 to scale along 120 (using standard notation and measuring anticlockwise) from the end of F_1.

5. Complete the triangle by joining F_2 to the origin. This represents the magnitude of C, the resultant cylinder. C is obtained by measurement: $C = +2.00\,D$

6. Measure the angle between F_1 and C. This is 2θ. To express the resultant cylinder's axis in standard notation, θ must be added to the axis of F_1.

By measurement, $2\theta = 60$ and $\theta = 30$

the axis of C is therefore $\theta + 90 = 120$

Alternatively, bisect the angle between F_1 and C; the angle between the bisector and the horizontal axis of the origin is the resultant cylinder's axis expressed in standard notation by measurement, the axis of C = **120**.

7. Find the sphere from:

$$S = \frac{(F_1 + F_2 - C)}{2}$$

$$S = \frac{+2.00 + (+2.00) - (+2.00)}{2} = \textbf{+1.00 D}$$

8. Refer to Step 1 and add S to any sphere from Step 1:

total sphere = $+1.00 + (-4.00) + (-6.00) =$ **−9.00 DS**

the value of F_2 in sph–cyl form is:

$-9.00\,\text{DS}/+2.00\,\text{DC} \times 120$

and as a crossed-cylinder:

$-9.00\,\text{DC} \times 30/-7.00\,\text{DC} \times 120$

The above problem illustrates how Stoke's construction can be used to determine one of the two surface powers in the case of a bitoric lens. The construction is no different from the first example. The only variation between the two examples is that when the problem involves the calculation of a surface power, the construction is always preceded by simple algebraic manipulation, so that the two values in the problem can be added together. Remember, this is what Stoke's construction does – it *adds* plano-cylinders together.

Examples 23.1 and 23.2 (and all problems involving obliquely crossed cylinders) can be solved mathematically using the formulae given in the various textbooks for ophthalmic lenses. Students can obviously use whichever method they prefer, although most seem to favour the use of Stoke's construction.

The theory of obliquely crossed cylinders can be applied to two practical dispensing situations, namely the effect of *tilting* a spectacle lens and the effect of dispensing a spectacle lens *off-axis*. The theory of obliquely crossed cylinders is also used:

- in the more advanced automatic subjective refraction units
- in the manufacture of certain bifocal lenses, e.g. cemented bifocals that require a different astigmatic correction for the near portion compared with the distance portion.

However, in the context of this chapter, we now use Stoke's construction to determine the result of an over-refraction of a trial toric contact lens on an eye.

Example 23.3

An eye is fitted with a toric soft contact lens of BVP −5.00 DS/−2.50 DC × 20. An over-refraction is performed that produces the prescription plano/−1.00 DC × 60. What lens should now be ordered?

This problem can be solved by the application of Stoke's construction. The procedure outline in Example 23.1 is used:

1. State both the lens power and over-refraction in the plus cylinder form:

   $-7.50\,\text{DS}/+2.50\,\text{DC} \times 110$

   $-1.00\,\text{DS}/+1.00\,\text{DC} \times 150$

   at this point ignore the spheres.
2. Select the cylinder of the axis numerically nearer to zero as F_1. In this case, 110 is nearer to 0 than 140, so:

   $F_1 = +2.50\,\text{DC} \times 110$

   $F_2 = +1.00\,\text{DC} \times 150$
3. Construct F_1 to scale along its given axis direction (refer to **Figure 23.9**).
4. Calculate the angle 2α using:

   $2\alpha = 2(\text{axis}\,F_2 - \text{axis}\,F_1)$

   $2\alpha = 2(150 - 110) = 80$

   construct F_2 to scale along 80 (using standard notation and measuring anticlockwise) from the end of F_1 (refer to Figure 23.9).

5. Complete the triangle by joining F_2 to the origin. This represents the magnitude of C, the resultant cylinder, and is obtained by measurement. It is necessary to allow for any scale used (refer to Figure 23.9). By measurement, $C = +2.84\,D$

6. The angle between the bisector and the horizontal axis of the origin is the resultant cylinder's axis expressed in standard notation (refer to Figure 23.9). By measurement, the axis of C is 120.

7. Find the sphere from:

$$S = \frac{(F_1 + F_2 - C)}{2}$$

$$S = \frac{+2.50 + (+1.00) - (+2.84)}{2} = +0.33\,D$$

8. Refer to Step 1 and add S above to any sphere from Step 1:

total sphere =
$+0.33 + (-7.50) + (-1.00) = -8.17\,DS$

the resultant sph–cyl is:

$-8.17\,DS/+2.84\,DC \times 120$ or
$-5.33\,DS/-2.84 \times 30$

The likely soft toric lens to be ordered is:

$-5.25\,DS/-2.75\,DC \times 30$

Example 23.4

A trial soft toric contact lens of power +4.00 DS/ −2.00 DC × 100 is fitted to a right eye and, on settling, is found to rotate anti-clockwise by 10°. An over-refraction is performed and best visual acuity is obtained with the trial lens combination +0.75 DS/−1.50 DC × 80. What power would be ordered?

Again, this is a straightforward Stoke's construction but with a slight twist at the start. We are told that, after settling, the lens rotates anti-clockwise by 10 degrees. The on-eye cylinder axis is therefore oriented at 110 degrees and not 100 degrees. The power of the soft toric contact lens is effectively **+4.00 DS/−2.00 DC × 110**. This rotation needs to be considered at the start of the construction and at the very end when the final power is recorded. Stoke's construction is as follows and is illustrated in **Figure 23.10**.

1. State both the lens power and over-refraction in the plus cylinder form:

$+2.00\,DS/+2.00\,DC \times 20$

$-7.50\,DS/+1.50\,DC \times 170$

at this point ignore the spheres.

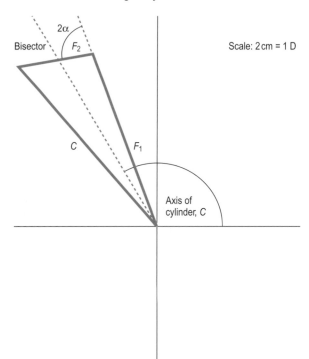

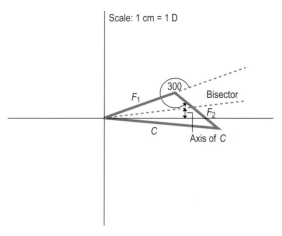

Figure 23.9 Diagram for example 23.3.

Figure 23.10 Diagram for example 23.4.

2. Select the cylinder of the axis which is numerically nearer to zero to be F_1. In this case, 20 is nearer to zero than 170 so:

$F_1 = +2.00 \, \text{DC} \times 20$

$F_2 = +1.50 \, \text{DC} \times 170$

3. Construct F_1 to scale along its given axis direction.
4. Calculate the angle 2α using:

$2\alpha = 2(\text{axis } F_2 - \text{axis } F_1)$

$2\alpha = 2(170 - 20) = 300$

Construct F_2 to scale along 300 (using standard notation and measuring anti-clockwise) from the end of F_1. Care needs to be taken when plotting large values for 2α on the diagram.

5. Complete the triangle by joining F_2 to the origin. This represents the magnitude of C, the resultant cylinder, and is obtained by measurement. It is necessary to allow for any scale used.

By measurement, $C = +3.04 \, \text{D}$

6. The angle between the bisector and the horizontal axis of the origin is the resultant cylinder's axis expressed in standard notation. By measurement the axis of C is 7°.

By measurement, the axis of $C = 7$ degrees

7. Find the sphere from:

$S = \dfrac{(F_1 + F_2 - C)}{2}$

$S = \dfrac{+2.00 + (+1.50) - (+3.04)}{2} = +0.23 \, \text{D}$

8. Refer to Step 1 and add S above to any sphere from Step 1:

total sphere $= +0.23 + (+2.00) + (-0.75) = +1.48 \, \text{DS}$

the resultant sph–cyl is:

$+1.48 \, \text{DS}/+3.04 \, \text{DC} \times 7$ or
$+4.52 \, \text{DS}/-3.04 \times 97$

Now allowing for the fact that, when the lens is placed on the eye, it rotated anti-clockwise by 10 degrees, the likely soft toric lens to be ordered is:

$+4.50 \, \text{DS}/-3.00 \, \text{DC} \times 87$

Example 23.5

A right eye is fitted with a toric soft contact lens of BVP $-2.00 \, \text{DS}/-1.50 \, \text{DC} \times 180$. On settling, the lens is found to have rotated 15 degrees clockwise. The spectacle prescription for this patient is $-5.00 \, \text{DS}/-1.50 \, \text{DC} \times 60$ at $15 \, \text{mm}$. What over-correction would be found at the same vertex distance?

This example is rather more complicated because it involves a spectacle refraction (F_{sp}) to ocular refraction (K) conversion, Stoke's construction and a step-back ray-trace.

First of all, the F_{sp} to K conversion: this has to be performed for each meridian in turn.

Along 60

$$K = \frac{F_{sp}}{1 - dF_{sp}} = \frac{-5.00}{1 - (0.015 \times -5.00)} = -4.6512 \, \text{D}$$

Along 150

$$K = \frac{F_{sp}}{1 - dF_{sp}} = \frac{-6.50}{1 - (0.015 \times -6.50)} = -5.9225 \, \text{D}$$

In sph–cyl form the ocular refraction is $-4.6512 \, \text{DS}/-1.2713 \, \text{DC} \times 60$.

After rotation, the cylinder axis of the soft toric contact lens on the eye lies along 165. The power of the soft toric contact lens on the eye is therefore $-2.00 \, \text{DS}/-1.50 \, \text{DC} \times 165$. The on-eye over-refraction (O_{Rx}) is simply the difference between the ocular refraction and the power of the soft toric contact lens and can be found using the Stoke's construction method described in Example 23.2.

$K = -4.6512 \, \text{DS}/-1.2713 \, \text{DC} \times 60$

$F_{cl} = -2.00 \, \text{DS}/-5.00 \, \text{DC} \times 165$

$O_{Rx} = K - F_{cl}$

$O_{Rx} = -4.6512 \, \text{DS}/-1.2713 \, \text{DC} \times 60 - (-2.00 \, \text{DS}/-1.50 \, \text{DC} \times 165)$

Now, as Stoke's construction *adds* plano-cylinders together, we need to apply some simple algebra to the above. If we change the *minus* sign between the two values to a *plus* we are then *adding* the powers together. We want to do this because Stoke's construction *adds* cylinders. If we do this, the signs of *all* the terms on the right-hand side have to be changed as well, and we obtain:

$$O_{Rx} = -4.6512\,DS/-1.2713\,DC \times 60 + (+2.00\,DS/+1.50\,DC \times 165)$$

Note that the axis has *not* changed. What we have done is simple algebra *not* transposition! We can now proceed with a *standard* Stoke's construction in order to add together the following:

$-4.6512\,DS/-2.2713\,DC \times 60$

$+2.00\,DS/+1.50\,DC \times 165$

The usual Stoke's procedure should now be followed. If you work through this problem, the value of over-refraction *on the eye* should be:

$-1.20\,DS/-2.68\,DC \times 68$

This now has to be converted to an over-refraction at 15 mm. As always, each meridian has to be taken in turn. The over-refraction at 15 mm is:

$-1.18\,DS/-2.48 \times 68$

Concluding points for Chapter 23

- Improvements in toric lens performance mean that practitioners should be confident in fitting theses lenses when poor vision is caused by astigmatism.

- Fitting is less reliable with oblique cylinders.
- Monocular patients are more likely to be disturbed by any instability of vision caused by temporary lens rotation on blinking.
- Assessing rotational stability is the key to success.
- Practitioners require a variety of lens designs from which to chose.
- Practitioners should allow for more than one lens per eye, in terms of both time and financial consideration, but attempt only a maximum of three lenses per eye!

Once again the mathematical material in this chapter relies on an understanding of basic optical principles and the ability to perform step-along and step-back ray-traces. Some questions involving soft toric lenses are complicated and there can be opportunities for error. A logical and organised approach is required. In practice, try to avoid over-refracting soft toric lenses, especially with cylinders!

References

Kerr C, Ruston D (2007) *The ACLM Contact Lens Year Book 2007.* Association of Contact Lens Manufacturers, London

Further recommended reading

Douthwaite W A (2006) *Contact Lens Optics and Lens Design.* Elsevier, Oxford
Efron N (2002) *Contact Lens Practice.* Butterworth-Heinemann, Oxford
Phillips A J, Speedwell L (2006) *Contact Lenses.* Elsevier, Oxford

Ray-tracing Through Soft Contact Lenses and Soft Contact Lens Hydration Factors

Andrew Keirl

Introduction

When a soft lens is fitted to an eye, the flexible nature of the lens material means that the geometry of the back surface of a soft lens conforms to that of the anterior corneal surface. The soft lens drapes to fit the cornea and the shape of the cornea is transmitted to the contact lens. This is particularly true for traditional hydrogel materials. Thus, any tear lens produced by a soft contact lens is small (negligible), unpredictable and often afocal. However, unlike a rigid gas permeable (RGP) lens, this cannot be reliably predicted and is generally assumed to remain unchanged at zero when changes to the back optic zone radius (BOZR) of a soft lens are made. Thus, the required back vertex power (BVP) of a soft contact lens measured in air is assumed to be equal to the patient's ocular refraction. This applies to both spherical and astigmatic soft contact lenses. If the front and back surface radii are known, along with the centre thickness and refractive index, simple two-surface ray-tracing techniques can be applied to soft lenses.

Chapter content

- The back vertex power of soft contact lenses
- Soft lens manufacture
- Calculations involving hydrated and dehydrated soft contact lenses

The back vertex power of soft contact lenses

Example 24.1

Find the BVP of the following hydrogel (soft) contact lens in air:

Front optic zone radius (FOZR or r_1) = 9.8489 mm

BOZR (r_2) = 8.60 mm

$n = 1.42$

$t_c = 0.15$ mm

All that is required here is a simple step-along ray-trace. The contact lens of course is considered to be a thick lens so the equivalent air distance (EAD) will be used when ray-tracing. The BVP can be calculated using a standard step-along ray-trace.

The first task is to calculate the two surface powers. The radius of curvature r is substituted in metres and **standard form** used as appropriate. To find surface powers use:

$$F_1 = \frac{n' - n}{r_1} \quad \text{and} \quad F_2 = \frac{n - n'}{r_2}$$

Sign convention tells us that both r_1 and r_2 are positive:

$$F_1 = \frac{1.42 - 1.00}{+9.8489^{-3}} = +42.6443\,\text{D}$$

$$F_2 = \frac{1.0000 - 1.4200}{+8.6^{-3}} = -48.8372\,\text{D}$$

The EAD (with t_{cl} substituted in metres) is given by:

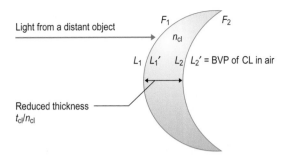

Figure 24.1 Diagram for Example 24.1.

$$EAD = \frac{t_{cl}}{n_{cl}} = \frac{1.5^{-4}}{1.42} = 1.0563^{-4}\,m$$

Now the EAD has been determined, the refractive index between the two surfaces can be assumed to be **equal to that of air** ($n = 1$). This value is used in the following ray-trace. A step-along ray-trace is used to find the BVP. As always, the computation takes place in two columns, one for vergences and the other for distances. The vergences for this problem are illustrated in **Figure 24.1**.

Vergences (D)   Distances (m)

$L_1 = 0.0000$

$F_1 = +42.6442\,D$

$L_1' = L_1 + F_1 = 42.6442\,D \rightarrow l_1' = \dfrac{n}{L_1'}$

$$l_1' = \frac{1}{+42.6442} = +0.02345\,m$$

$$l_2 = l_1' - \left(\frac{t_{cl}}{n_{cl}}\right)$$

$L_2 = \dfrac{n}{l_2} \leftarrow l_2 = +0.02345 - 1.0563^{-4} = +0.02334\,m$

$$L_2 = \frac{1}{+0.02344} = +42.8373\,D$$

$L_2' = L_2 + F_2$

$L_2' = +42.8372 + (-48.8372) = -6.0000\,D$

As $L_1 = 0.0000$, $L_2' = BVP$ and $F_v' = -6.0000\,D$

The BVP of this soft contact lens in air is therefore −6.0000 D.

Soft lens manufacture

Cast moulding, hydrated wet moulding, spin casting and lathe cutting have been the usual methods of soft lens manufacture. Wet moulding can produce both fully and partially hydrated lenses whereas current silicone hydrogel lenses are cast moulded. If a soft lens is manufactured using the techniques of cast moulding, spin casting or lathing, the lens needs to be hydrated after manufacture. The hydration of a dry or dehydrated soft contact lens alters the physical parameters of the finished lens because the refractive index of a soft lens material *reduces* when hydrated. A lens manufacturer must therefore assume a particular 'swell' or 'expansion' factor for a given material and produce a lens that, on hydration, swells to the desired specification:

$$\text{Expansion factor} = \frac{\text{Wet parameter}}{\text{Dry parameter}}$$

and

$$\text{Dry parameter} = \frac{\text{Wet parameter}}{\text{Expansion factor}}$$

For example, the expansion factor for a 58% water content hydrogel lens is 1.40. This means that a hydrogel lens with a 58% water content, a BOZR of 8.60 mm and a total diameter of 14.40 mm would have corresponding dehydrated or dry dimensions of BOZR 6.143 mm and total diameter 10.286 mm. The dimensions of the dry lens are therefore based on the expansion factor of the material being used.

Although lathe cutting was the traditional method of soft lens manufacture, some commercially available soft lenses are now moulded in either a fully or partially hydrated state.

Lathe cutting

This conventional method of producing soft lenses works exactly the same as the process of lathing RGP lenses. The lens material is dry and in button form. First, the lathe cuts the back surface to the desired design followed by the front surface. After hydration and sterilisation, the finished soft lens is ready for use. Lathe cutting is suitable for all soft lens materials and gives total flexibility of lens design and available parameters. It is, however, time-consuming, relatively expensive, and does not lend itself to mass production. Examples are Menicon **Soft 72** and Coopervision **Lunelle ES70**.

Spin casting

Introduced by Bausch & Lomb in 1971, this method used a spinning mould, into which a mixture of monomers was injected. The centripetal force caused the mixture to extend up the wall of the mould to form the required shape. The curvature of the mould determined the front surface of the lens and the back surface was determined by the speed of the spin and the volume of monomers. The formed dehydrated lens had to be edged before being hydrated and released from its mould. This method of manufacture was the earliest example of mass production of soft lenses. However, it employed low water content materials only. Examples were the Bausch & Lomb **U3/U4** series of lenses, which are no longer available.

Conventional cast moulding

With this technique, male and female mould components are formed from polypropylene using stainless steel tools. Liquid monomer is dispensed into the female mould and the male mould is clamped to the female mould under carefully controlled pressure. The monomer is cured with ultraviolet (UV) light, forming the dehydrated lens. The lens then has to be fully hydrated before packaging. Any tiny flaws in the dehydrated lens are greatly magnified when hydrated and, with modern high-water materials, it is harder to control parameter reproducibility and lens quality. Examples are Ciba Vision **Precision UV** and Coopervision **Biomedics 55**.

Award technology

Lenses manufactured using this technique began production in 1993 and the process was acquired by Bausch & Lomb in 1996. It is a novel process because one half of the lens mould forms part of the final packaging, which greatly reduced manufacturing costs. An example is Bausch & Lomb **SofLens One Day**.

Stabilised soft moulding

This technique was developed by Johnson & Johnson in 1988 to allow the production of lenses with higher water content and minimal expansion on hydration. The monomer, diluents, tints and UV inhibitor are injected into the front curve of high-precision polystyrene moulds. The inert diluents that are added to the liquid monomer formulation take up the space that is later occupied by water. Polymerisation by visible light takes place in a curing tunnel. After moulding the diluents are replaced by de-ionised water, which results in minimal expansion of the lens. An example is Johnson & Johnson **ACUVUE 2**.

Stabilised soft moulding (maximised technology)

This process has lead to a fully automated, continuous flow production operation with no human intervention, thus reducing unit costs essential for daily disposable production. An example is the Johnson & Johnson **ACUVUE 1-Day**.

Lightstream technology

This technique was developed by Ciba Vision in 1997 for the production of daily disposable lenses. Reusable moulds are employed which are manufactured from high-quality quartz. The desired lens geometry is incorporated into the mould surface. Fully hydrated moulding is achieved using a water-soluble modified polyvinyl alcohol (PVA) material and no further diluents extraction is required. An example is the Ciba Vision **Focus Dailies**.

Example 24.2
A soft contact lens is specified as follows:
8.70/14.20/+5.00 D.
The centre thickness is 0.2 mm and the lens material has an expansion factor of 20% with

refractive indices of 1.4 (wet) and 1.5 (dry). Find the complete specification of the lens in its dehydrated (dry) state.

For the hydrated (wet) lens we know the following:

- BOZR (r_2)
- Diameter
- Centre thickness
- BVP
- Refractive index

We are missing the FOZR (r_1). This can be easily found using a step-back ray-trace in the usual way. Once we know the FOZR, we can use the stated expansion factor to find all the dehydrated parameters. The BVP of the dehydrated lens can then be found using a step-along ray-trace and the refractive index of the dehydrated lens. In this example, the unknown value is the front surface radius, so we have to use a step-back ray-trace to find F_1 and hence the FOZR. We are given the BOZR (r_2), the centre thickness (t_{cl}), the refractive index of the lens material (n_{cl}) and the BVP (L_2'). Using what we know, we can calculate the back surface power (F_2) and the reduced thickness (t_{cl}/n_{cl}). The back surface power is given by:

$$F_2 = \frac{n - n'}{r_2}$$

Sign convention tells us that r_2 is positive:

$$F_2 = \frac{1.00 - 1.40}{+8.7^{-3}} = -45.9770\,D$$

The equivalent air distance (with t_{cl} substituted in metres) is given by:

$$EAD = \frac{t_{cl}}{n_{cl}} = \frac{2^{-4}}{1.40} = 1.4286^{-4}\,m$$

Now that the EAD has been determined, the refractive index between the two surfaces can be assumed to be **equivalent to that of air** $(n = 1)$. This value is used in the following ray-trace. A step-back ray-trace is used to find the front surface power. As always, the computation takes place in two columns, one for vergences and the other for distances.

Vergences (D)	Distances (m)

$$L_2' = +5.0000\,D$$

$$L_2 = L_2' - F_2$$

$$L_2 = +5.0000 - (-45.9770) = +50.9770\,D$$

$$L_2 = +50.9770\,D \rightarrow l_2 = \frac{1}{+50.9770} = +0.01962\,m$$

$$l_1' = l_2 + \left(\frac{t_{cl}}{n_{cl}}\right)(\text{step-back})$$

$$\leftarrow l_1' = +0.01962 + 1.4286^{-4} = +0.01976\,m$$

$$L_1' = \frac{1}{+0.01976} = +50.6084\,D$$

As $L_1 = 0.0000$, $L_1 = F_1$

$$F_1 = +50.6084\,D$$

The FOZR can be found using:

$$r_1 = \frac{n' - n}{F_1}$$

$$r_1 = \frac{1.4000 - 1.0000}{+50.6084} = +7.9038^{-3}\,m$$

The FOZR of the contact lens is therefore +7.9038 mm.

We now know all the hydrated (wet) parameters:

- FOZR $(r_1) = 7.9038\,mm$
- BOZR $(r_2) = 8.7000\,mm$
- Diameter $= 14.20\,mm$
- Centre thickness $= 0.20\,mm$

To find the dehydrated parameters we simply need to divide all of the above by the expansion factor. The given expansion factor is 20% which is the same as 1.20.

$$\text{Dry parameter} = \frac{\text{Wet parameter}}{\text{Expansion factor}}$$

$$\text{Dry parameter} = \frac{\text{Wet parameter}}{1.20}$$

The details of the two lenses are given in Table 24.1. The only unknown dehydrated value is the back vertex power. This can be found using a step-along ray-trace, the above surface radii and the refractive index of the dehydrated lens $(n = 1.50)$.

Table 24.1 The hydrated and dehydrated details of the lens in Example 24.2

	Hydrated	Dehydrated
FOZR (mm)	+7.9038	+6.5865
BOZR (mm)	+8.7000	+7.2500
Diameter (mm)	14.20	11.83
Centre thickness (mm)	0.20	0.1667

BOZR, back optic zone radius; FOZR, front optic zone radius.

Table 24.2 The full details of the hydrated and dehydrated lenses in Example 24.2

	Hydrated	Dehydrated
FOZR (mm)	+7.9038	+6.5865
BOZR (mm)	+8.7000	+7.2500
Diameter (mm)	14.20	11.83
Centre thickness (mm)	0.20	0.1667
Refractive index	1.40	1.50
Back vertex power (D)	+5.0000	+7.7235

BOZR, back optic zone radius; FOZR, front optic zone radius.

The first task is to calculate the two surface powers of the dehydrated lens. The radius of curvature is substituted in metres and **standard form** used as appropriate. To find surface powers use:

$$F_1 = \frac{n'-n}{r_1} \quad \text{and} \quad F_2 = \frac{n-n'}{r_2}$$

Sign convention tells us that both r_1 and r_2 are positive:

$$F_1 = \frac{1.50-1.00}{+6.5865^{-3}} = +75.9128\,D$$

$$F_2 = \frac{1.00-1.50}{+7.25^{-3}} = -68.9655\,D$$

The EAD (with t_{cl} substituted in metres) is given by:

$$EAD = \frac{t_{cl}}{n_{cl}} = \frac{2.0^{-4}}{1.50} = 1.3333^{-4}\,m$$

Now that the EAD has been determined, the refractive index between the two surfaces can be assumed to be **equal to that of air** ($n = 1$). This value is used in the following ray trace. A step-along ray-trace is used to find the BVP. As always, the computation takes place in two columns, one for vergences and the other for distances. The vergences for this problem are illustrated in Figure 24.1.

Vergences (D)   Distances (m)

$L_1 = 0.0000$

$F_1 = +75.9128\,D$

$L_1' = L_1 + F_1 = 75.9128\,D \rightarrow l_1' = \dfrac{n}{L_1'}$

$$l_1' = \frac{1}{+75.9128} = +0.01317\,m$$

$$l_2 = l_1' - \left(\frac{t_{cl}}{n_{cl}}\right)$$

$$L_2 = \frac{n}{l_2} \leftarrow l_2 = +0.01317 - 1.3333^{-4} = +0.01304\,m$$

$$L_2 = \frac{1}{+0.01304} = +76.6890\,D$$

$$L_2' = L_2 + F_2$$

$$L_2' = +76.6890 + (-68.9655) = +7.7235\,D$$

As $L_1 = 0.0000$, $L_2' = BVP$ and $F_v' = +7.7235\,D$

The BVP of the dehydrated soft contact lens in air is therefore +7.7235 D.

The full details of the hydrated and dehydrated lenses are given in Table 24.2. As the dehydrated lens has a higher refractive index and steeper surface curves, inspection of Table 24.2 provides few surprises. It will be noted that the dehydrated lens is stronger than the hydrated lens. However, on hydration, the back vertex power reduces to its required clinical value.

Example 24.3

A front surface toric hydrogel contact lens is specified as follows:

8.60/14.40/−6.00 DS/−2.00 DC × 180

The centre thickness is 0.10 mm. The lens material has an expansion factor of 1.25 and refractive indices of 1.42 (hydrated) and 1.54

257

(dehydrated). Determine all the dehydrated parameters of the lens.

Although the lens in Example 24.3 is toric, it presents no new challenges. It is in fact the same as Example 24.2 but longer because we have to do everything twice because each meridian of the toric lens is treated as a spherical lens. For the hydrated lens we know:

- BOZR (r_2)
- Diameter
- Centre thickness
- BVP
- Refractive index

Again, we are missing the FOZR (r_1). As the lens is toric there are two FOZRs, one for each principal meridian of the lens. The principal radii of curvature of the front (toroidal) surface can be found using *two* step-back ray-traces along each principal meridian. When we know both FOZRs, we can use the stated expansion factor to find all the dehydrated parameters. The BVPs of the dehydrated lens can then be found using a step-along ray-trace and the refractive index of the dehydrated lens. In this example, the unknown value is the front surface radius. The back surface power is given by:

$$F_2 = \frac{n - n'}{r_2}$$

Sign convention tells us that r_2 is positive:

$$F_2 = \frac{1.00 - 1.42}{+8.6^{-3}} = -48.8372\,D$$

The EAD for the hydrated lens (with t_{cl} substituted in metres) is given by:

$$EAD = \frac{t_{cl}}{n_{cl}} = \frac{1^{-4}}{1.42} = 7.0422^{-5}\,m$$

The BVP of this hydrated toric lens is −6.00 DS/ −2.00 DC × 180. This means that the principal powers are −6.00 D *along* 180 and −8.00 D *along* 90. If step-back ray-tracing is used for each meridian of the lens, the radius of curvature of the front toroidal surface is found to be:

1. r_1 along 180 = +9.8341 mm
2. r_1 along 90 = +10.3143 mm

We now know all the hydrated (wet) parameters:

- FOZR (r_1) = +9.8341 mm along 180
- FOZR (r_1) = +10.3143 mm along 90
- BOZR (r_2) = +8.60 mm (spherical)
- Diameter = 14.40 mm
- Centre thickness = 0.10 mm

To find the dehydrated parameters we simply need to divide all of the above by the expansion factor. The given expansion factor is 1.25:

$$Dry\ parameter = \frac{Wet\ parameter}{Expansion\ factor}$$

$$Dry\ parameter = \frac{Wet\ parameter}{1.25}$$

The details of the two lenses so far are given in Table 24.3. The only unknown dehydrated value is the BVP, which can be found using two step-along ray-traces, one for each principal meridian of the lens, using the above surface radii and the refractive index of the dehydrated lens ($n = 1.54$).

The first task is to calculate the *three* surface powers of the dehydrated lens. There are two powers on the front surface and one on the back. The radius of curvature r is substituted in metres and **standard form** used as appropriate. To find surface powers use:

$$F_1 = \frac{n' - n}{r_1} \quad and \quad F_2 = \frac{n - n'}{r_2}$$

Sign convention tells us that both r_1 and r_2 are positive.

Table 24.3 The hydrated and dehydrated details of the lens in Example 24.3

	Hydrated	Dehydrated
FOZR along 180 (mm)	+9.8341	+7.8673
FOZR along 90 (mm)	+10.3143	+8.2514
BOZR (spherical) (mm)	+8.6000	+6.8800
Diameter (mm)	14.40	11.52
Centre thickness (mm)	0.10	0.08

BOZR, back optic zone radius; FOZR, front optic zone radius.

Along 180

$$F_1 = \frac{1.54 - 1.00}{+7.8673^{-3}} = +68.6385\,\text{D}$$

Along 90

$$F_1 = \frac{1.54 - 1.00}{+8.2514^{-3}} = +65.4434\,\text{D}$$

Back surface (spherical)

$$F_2 = \frac{1.00 - 1.54}{+6.88^{-3}} = -78.4884\,\text{D}$$

The equivalent air distance for the dehydrated lens (with t_{cl} substituted in metres) is given by:

$$\text{EAD} = \frac{t_{cl}}{n_{cl}} = \frac{8^{-5}}{1.54} = 5.1948^{-5}\,\text{m}$$

Two step-along ray-traces can now be performed along each principal meridian of the lens, which give BVPs of:

BVP along 180 = −9.6042 D

BVP along 90 = −12.8217 D

The full details of the hydrated and dehydrated lenses are given in Table 24.4.

Concluding points for Chapter 24

Once again the material in this relatively straightforward chapter relies on an understanding of basic optical principles and the ability to perform step-along and step-back ray traces. Most questions of this type are academically undemanding. However, they are somewhat tedious and long-winded. This is particularly true for problems involving toric lenses, which means that everything has to be done twice!

Table 24.4 The full details of the hydrated and dehydrated lenses in Example 24.3

	Hydrated	Dehydrated
FOZR along 180 (mm)	+9.8341	+7.8673
FOZR along 90 (mm)	+10.3143	+8.2514
BOZR (spherical) (mm)	+8.6000	+6.8800
Diameter (mm)	14.40	11.52
Centre thickness (mm)	0.10	0.08
Refractive index	1.42	1.54
Back vertex power	−6.00 DS/ −2.00 DC × 180	−9.6042 DS/ −3.2175 DC × 180

BOZR, back optic zone radius; FOZR, front optic zone redius.

Further recommended reading

Douthwaite W A (2006) *Contact Lens Optics and Lens Design*. Elsevier, Oxford

Efron N (2002) *Contact Lens Practice*. Butterworth-Heinemann, Oxford

The Correction of Presbyopia with Contact Lenses

Caroline Christie and Andrew Keirl

Introduction

The onset of presbyopia creates a wealth of patients looking for alternatives to spectacles. Reading glasses are seen as an admission of aging and a step back from what and where patients used to be. It is a misconception that bifocal/multifocal contact lenses do not work and are difficult to fit. However, it is challenging and sometimes confusing to manage all the options for the correction of presbyopia with the seemingly vast range of contact lenses available today. The secret of success is to find the right design for the right patient.

The choices available to correct the presbyopic contact lens wearer include:

- Modified distance correction
- Use of reading spectacles together with distance contact lenses
- Monovision
- Alternating (translating) bifocals
- Simultaneous multifocals
  - spherical
  - aspherical

Chapter content

- The correction of early presbyopia by modifying the distance correction
- The correction of presbyopia using reading spectacles with distance contact lenses
- Monovision
- Alternating lens designs (translating bifocals)
- Simultaneous designs (bifocals and multifocals)
- Materials, design and modality
- Calculations involving multifocal contact lenses

The correction of early presbyopia by modifying the distance correction

By under-correcting the myopic patient or over-correcting the hypermetropic patient for distance, a marginal improvement in near vision may be obtained. This is clearly helpful only for the person with very early presbyopia and proves of limited value for the newly fitted myopic patient who is already exerting more accommodation and convergence with contact lenses for near vision than with spectacles. In any event, the option is a short-term solution only and at best may serve as an interim measure until a more permanent solution can be found.

The correction of presbyopia using reading spectacles with distance contact lenses

Although this is the simplest option for the correction of the presbyopic contact lens wearer, it does not address the problem for the patient who does not wish to wear spectacles. Nevertheless, the quality and stability of vision in this mode of correction are such that it may prove to be a necessary additional method of correction for those with visual demands that cannot be satisfied using other methods of correction.

Monovision

This is the least complicated method for providing a presbyopic correction using contact lenses. As the lenses employed are of the single vision type, fitting is by conventional means and the addition for near is provided in the non-dominant eye, using the minimum plus that affords adequate near correction. Many practitioners still use sighting dominance tests (hole in the card, pointing finger, etc.) to confirm the dominant eye; however, there is no scientific literature to relate sighting dominance to blur suppression. Sensory dominance tests (fogging techniques) are far more likely to be predicative because they are a direct measure of the patient's ability to suppress an out-of-focus image.

The +2.00 blur test is acknowledged as being one of the best and simplest clinical methods of identifying the dominant eye. The test is carried out with the patient fully corrected for distance vision. The patient is asked to observe a distance target binocularly and a +2.00DS trial lens is held in front of each eye in turn. The patient is asked to decide under which distance-viewing option the vision is least disturbed and the eyes feel most comfortable. Once the preference is decided the practitioner should record in front of which eye the +2.00DS lens is placed. This is defined as the non-dominant eye to be corrected for near vision. Where there is no particular preference, the decision as to which eye to select is made bearing in mind other factors such as residual astigmatism and corrected acuity. Monovision is of course totally unsuitable for amblyopic patients.

Monovision is an extremely successful option for early presbyopia and quoted success rates vary from 67% to 86%. However, there is usually a decreased level of stereoacuity at near and slightly decreased contrast sensitivity. The degree of interocular blur suppression (which varies between individuals) may be linked to the final success of monovision. The most commonly reported problem with monovision is glare when driving at night. For most patients, once the addition reaches +2.25D, the disparity between the two eyes will be difficult to accept. The lowest acceptable near addition possible allows quicker adaptation to the concept and better intermediate vision.

Partial monovision, in which a low add of +0.50 or +0.75D gives sufficient convenience for intermittent near vision (price tags, menus, headlines) and when patients are happy to use additional reading spectacles for prolonged close work, can often prove successful where full monovision has failed. The over-reading spectacles in this instance however, need to be properly balanced and should not be ready-made reading spectacles with the same prescription in both lenses. The authors feel that the most important indicator of success with monovision is the patient's initial impression; if the patient does not feel disoriented in the consulting room environment, it is worth proceeding with a monovision trial. Monovision is also discussed in Chapter 31.

Alternating lens designs (translating bifocals)

The basic principle underlying the alternating design is that as the patient looks down, the lens rides up and the near segment covers part of the pupil, allowing the patient to read. When the patient looks straight ahead the lens re-centres and the visual axis passes through the distance portion. Alternating designs have the best potential for crisp vision at distance and near, but, as they are gaze dependent, they may be unsuitable for prolonged computer use. In crossover trials they have been shown to be as successful as monovision, but the fitting of an alternating design requires skill, time and experience. There is minimal contrast loss but some flare with larger pupils, and it can be difficult to achieve inferior lens positioning, particularly in myopic patients. Alternating designs employ no-jump or monocentric optics.

Until recently alternating designs have almost exclusively been available as rigid gas permeable (RGP) lenses because no soft alternating lens had been found to work reliably and be comfortable and physiologically acceptable for long periods of wear.

Alternating designs are similar in construction to a bifocal spectacle lens and both solid and fused constructions have been available. Solid alternating lenses have a lower segment and perform in a similar fashion to a bifocal spectacle lens. Prism ballast is used to orient the segment in the correct position and may be stabilised further with truncation inferiorly.

The **Fluoroperm ST** bifocal had a straight-top, high refractive index segment encapsulated within the lens. The anterior and posterior surfaces are smooth and the segment monocentric. The Fluoroperm ST bifocal is now discontinued.

The **Tangent Streak** Bifocal (also available as a trifocal) consists of two wide segments meeting at a straight horizontal junction, similar to an E-type bifocal spectacle lens. The Tangent Streak bifocal is illustrated in **Figure 25.1**.

The segment of the **PresbyLite** bifocal is triangular in shape. This makes the design more independant of pupil diameter compared with straight-top bifocals, and results in similar and consistent distance and reading performance even in varying levels of illumination. Another advantage of the triangular segment is that less translation is necessary for optimal visual performance. The segment is only a few micrometres in depth and is cut from the front surface of the lens with no noticeable transition. This aids wearing comfort, reduces reflections and shortens adaptation time. The triangular near area allows up to 30° rotation without visual disturbance. The initial designs (first generation) had a spherical back surface, included the use of prism ballast and truncation and offered two segment variants. This made fitting the lens complex and large amounts of prism were often required to prevent high riding fits. PresbyLite 2 (second generation) has a new aspheric back surface design. This provides easier translation, needs only a minimal (standard) prism for stabilisation and rarely requires truncation. This, combined with the use of larger diameters, is a benefit for this type of alternating design. The PresbyLite lens is shown in **Figure 25.2**.

Concentric lenses, as the name suggests, have concentric rings of power that may involve a centre distance or centre near configuration. The **Menifocal Z** is a concentric design bifocal with a distance optical zone, a unique transition zone that helps to reduce the degree of blur and doubling, plus a peripheral optic zone for near. As the add increases, both distance and transition zones reduce, allowing the reading area to become larger. The front surface is aspheric, which contributes to the effective addition,

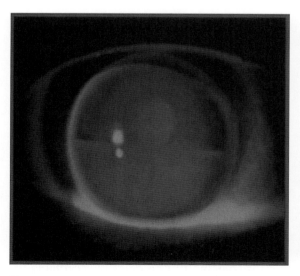

Figure 25.1 Fluorescein pattern of the Tangent Streak RGP bifocal (translating design). (See also colour plates.)

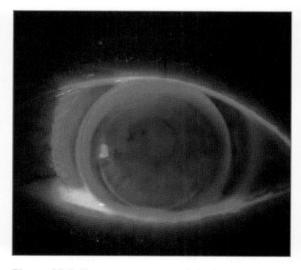

Figure 25.2 Fluorescein pattern of the PresbyLite RGP bifocal (translating design). (See also colour plates.)

whereas a spherical back surface is used to simplify fitting. This bifocal requires translation of the lens on downward gaze. The fit requires the lens to position centrally with free and rapid movement.

Soft translating designs

Until recently, sucessful soft translating designs did not exist! Problems with earlier attempts at producing such lenses included poor comfort and long-term physiological problems as a result of the low oxygen transmissibility, truncation and use of conventional low water content materials.

A new approach is the **Royal** bifocal contact lens (Figure 25.3). A triangular reading segment (based on the same principle as the PresbyLite RGP bifocal lens) is located on the front surface in the inferior part of the lens. Two segment types are available: the full triangle (distance zone 0) and a 'topless' variant (distance zone 2), which is the standard version. Situated just below the reading segment on the front surface is an arched zone. This zone, when aligned with the lower eyelid, enables the lens to be pushed upward when the gaze is directed downwards to make reading possible. The specially designed back surface geometry aids in easy translation. Stabilisation is achieved by a combination of prism ballast and dynamic stabilisation. These features result in a relatively thin design, which is beneficial in terms of comfort and corneal physiology. The lens is also available in a toric option with cylinders from 0.75 DC to 3.00 DC.

Tips

- All translating lenses must move freely upwards on downward gaze to bring the near portion in front of the pupil.
- The lens (segment) needs to sit slightly low in the primary position and, although good translation (movement) on downgaze is required, the lens must recover quickly after each blink so as not to disturb distance vision (particularly during driving).
- The patient's lower lid must have sufficient

Figure 25.3 Royal soft bifocal (translating design).

tone to hold the lens stationary when the eye rotates downwards.
- The lower lid ideally needs to be positioned at a point that is no lower than the lower edge of the visible iris. To assist the lid in locating against the lower edge of the contact lens, the edge profile of the contact lens may need to be relatively square in cross-section rather than the normal tapered and rounded edge.

Translating bifocals: a summary

Advantages

- Excellent visual acuity (VA) for distance and near
- Good contrast
- High near adds available
- Wide range of parameters
- Toric options available
- Normal stereopsis

Disadvantages

- Stabilisation necessary e.g. prism ballast
- Gaze dependent
- Intermediate distances can be a problem
- Lens needs to be very mobile on eye
- Comfort can be an issue

Simultaneous designs (bifocals and multifocals)

There is potential for some confusion in using the term multifocal as this implies a similar mechanism to a multifocal spectacle lens. Simultaneous vision bifocals produce an optical system that places two images on the retina at the same time and then relies on the visual system to select the clearer picture. The basic principle of a simultaneous design however remains the same irrespective of whether the power of the lens varies in a discrete or a progressive manner across the lens surface.

Good centration and stability of fit is required for optimal performance. One of the main problems with these lenses is a loss of contrast sensitivity within the superimposed retinal images. The problem is worse in low illumination and can give particular difficulties with near vision. Despite many attempts to improve the visual performance some loss of contrast always occurs with simultaneous designs. Whilst they are gaze independent, the balance between distance and near is almost always influenced by the level of ambient light (pupil diameter). Some designs on the market exploit or control spherical aberration in an attempt to enhance the refractive effect. Since there is no need to depress the gaze in order to read. This has an obvious benefit for certain occupations (prolonged computed use). For advanced presbyopes 'top-up' spectacles will often be required, although only for certain specific tasks.

With aspheric simultaneous designs, a gradual change in curvature on either the front or the back surface provides the required reading addition. These designs may be of the centre near (CN) type (maximum plus at the centre of the lens decreasing peripherally and employing front or back surface asphericity) or the centre distance (CD) type (minimum plus at the centre of the lens increasing peripherally and again employing front or back surface asphericity). Both designs are shown in Figure 25.4.

These lenses are available in RGP, soft and more recently silicone hydrogel materials. All share one essential drawback, as the asphericity is the only means of providing the addition,

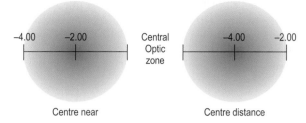

Figure 25.4 Schematic diagram of a centre near and a centre-distance aspheric simultaneous design.

there are implications for the range and quality of vision. The greater the eccentricity (or rate of flattening), the higher the reading power in relation to distance power. However, the higher the reading addition, the more likely it is that distance vision will be adversely affected, especially in low contrast and/or low light conditions. For this reason, aspheric lenses often work well in individuals with early presbyopia where the required addition is low and the compromises on distance vision are minimised. It is essential that any aspheric lens centres well on the eye. If it does not, the visual performance is automatically substandard.

Front surface RGP aspheric simultaneous designs produce a power curve that increases from the centre out to the periphery. This means that most lenses are CD. By controlling the spherical aberration within the power gradient, there is a smooth transition from distance to near, which is why they are often referred to as having a true 'multifocal' effect.

Concentric simultaneous designs employ two distinct zones, with the central zone being used for either reading (CN) or distance (CD). With some of these designs practitioners can select the size of the central zone according to specific requirements, although others have fixed zone diameters related to the amount of addition required.

RGP aspheric multifocals are available with high and low eccentricities. High eccentricity designs are typically fitted very steep, as much as 0.60 mm steeper than the flattest keratometry reading. Good centration is essential and a central fluorescein pool is observed. Posterior aspheric designs can provide an effective near addition up to +2.50 DS. It is important to note that higher additions produce smaller distance zones.

The No7 **Quasar Plus** is an example of a high eccentricity design lens. It is a centre distance bifocal created via a back-surface RGP aspheric design, with the back surface of the lens being made significantly steeper than the cornea and the add power increasing with eccentricity. It is generally fitted 0.4 mm steeper than flattest K and good centration is obtained with the help of the steep back optic zone radius (BOZR). The fluorescein pattern exhibits apical clearance with midperipheral alignment. Ideal movement is 1.0–1.5 mm, with 2 mm on downward gaze. An increase in the reading addition is achieved by fitting a steeper BOZR. The distance zone however, is reduced and this may compromise distance acuity. For high additions of up to +4.00 DS, +2.50 D of the addition is generated on the back surface, with the remaining power worked as an aspheric front surface. By incorporating the asphericity into the back surface, the profile of the tear film is altered when the patient is looking down. As the near portion of the lens positions itself, the pressure of the fitting relationship flattens the central cornea and creates a steep configuration inferiorly, where there is a clearance between the cornea and the contact lens. This can lead in some cases to corneal moulding and although this produces no effect on the vision obtained with the contact lens, it may cause spectacle blur in some patients. **Figure 25.5** shows a back surface aspheric design and **Figure 25.6** how topography can be used to demonstrate corneal moulding resulting in spectacle blur.

Low eccentricity designs have a lower range of posterior flattening and the BOZR is usually only 0.05–0.10 mm steeper than flattest K. These lenses need to centre well or slightly superiorly and show an alignment fluorescein pattern. Some translation on downwards gaze helps with near vision. The **Aqualine MF200** (Cantor & Nissel) is an example of a low eccentricity aspheric front surface lens designed to produce a progressive power curve that increases from the centre outwards. This lens is a centre distance design with an intermediate to near addition of up to +2.00 D within the pupil area. Plus power is further enhanced on upward or

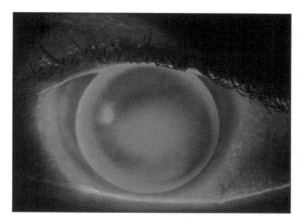

Figure 25.5 Fluorescein pattern of a well fitted RGP, back surface, aspheric multifocal (fitted typically 0.4 mm steeper than flattest K). (See also colour plates.)

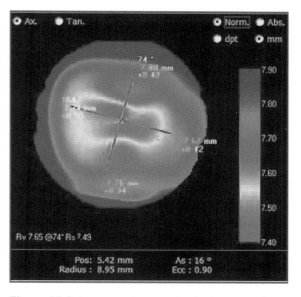

Figure 25.6 A topography map clearly demonstrating corneal moulding after wear of an RGP back surface aspheric multifocal. (See also colour plates.)

downward gaze. Control of spherical aberration within the power gradient gives a smooth transition from distance to near vision. The reading addition cannot work effectively if the lens is fitted too steep or too flat. Aqualine MF200 is generally fitted in alignment. If a higher add is required, the non-dominant eye can be over-plussed by +0.50 D, but care should be taken not to over-minus at distance. For best near vision while reading, the patient should hold the chin up to use the intermediate zone.

Pupil size dependency

All the above designs exhibit variation in performance depending on the size of pupil. It is barely of significance in segment alternating lenses but is most noticeable with concentric and aspheric simultaneous vision types. Pupil size varies with light intensity, age and visual task. Even with careful measurements of pupil size in the consulting room, there is no guarantee that the lenses will be worn in similar conditions of illumination in the 'real world'. In CN types low light levels favour distance vision at the expense of near vision whereas bright sunlight favours near vision at the expense of distance vision. Thus wearers of CN concentric bifocals should be advised to wear sunglasses when driving in sunlight and be warned that there might be difficulties when reading in poor light. The reverse is true of CD types, where distance vision is best in bright light and near vision is best in poor light. In addition there is the advantage that some translation during inferior gaze will benefit near vision when CD lenses are used.

The **ACUVUE Bifocal** (J & J) is listed in the literature as a multi-zone concentric lens and is also referred to as 'pupil intelligent'. The design incorporates five alternating distance and near rings over an 8 mm optic zone. Based on research of different pupil sizes within the presbyopic population, the size and spacing of each zone are designed to optimise vision for various lighting and viewing conditions. There are four add powers available, which allow the practitioner to modify the prescription to the exact individual requirements of the patient. However, near vision is biased with small pupils and distance vision with larger pupils. The **ACUVUE Bifocal** is illustrated in **Figure 25.7**.

When using aspheric lenses, the same basic factors apply. In centre distance lenses with limited addition power, such as the Bausch & Lomb **Occasions** lens, the maximum amount of add available depends on the pupil size. As the pupil decreases in size, so the amount of avail-

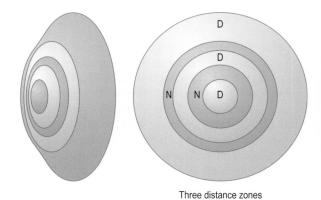

Three distance zones
Two near zones

Figure 25.7 Schematic of the ACUVUE Bifocal (multizone bifocal).

able add decreases. Unfortunately, pupil size tends to decrease with age so that as more addition is required, less addition is available! In centre near aspheric designs such as the Ciba **Focus Dailies Progressives** and the Bausch & Lomb **Purevision Multifocal**, the lens front surface progressively changes from the centre to the periphery, creating a gradual power gradient. Aspheric multifocal lenses do not have a specific addition power but a span of power from a pre-set near point out to distance, created by the aspheric optics. Distance vision is limited by pupil size and is better with larger pupils.

The **Proclear Multifocal** is an amalgamation of spherical and aspheric optics with unique zone sizes to produce two complementary but inverse geometry lenses. A centre distance **D lens** is used for the dominant eye and a centre near **N lens** for the non-dominant eye. These lenses are multi-zonal with a spherical central area of different sizes for the two designs (2.3 mm for the distance and 1.7 mm for the near). Surrounding the central zone is a 5 mm diameter aspheric annulus and surrounding this, a final spherical band that completes the total optic zone of 8.5 mm. Although this may sound a bit like monovision, each lens is a true multifocal in its own right and bifocal summation is maintained. Four specific additions allow the practitioner to independently prescribe the distance and near powers for the individual patient's needs. The Proclear Multifocal is shown in **Figure 25.8**.

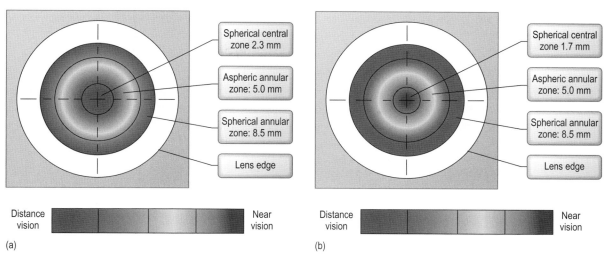

Figure 25.8 Schematic of the Proclear Multifocal (multi-zone multifocal). (a) D lens: dominant eye; (b) N lens: non-dominant eye. (See also colour plates.)

Simultaneous multifocals: a summary

Advantages

- Independent of lens rotation
- Comfort equivalent to that of a single vision lens
- Good VA at any distance
- Multifocal effect (computer workers)
- Easy to fit
- Normal stereopsis

Disadvantages

- Some compromise in vision may occar
- Lost of contrast
- Near additions limited
- Limited parameters

Diffractive designs

Diffractive (zone plate) lens designs allow the distance image to be formed using refractive optics whereas the near image formation is achieved by central diffractive echelettes. This lens type is termed 'pupil independent' because the ratio split between distance and near image formation is equal regardless of pupil size. As a result of the 50/50 split in light, there is a correlating drop in contrast and limited optical performance, especially with higher additions. Another disadvantage is the build-up of debris in the prism rings, which are located on the back surface of the lens. The major positive effect of diffractive bifocals is that unlike all other designs, they are pupil independent.

This diffractive technique was employed on both the soft **Echelon** and the RGP **Diffrax** lenses with limited success. Neither product is now available.

Modified monovision

In this technique, multifocals are used with the lens selected for the dominant eye biased for distance vision (relatively under-plussed or over-minused) and the non-dominant eye biased for near (relatively over-plussed or under-minused). In this way the patient theoretically enjoys some of the benefits of bifocal lenses with the added advantages of monovision. This may be essential with back surface aspheric lenses with limited reading addition, because the patient with advanced presbyopia requires higher reading additions.

Enhanced monovision

With this, one multifocal is incorporated into the monovision system. To improve (enhance) either distance *or* near vision:

- **Near**: use a near-biased multifocal in the dominant eye and a full near-vision correcting single vision contact lens in the non-dominant eye.

- **Distance**: use a distance-biased multifocal in the non-dominant eye and a full distance vision-correcting single vision contact lens in the dominant eye.
- **Intermediate**: use an intermediate-biased multifocal in the non-dominant eye and a full distance vision or slightly over-plussed single vision contact lens in the dominant eye.

The technique of enhancing monovision may improve the visual comfort by affording a degree of binocular summation at the most critical viewing distance.

Design choice

Early presbyopia (up to +1.00 DS)

- Simultaneous: full correction to both eyes
- Monovision: distance and *full* near

Mid presbyopia (+1.25 DS to +2.00 DS)

- Simultaneous: full correction to both eyes
- Translating: full correction to both eyes (do not over-correct addition)
- Monovision: distance and *full* near

Late presbyopia (+2.25 DS to +3.00 DS)

- Translating: full correction to both eyes
- Simultaneous:
  – modified monovision
  – enhanced monovision
- Monovision:
  – distance and *partial* near
  – consider top-up spectacles to provide additional plus

Establishing patient expectations

Given that all the corrections for presbyopia (including spectacle lenses) produce some limitations on visual quality and/or range of clear vision, it is critically important that patients seeking contact lens correction for presbyopia are made fully aware of the limitations of this mode of correction. Multifocal lenses may demonstrate lost lines of distance acuity, reduction in distance and near stereopsis, and mild-to-moderate difficulties with night driving, ghost images and haloes. In some cases auxiliary spectacles may be

a solution to occasional visually demanding tasks. Patients need to appreciate that there is no perfect solution to presbyopia and need to be counselled in advance, otherwise they often perceive the outcome as a failure. Although the ultimate goal is to provide a full binocular correction at distance and near, it is important to ascertain from patients whether their priority is good distance vision or good near vision. This should be determined early on in the initial consultation.

Remember that although the practitioner may be obsessed about the patient achieving 6/6 and N5, the patient is more interested in visual quality, a function that cannot be measured using a high-contrast letter chart. It is thus incumbent on the practitioner to act on patient feedback after a lens trial rather than on acuity measurement alone. However, low-contrast acuity charts are often a useful way of assessing the 'real-world' vision obtained with a presbyopic lens design. Even with modern and sophisticated lens designs, monovision still has a higher success rate when compared with other methods of correcting presbyopia with contact lenses.

The fitting of individuals with presbyopia is challenging and rewarding. However, the practitioner needs to understand patients' visual needs and act on their subjective feedback as a means of refining the correction.

Materials, design and modality

As with single vision lenses, it is important to select a suitable material, replacement frequency and wearing schedule (modality) for presbyopic patients, taking into account their visual and physiological demands as well as their lifestyle. Examples of contact lenses for the correction of presbyopia currently available include: Soft Daily Disposables (CIBA **Focus Dailies Progressives**), soft disposable 2-weekly and monthly lenses (J & J **ACUVUE Bifocal**, Bausch & Lomb **Soflens 66** Multifocal, Coopervision **Proclear Multifocal**) and silicone hydrogel monthly disposable or 30-day continuous wear lenses (Bausch & Lomb **Purevision Multifocal**). Most soft planned replacement and conventional lenses are made to order and the lists of materials, designs and products available constantly change, with many products being withdrawn due to lack of use. Two new interesting products allowing the correction of both presbyopia and astigmatism include the **Royal Toric Bifocal** from No7 and CIBASOFT **Progressive Toric**, the latter being made in a low water content (37%) materical.

In the RGP market, the possibility of lenses being made to order makes the list almost endless. Most lenses are available in a range of materials and *Dk* values, allowing extended-wear options where required. Key players in the UK market include No7 **Quasar** (back and front surface aspheric designs), David Thomas **Essentials** (front surface aspheric S-form technology), CIBA Vision **Astrocon MF** (front surface aspheric), Menicon **Menifocal Z** (concentric) No7 & Scotlens, **PresbyLite2** (translating triangle segment) and No7 **Tangent Streak** (translating long line).

The number of presbyopic lens designs available is constantly growing, thus it is impossible to cover them all in this text. The *ACLM Contact Lens Year Book* or website (www.aclm.org.uk) contains a current and comprehensive list of products currently available in the UK.

Clinical pearls

- Given that all corrections for presbyopia produce some limitations on visual quality, it is critically important that patients seeking contact lens correction for presbyopia are made fully aware of the limitations of this mode of correction.
- Pre-determine goals for patients, stress that there will be some compromise when lenses are used for full-time wear and attempt to gain patient acceptance of set goals.
- It is extremely important to be familiar with the manufacturer's fitting advice for any particular design of lens, because they all differ slightly in design characteristics and optimal fitting.
- Alternating lenses are ideal for the rigorous demands of higher additions whereas simultaneous lenses are better suited for the correction of the lower additions. It follows that the additions should be ≥1.00 D for alternators and <2.00 D for the simultaneous candidates. However, where patients with a high additions are prepared to use 'top-up' spectacles from time to time, they may be very happy with the results provided by simultaneous designs.
- In advanced presbyopia (additions >2.25 D), pupils tend to be smaller and alternating vision tends to work better. In earlier presbyopia, the pupils are larger and simultaneous designs tend to be indicated.
- Success is generally lower with patients having excellent uncorrected distance vision. Given that all contact lens systems

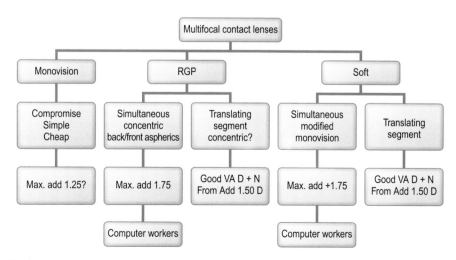

Figure 25.9 Simple flow chart to aid in the selection of the most appropriate presbyopic correction system for a particular patient. D, distance; GP, gas permeable; N, normal; VA, visual acuity.

Table 25.1 Terminology applicable to fused bifocal spectacle and contact lenses

Symbol	Explanation
F_1	Power of the front surface of the main lens
F_2	Power of the back surface of the main lens
r_1	Radius of curvature of the front surface of the main lens
r_2	Radius of curvature of the back surface of the main lens
F_3	Power of the front surface of the segment (if segment is on the front surface) or power of the back surface of the segment (if segment is on the back surface)
F_4	Power of the back surface of the segment in air (if segment is on the front surface) or power of the front surface of the segment in air (if segment is on the back surface)
r_c	Radius of curvature of the depression curve that is ground into the main lens in order to accept the segment
F_c	Power of the depression curve in air, i.e. segment removed from the main lens
F_{con}	Power of the contact surface when the segment is in contact with depression curve
F_s	Power of the segment in isolation, i.e. in air (segment assumed to be thin)
K	The fused blank ratio
n	Refractive index of the main lens
n_s	Refractive index of the segment
A	Reading addition

for presbyopia correction impact to a greater or lesser degree on the quality of distance vision, it follows that those patients who are accustomed to relatively good uncorrected distance vision may be less able to cope with the relative impairment in their distance vision when a contact lens is worn to correct presbyopia. It is generally better to look for uncorrected vision of 6/9 or worse. However, this should be seen only as a relative contraindication and a well-motivated patient can be surprisingly successful.

- Use the Snellen chart only to record VA for legal purposes and as a benchmark for future lens changes. Asking the patient to look out of the consulting room window can often be a better way to assess patient satisfaction.
- If patients are to be successful they will generally decide within the first two aftercare appointments. The more appointments made the less likely they are to be successful.

Figure 25.9 is a flow chart that may be helpful when selecting a correction system for a particular patient.

Calculations involving multifocal contact lenses

Problems involving fused bifocal contact lenses can be tackled in exactly the same way as problems that involve a fused bifocal spectacle lens. There are two ways of addressing such problems. One method uses first principles whereas the other uses the fused blank ratio. The segment of a fused bifocal spectacle lens is usually incorporated into the front surface of the main lens whereas the segment of a fused bifocal contact lens is usually incorporated into the back surface of the main lens. The usual questions involve the determination of the radius of curvature of the depression curve. It is important to understand the terminology used with fused bifocals. The common terms are given in Table 25.1 and can be applied to both a contact lens and a spectacle lens.

Figures 25.10–25.12 illustrate the common terminology used with front and back surface segment fused bifocal lenses. They apply to both spectacle lenses and contact lenses.

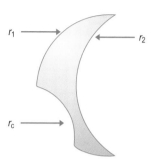

Figure 25.10 Fused bifocal terminology: front surface segment: r_1 is the radius of curvature of the front surface of the main lens; r_c radius of curvature of the depression curve. The power of this surface in air would be F_c; r_2 is the radius of curvature of the back surface of the main lens.

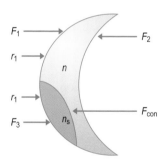

Figure 25.11 Fused bifocal terminology: front surface segment: F_1 is the power of the front surface of the main lens; F_3 is the power of the front surface of the segment; F_{con} is the power of the interface between the depression curve and the main lens; F_2 is the power of the back surface of the main lens.

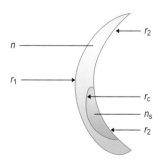

Figure 25.12 A back surface segment fused bifocal contact lens.

Expressions applicable to both fused bifocal spectacle and contact lenses

Front surface segment: fused blank ratio method

$$F_1 = \frac{n-1}{r_1}$$

$$K = \frac{n-1}{n_s - n}$$

$$F_c = F_1 - AK$$

$$r_c = \frac{n-1}{F_c}$$

$$F_{con} = \frac{n - n_s}{r_c}$$

$$Fs = A(K+1)$$

Back surface segment: fused blank ratio method

$$F_2 = \frac{1-n}{r_2}$$

$$K = \frac{n-1}{n_s - n}$$

$$F_c = F_2 - AK$$

$$r_c = \frac{1-n}{F_c}$$

$$F_{con} = \frac{n_s - n}{r_c}$$

$$Fs = A(K+1)$$

Front surface segment: first principles method

$$F_1 = \frac{n-1}{r_1}$$

$$r_1 = \frac{n-1}{F_1}$$

$$F_3 = \frac{n_s - 1}{r_1}$$

Add due to the front surface $= F_3 - F_1$

Total add $= (F_3 - F_1) + F_{con}$

$$F_{con} = \text{Total add} - (F_3 - F_1)$$

271

$$r_c = \frac{n - n_s}{F_{con}}$$

$$F_s = F_3 + F_4$$

$$F_4 = \frac{1 - n_s}{r_c}$$

Back surface segment: first principles method

$$F_2 = \frac{1 - n}{r_2}$$

$$r_2 = \frac{1 - n}{F_2}$$

$$F_3 = \frac{1 - n_s}{r_2}$$

Add due to the back surface $= F_3 - F_2$

Total add $= (F_3 - F_2) + F_{con}$

$$F_{con} = \text{Total add} - (F_3 - F_2)$$

$$r_c = \frac{n_s - n}{F_{con}}$$

$$F_s = F_3 + F_4$$

$$F_4 = \frac{n_s - 1}{r_c}$$

Example 25.1

A front surface fused bifocal spectacle lens is made as follows:

$F = +3.00\,D$, $F_2 = -5.00\,D$, Add $= +2.50\,D$, $n = 1.52$, $n_s = 1.65$. Find the radius of curvature of the depression curve and the power of the segment in isolation. The lens is assumed to be thin.

Both methods are used in this example. We start with the first principles method. As the lens is thin, the power of the front surface can be found using:

$$F = F_1 + F_2$$

This rearranges to give:

$$F_1 = F - F_2 = +3.00 - (-5.00) = +8.00\,D$$

Next, find the radius of curvature of the front surface of the main lens r_1:

$$r_1 = \frac{n - 1}{F_1} = \frac{1.52 - 1.00}{+8.00} = +0.065\,m$$

The physical construction of a fused bifocal means that the front surface of the segment has exactly the same radius of curvature (r_1) as the front surface of the main lens. However, its front surface power (F_3) is greater than the power of the front surface of the main lens (F_1) because its refractive index is higher:

$$F_3 = \frac{n_s - 1}{r_1} = \frac{1.65 - 1.00}{+0.065} = +10.00\,D$$

The addition due to the front surface is the difference between F_3 and F_1:

$$F_3 - F_1 = +10.00 - (+8.00) = +2.00\,D$$

As the total addition of the segment is given by:

$$(F_3 - F_1) + F_{con}$$

F_{con} can be found using:

$$F_{con} = \text{Total addition} - (F_3 - F_1) = +2.50 - (+2.00)$$
$$= +0.50\,D$$

Now we know F_{con} we can find the radius of curvature of the depression curve using:

$$r_c = \frac{n - n_s}{F_{con}} = \frac{1.52 - 1.65}{+0.50} = -0.26\,m$$

If the segment element was removed, the power of the depression curve *in air* would be:

$$F_c = \frac{n - 1}{r_c} = \frac{1.52 - 1.00}{-0.26} = -2.00\,D$$

As the segment is assumed to be thin, we can find the power of the segment in isolation using:

$$F_s = F_3 + F_4$$

We already know F_3, F_4 can be found using:

$$F_4 = \frac{1 - n_s}{r_c} = \frac{1.00 - 1.65}{-0.26} = +2.50\,D$$

$$F_s = F_3 + F_4 = +10.00 + (+2.50) = +12.50\,D$$

Example 25.1 is now re-worked using the fused blank ratio method. As before, the power of the front surface can be found using:

$$F = F_1 + F_2$$

$$F_1 = F - F_2 = +3.00 - (-5.00) = +8.00\,D$$

Now find the blank ratio K

$$K = \frac{n-1}{n_s - n} = \frac{1.52 - 1.00}{1.65 - 1.52} = 4.00$$

The *power* of the depression curve in air can now be found using:

$$F_c = F_1 - AK = +8.00 - (+2.50 \times 4.00) = -2.00\,D$$

The *radius of curvature* of the depression curve now be found using:

$$r_c = \frac{n-1}{F_c} = \frac{1.52 - 1.00}{-2.00} = -0.26\,m$$

The power produced when the segment is placed in contact with the depression curve (F_{con}) is given by:

$$F_{con} = \frac{n - n_s}{r_c} = \frac{1.52 - 1.65}{-0.26} = +0.50\,D$$

The power of the segment in isolation and in air can be found using:

$$F_s = A(K+1)$$

$$F_s = +2.50(4.00 + 1.00) = +12.50\,D$$

Example 25.2

A back surface fused bifocal is made as follows:

$F = -3.00\,DS$, Add $= +3.00\,D$, $F_1 = +5.00\,D$,
$n = 1.525$, $n_s = 1.675$.

Find the radius of curvature of the depression curve and the power of the segment element in isolation. Assume a thin lens.

Once again, both methods are used to solve this problem. We start with the first principles method. We need the power of the back surface of the main lens, F_2, so we can use:

$$F = F_1 + F_2$$

This becomes:

$$F_2 = F - F_1 = -3.00 - (+5.00) = -8.00\,D$$

We can now find r_2 using:

$$r_2 = \frac{1-n}{F_2} = \frac{1.000 - 1.525}{-8.00} = +0.06562\,m$$

As r_2 is constant over the whole of the back surface of the lens, F_3 can be found using:

$$F_3 = \frac{1 - n_s}{r_2} = \frac{1.000 - 1.675}{+0.06562} = -10.28\,D$$

The addition due to the back surface is therefore:

$$F_3 - F_2 = -10.28 - (-8.00) = -2.28\,D$$

As the total addition of the segment is given by:

$$\text{Total addition} = (F_3 - F_2) + F_{con}$$

$$F_{con} = \text{Total addition} - (F_3 - F_2)$$

$$F_{con} = +3.00 - (-2.28) = +5.28\,D$$

F_{con} is the power produced when the power of the depression curve ground into the main lens is in contact with the segment. We can use F_{con} to find the radius of curvature of the depression curve r_c

$$r_c = \frac{n_s - n}{F_{con}} = \frac{1.675 - 1.525}{+5.28} = +0.02838\,m$$

If the segment element was removed, the power of the depression curve *in air* would be:

$$F_C = \frac{1-n}{r_c} = \frac{1.000 - 1.525}{+0.02838} = -18.50\,D$$

The power of the segment element in isolation and in air is given by:

$$F_s = F_3 + F_4$$

We have already found F_3. The front surface power of the segment in air (F_4) is given by:

$$F_4 = \frac{n_s - 1}{r_c} = \frac{1.675 - 1.000}{+0.02838} = +23.78\,D$$

$$F_3 = -10.28\,D$$

$$F_s = -10.28 + (+23.78) = +13.50\,D$$

So the power of the segment element in isolation and in air is +13.50 D. We now use the fused blank ratio method. We already know that $F_2 = -8.00\,D$.

The blank ratio is given by:

$$K = \frac{n-1}{n_s - n} = \frac{1.525 - 1.000}{1.675 - 1.525} = 3.50$$

The power of the depression curve in air is given by:

$$F_c = F_2 - AK = -8.00 - (+3.00 \times 3.50) = -18.50\,D$$

The radius of curvature of the depression curve in air is given by:

$$r_c = \frac{1-n}{F_c} = \frac{1.000-1.525}{-18.50} = +0.02838\,m$$

The power produced when the segment is placed in contact with the depression curve (F_{con}) is given by:

$$F_{con} = \frac{n_s - n}{r_c} = \frac{1.675 - 1.525}{+0.02838} = +5.28\,D$$

The power of the segment element in isolation in air (F_s) is found using:

$$F_s = A(K+1) = +3.00(3.50+1) = +13.50\,D$$

Example 25.3

The prescription −3.00 DS, Add +2.00 D is to be made as a back surface fused bifocal contact lens. The BOZR (r_2) is 8.00 mm, centre thickness 0.2 mm, and the refractive index of the main lens is 1.50 and of the segment is 1.60. Find all the surface radii.

The surface radii that need to be found are:

- The radius of curvature of the front surface of the main lens
- The radius of curvature of the depression curve.

As the contact lens is thick, a step-back ray trace is needed to find r_1. The first task is to find the power of the back surface of the main lens F_2 and the equivalent air distance (EAD) for the contact lens:

$$F_2 = \frac{1-n}{r_2} = \frac{1.00-1.50}{+8.00^{-3}} = -62.50\,D$$

The EAD (with t_{cl} substituted in metres) is given by:

$$EAD = \frac{t_{cl}}{n_{cl}} = \frac{2^{-4}}{1.50} = 1.3333^{-4}\,m$$

Now that the EAD has been determined, the refractive index between the two surfaces can

be assumed to be **equivalent to that of air** ($n = 1$). This value is used in the following ray-trace. A step-back ray-trace is used to find the front surface power. As always, the computation takes place in two columns, one for vergences and the other for distances. The starting point is L'_2, which is equivalent to the back vertex power (BVP) of the contact lens in air. In this example, this is −3.00 D.

Vergences (D)	Distances (m)

$$L'_2 = -3.0000\,D$$

$$L_2 = L'_2 - F_2$$

$$L_2 = -3.0000 - (-62.5000)$$

$$L_2 = +59.5000\,D \rightarrow l_2 = \frac{1}{+59.5000} = +0.01681\,m$$

$$l'_1 = l_2 + \left(\frac{t_{cl}}{n_{cl}}\right)\ \text{(step-back)}$$

$$\leftarrow l'_1 = +0.01681 + 1.3333^{-4} = +0.01694\,m$$

$$L'_1 = \frac{1}{+0.01694} = +59.0317\,D$$

As $L_1 = 0.00$, $L_1 = F_1$

$$F_1 = +59.0317\,D$$

The FOZR can be found using:

$$r_1 = \frac{n'-n}{F_1}$$

$$r_1 = \frac{1.50-1.00}{+59.0317} = +8.4700^{-3}\,m$$

The front surface radius of the contact lens is therefore +8.4700 mm.

First principles are used to find the radius of curvature of the depression curve. As r_2 is constant over the whole of the back surface of the lens, F_3 can be found using:

$$F_3 = \frac{1-n_s}{r_2} = \frac{1.00-1.60}{+8.00^{-3}} = -75.00\,D$$

The addition due to the back surface is therefore:

$$F_3 - F_2 = -75.00 - (-62.50) = -12.50\,D$$

As the total addition of the segment is given by:

$$\text{Total add} = (F_3 - F_2) + F_{con}$$

$$F_{con} = \text{Total add} - (F_3 - F_2)$$

$$F_{con} = +2.00 - (-12.50) = +14.50\,\text{D}$$

F_{con} is the power produced when the depression curve ground into the main lens is in contact with the segment. We can use F_{con} to find the radius of curvature of the depression curve r_c:

$$r_c = \frac{n_s - n}{F_{con}} = \frac{1.60 - 1.50}{+14.50} = +6.8965^{-3}\,\text{m}$$

The radius of curvature of the depression curve is therefore **+6.8965 mm**.

Alternatively, we can use the fused blank ratio method. We know that $F_2 = -62.50\,\text{D}$ The blank ratio is given by:

$$K = \frac{n-1}{n_s - n} = \frac{1.50 - 1.00}{1.60 - 1.50} = 5.00$$

For a back surface segment, the power of the depression curve in air is given by:

$$F_c = F_2 - AK = -62.50 - (+2.00 \times 5.00)$$

$$= -72.50\,\text{D}$$

The radius of curvature of the depression curve in air is given by:

$$r_c = \frac{1-n}{F_c} = \frac{1.00 - 1.50}{-72.50} = +6.8965^{-3}\,\text{m}$$

The radius of curvature of the depression curve is therefore **+6.8965 mm**, which agrees with the answer using the first principles method.

In this example, the bifocal segment is incorporated into the back surface of the main lens. It is interesting to note that, if the segment of a fused bifocal contact lens is manufactured on the *front* surface, the radii r_1 and r_c are often very similar. This means that the lens is difficult to produce and results in a segment that is very thin. Consequently, fused bifocal contact lenses are usually produced with back surface segments.

Example 25.4

A concentric front surface CD RGP bifocal contact lens is to be made with an addition of +2.50 D. If the radius of the central distance portion is 8.333 mm and the refractive index of the lens material is 1.50, find the radius of curvature required to produce the required addition.

As this lens is a front surface CD design, the central area of the lens is used for distance vision and the peripheral for near (**Figure 25.13**). Two separate zones and therefore two separate radii need to be worked on the front surface.

The power of the central area of the front surface can be calculated using:

$$F_1 = \frac{n_{cl} - 1}{r_1}$$

where $n_{cl} = 1.50$ and $r_1 = +8.333\,\text{mm}$. Substitution into the above expression gives:

$$F_1 = \frac{1.50 - 1.00}{+8.333^{-3}} = +60.0024\,\text{D}$$

To produce an addition of +2.50 D the front peripheral curve must have a power of:

$$+60.0024 + (+2.5000) = +62.5024\,\text{D}$$

The front peripheral radius can be found using:

$$r_1 = \frac{n_{cl} - 1}{F_1} = \frac{1.500 - 1.000}{+62.5024} = +7.9997^{-3}\,\text{m}$$

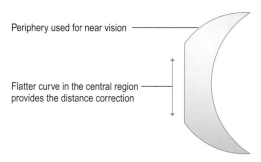

Periphery used for near vision

Flatter curve in the central region provides the distance correction

Figure 25.13 Front surface centre–distance solid bifocal contact lens. Central region has flatter curve.

The front surface of this concentric bifocal contact lens therefore has a central radius of curvature of +8.333 mm and peripheral radius of curvature of +7.9997 mm. It is the steeper peripheral curve worked on the front surface that is responsible for the generation of the reading addition.

Summary

- Flatter curve in central region provides distance correction
- Steeper curve in peripheral region provides near correction
- Periphery of lens is more positive in power

Example 25.5

A concentric front surface CN RGP bifocal contact lens is to be made with an addition of +2.00 D. If the radius of the central near portion is 7.85 mm and the refractive index of the lens material is 1.49, find the radius of curvature required to produce the necessary distance correction.

As this lens is a front surface CN design, the central area of the lens is used for near vision and the peripheral area for distance vision (**Figure 25.14**). Two separate zones and therefore two separate radii need to be worked on the front surface. The power of the central area of the front surface can be calculated using:

$$F_1 = \frac{n_{cl} - 1}{r_1}$$

where $n_{cl} = 1.49$ and $r_1 = +7.85$ mm. Substitution into the above expression gives:

$$F_1 = \frac{1.49 - 1.00}{+7.85^{-3}} = +62.4204 \, \text{D}$$

The above value is providing the near correction. If the reading addition is +2.00 D, the front peripheral curve must have a power that is 2.00 D less than this.

$$+62.4204 - (+2.00) = +60.4204 \, \text{D}$$

The front peripheral radius can be found using:

$$r_1 = \frac{n_{cl} - 1}{F_1} = \frac{1.49 - 1.00}{+60.4204} = +8.1098^{-3} \, \text{m}$$

The front surface of this concentric bifocal contact lens therefore has a central radius of curvature of +7.8500 mm and peripheral radius of curvature of +8.2753 mm. It is the steeper central curve worked on the front surface that is responsible for the generation of the reading addition.

Summary

- Flatter curve in peripheral region provides distance correction
- Steeper curve in central 'segment' region provides near correction
- Central area of the lens is more positive in power

There are two general problems with front surface designs that are worth mentioning:

1. The small difference in radius of curvature between the two portions of the front surface makes this type of lens difficult to manufacture. However, sophisticated CNC (computer numerically controlled) production techniques have overcome this difficulty.
2. Tear fluid accumulating on the front surface of the lens may alter the power of the near addition by 'filling in' the transition between the curves.

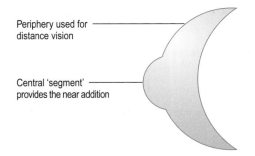

Figure 25.14 Front surface centre–near solid bifocal contact lens. Central region has steeper curve (NV) peripheral region has flatter curve (DV). DV, distance vision; NV, near vision.

Periphery used for distance vision

Central 'segment' provides the near addition

Example 25.6

A back surface CD solid bifocal is made using a material of refractive index 1.51. The BOZR of the near portion is 7.80 mm, which is fitted in alignment with the cornea. The BVP of the distance portion is to be −4.00 D and the reading addition +1.50 D. Find the radius of curvature of the distance portion and the power of the addition if measured in air.

This lens is illustrated in Figure 25.15. It is important to appreciate that this lens must produce an *on-the-eye* reading addition of +1.50 D. The first task is therefore to calculate the on-the-eye power of the near (peripheral) portion. This can be done using:

$$F = \frac{n_{\text{tears}} - n_{\text{cl}}}{r_2}$$

where n_{tears} is assumed to be 1.336, $n_{\text{cl}} = 1.51$ and $r_2 = 7.80$ mm. Substitution gives:

$$F = \frac{1.336 - 1.510}{+7.80^{-3}} = -22.3077 \text{ D}$$

As the above power is being generated by the peripheral (near) portion, it follows that the *on-the-eye-power* of the distance portion must be 1.50 D more negative in order to generate an on the eye reading addition of +1.50 D. The *on-the-eye* power on the distance portion is therefore −23.8077 D. The radius of curvature of this surface can be found using:

$$r_2 = \frac{n_{\text{tears}} - n_{\text{cl}}}{F_2}$$

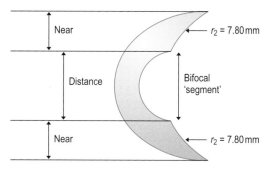

Figure 25.15 Centre distance back surface solid bifocal.

Substitution gives:

$$r_2 = \frac{1.336 - 1.510}{-23.8077} = +7.3086^{-3} \text{ m}$$

The radius of curvature of the back peripheral curve is +7.8500 mm. This gives the required on-the-eye reading prescription. The radius of curvature of the back central curve is +7.3086 mm, which gives the required on-the-eye distance prescription. The difference in radius of curvature between the distance and near portions is much greater with the back surface solid bifocals compared with the front surface solid bifocals. This makes the manufacturing of such a lens an easier prospect. To find the power of the addition in air, we need to find the power of both back surfaces in air. To do that we can use the expression:

$$F_2 = \frac{1 - n_{\text{cl}}}{r_2}$$

For the central (distance) portion:

$$F_2 = \frac{1.000 - 1.510}{+7.3086^{-3}} = -69.7808 \text{ D}$$

For the peripheral (near) portion:

$$F_2 = \frac{1.000 - 1.510}{+7.8500^{-3}} = -64.9681 \text{ D}$$

The addition in air is simply the difference between the above values. The addition is **+4.8127 D**. The large difference between the two powers makes the verification of the addition easier than with front surface solid bifocals. It should be noted that the addition in air is approximately three times that on the eye. The addition produced in air can be calculated using the expression:

$$\text{Add in air} = \text{Add on eye} \times \frac{n_{\text{cl}} - 1}{n_{\text{cl}} - n_{\text{tears}}}$$

If the refractive indices of the contact lens and tears are taken to be 1.490 and 1.336, respectively, the power of the addition in air is **3.18×** the power of the addition on the eye.

Summary

- Segment (central) area of the lens provides the distance correction.

- Periphery of the lens provides the near correction.
- As the surface generating the segment is in contact with the tear fluid and not air, there is a greater difference in surface radii, which makes lens manufacture easier.
- The large difference between the distance and near powers makes lens checking much easier than with front surface bifocals.
- The power of the addition measured in air is approximately 3.18× the power of the addition required on the eye.

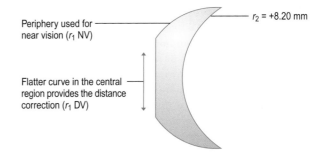

Figure 25.16 Diagram for Example 25.7. Front surface centre–distance solid bifocal contact lens. DV, distance vision; NV, near vision.

Example 25.7

A CD front surface solid bifocal contact lens is made to the prescription −1.00 DS, Add +2.25 DS. The lens has a constant thickness of 0.20 mm and is manufactured using a material of refractive index 1.490. The lens is to be fitted to a patient with keratometry readings of 8.20 mm along all meridians and in alignment with the cornea with no apical clearance. Calculate all the surface radii.

This question asks for all the surface radii, which means the BOZR, the front central radius that generates the distance prescription and the front peripheral radius that generates the near prescription. As the lens is to be fitted in alignment with no corneal clearance (and therefore no tear lens), the optical effect of the tears can be ignored. An alignment fit also means that the BOZR is the same as the corneal curvature given in the question. The BOZR is therefore 8.20 mm. To find the two front surface radii, we must perform two step-back ray-traces through this *thick* contact lens: one is through the central distance portion, the other through the peripheral near portion. Remember that we are ignoring the optical effect of the tear lens. The lens is illustrated in **Figure 25.16**.

So far we know r_2 (the BOZR), n_{cl}, t_{cl}, the BVP and the addition. We can easily determine the back surface power of the contact lens (F_2) and the EAD:

$$F_2 = \frac{1-n}{r_2} = \frac{1.00 - 1.49}{+8.20^{-3}} = -59.7561\,\text{D}$$

The EAD (with t_{cl} substituted in metres) is given by:

$$\text{EAD} = \frac{t_{cl}}{n_{cl}} = \frac{2^{-4}}{1.49} = 1.3423^{-4}\,\text{m}$$

Now that the EAD has been determined, the refractive index between the two surfaces can be assumed to be **equivalent to that of air** ($n = 1$). This value is used in the following ray-trace. A step-back ray-trace is used to find the front surface power. As always, the computation takes place in two columns, one for vergences and the other for distances. The first ray-trace finds the radius of curvature of the central distance portion. The starting point is L_2' which is equivalent to the BVP of the contact lens in air. In this example, this is −1.00 D.

Vergences (D)	Distances (m)
$L_2' = -1.0000\,\text{D}$	
$L_2 = L_2' - F_2$	
$L_2 = -1.0000 - (-59.7561)$	
$L_2 = +58.7561\,\text{D} \rightarrow l_2 = \dfrac{1}{+58.7561} = +0.01702\,\text{m}$	

$$l_1' = l_2 + \left(\frac{t_{cl}}{n_{cl}}\right) \text{(step-back)}$$

$$\leftarrow \quad l_1' = +0.01702 + 1.3423^{-4} = +0.01715\,\text{m}$$

$$L_1' = \frac{1}{+0.01715} = +58.2963\,\text{D}$$

As $L_1 = 0.00$, $L_1 = F_1$

$$F_1 = +58.2963\,\text{D}$$

The surface power of the front CD portion is therefore +58.2963 D. The corresponding radius of curvature can be found using:

$$r_1 = \frac{n' - n}{F_1}$$

$$r_1 = \frac{1.49 - 1.00}{+58.2963} = +8.4053^{-3} \text{ m}$$

The front surface radius of the central distance portion of the contact lens is therefore +8.4053 mm.

We now have to do the same again for the peripheral (near) portion. If the distance prescription is −1.00 D and the near addition +2.25, the BVP of the near portion is +1.25 D. This value is taken be equal to L_2' which is the starting point for the step-back ray-trace. Both F_2 and the EAD remain unchanged.

Vergences (D)   Distances (m)

$L_2' = +1.2500 \text{ D}$

$L_2 = L_2' - F_2$

$L_2 = +1.2500 - (-59.7561)$

$L_2 = +61.0061 \text{ D} \quad \rightarrow \quad l_2 = \dfrac{1}{+61.0061} = +0.01639 \text{ m}$

$$l_1' = l_2 + \left(\frac{t_{cl}}{n_{cl}}\right) \text{ (step-back)}$$

$\leftarrow l_1' = +0.01639 + 1.3423^{-4} = +0.01653 \text{ m}$

$$L_1' = \frac{1}{+0.01653} = +60.5106 \text{ D}$$

As $L_1 = 0.00$, $L_1 = F_1$

$F_1 = +60.5106 \text{ D}$

The surface power of the front peripheral near portion is therefore +60.5106 D. The corresponding radius of curvature can be found using:

$$r_1 = \frac{n' - n}{F_1}$$

$$r_1 = \frac{1.49 - 1.00}{+60.5106} = +8.0977^{-3} \text{ m}$$

The front surface radius of the peripheral near portion of the contact lens is therefore +8.0977 mm.

The question asks for all the surface radii so:

- r_1 distance vision (DV) = +8.4053 mm
- r_1 near vision (NV) = +8.0977 mm
- r_2 = +8.2000 mm

Example 25.8

A CD back surface solid bifocal contact lens is made to the prescription +5.00 DS, Add +2.25 DS. The lens has a uniform thickness of 0.25 mm and is manufactured using a material of refractive index 1.490. The lens is to be fitted to a patient with keratometry readings of 7.80 mm along all meridians and in alignment with the cornea, with no corneal clearance. Calculate all the surface radii if the near portion is fitted in alignment with the cornea. The refractive index of the tears is assumed to be 1.336.

Our starting point for this question is the fact that the near portion is in alignment with the cornea. Once again, as this is an alignment fit we can ignore the optical effect of the tears in the near portion. As the near portion is fitted in alignment with the contact lens, the BOZR of the near portion (r_2) is the same as the corneal radius. We therefore know r_2 for the near portion. The unknown radii are r_1 (the FOZR) and the back surface radius (r_2) of the central distance portion. As usual, we must start with values that we know. The first task is therefore to step back through the peripheral (near) portion of the contact lens to find r_1. The lens of course is a *thick* lens and is illustrated in **Figure 25.17**. We can

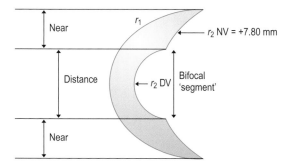

Figure 25.17 Diagram for Example 25.8. Centre distance back surface solid bifocal. The radius r_1 is a constant over the whole of the front surface and is found by step-back ray tracing right to left through the peripheral near portion. The curve r_2 distance vision (DV) is found by step-along ray-tracing left to right from r_1. NV near vision.

determine the back surface power of the contact lens (F_2) in the near portion and find the EAD:

$$F_2 = \frac{1-n}{r_2} = \frac{1.00-1.49}{+7.80^{-3}} = -62.8205\,\text{D}$$

The EAD (with t_{cl} substituted in metres) is given by:

$$\text{EAD} = \frac{t_{cl}}{n_{cl}} = \frac{2.5^{-4}}{1.49} = 1.6778^{-4}\,\text{m}$$

Now that the EAD has been determined, the refractive index between the two surfaces can be assumed to be **equivalent to that of air** ($n = 1$). This value is used in the following ray-trace. A step-back ray-trace is used to find the front surface power. As always, the computation takes place in two columns, one for vergences and the other for distances. The first ray-trace finds the radius of curvature of the front surface of the contact lens. As we are ray-tracing through the near portion, we must use the near prescription for L_2'. The near prescription is the distance prescription plus the addition. In this example, this value is +7.25 D.

Vergences (D)   Distances (m)

$L_2' = +7.2500\,\text{D}$

$L_2 = L_2' - F_2$

$L_2 = +7.2500 - (-62.8205)$

$L_2 = +70.0705\,\text{D} \rightarrow l_2 = \dfrac{1}{+70.0705} = +0.01427\,\text{m}$

$$l_1' = l_2 + \left(\frac{t_{cl}}{n_{cl}}\right) \text{(step-back)}$$

$\leftarrow l_1' = +0.01427 + 1.6778^{-4} = +0.01444\,\text{m}$

$L_1' = \dfrac{1}{+0.01444} = +69.2563\,\text{D}$

As $L_1 = 0.0000$, $L_1 = F_1$

$F_1 = +69.2563\,\text{D}$

The surface power of the front surface of the contact lens is therefore +58.2963 D. The corresponding radius of curvature can be found using:

$$r_1 = \frac{n'-n}{F_1}$$

$$r_1 = \frac{1.4900-1.0000}{+69.2563} = +7.0752^{-3}\,\text{m}$$

The front surface radius of the contact lens is therefore +7.0752 mm.

The above value has been obtained by ray-tracing through the *near* portion using the full *near* prescription. We must now find the radius of curvature of the back surface of the distance portion. There are two ways of approaching this: method 1 ignores the tears and assumes that the lens is off the eye and in air, whereas method 2 assumes that the lens is on the eye and therefore considers the optical effect of the tears.

Method 1

We must first find the addition provided by the lens when it is off the eye and in air. This is achieved using:

$$\text{Add in air} = \text{Add on eye} \times \frac{n_{cl}-1}{n_{cl}-n_{tears}}$$

Substitution into the above expression gives:

$$\text{Add in air} = +2.25 \times \frac{1.490-1}{1.490-1.336} = +7.1591\,\text{D}$$

The distance prescription in air is therefore the full near prescription minus the addition in air:

Distance Rx = Near Rx − Addition in air

Distance Rx = +7.2500 − (+7.1591) = +0.0909 D

We now perform a step-along ray-trace through the central zone (left to right) to find the radius of curvature of the back surface of the distance portion. We of course already know the power of the front surface of the lens (F_1) and the EAD from the previous part of this calculation. The above distance Rx is taken to be equal to L_2'. As always, the computation takes place in two columns, one for vergences and the other for distances.

Vergences (D)   Distances (m)

$L_1 = 0.0000$

$F_1 = +69.2563\,D$

$L_1' = L_1 + F_1 = 69.2563\,D \quad \rightarrow \quad l_1' = \dfrac{n}{L_1'}$

$$l_1' = \dfrac{1}{+69.2563} = +0.01444\,m$$

$$l_2 = l_1' - \left(\dfrac{t_{cl}}{n_{cl}}\right)$$

$L_2 = \dfrac{n}{l_2} \quad \leftarrow \quad l_2 = +0.01444 - 1.6778^{-4} = +0.01427\,m$

$L_2 = \dfrac{1}{+0.01427} = +70.0705\,D$

$F_2 = +0.0909 - (+70.0705) = -69.9796\,D$

Finally, the radius of curvature of the back surface of the central distance portion can be found using:

$$r_2 = \dfrac{1 - n_{cl}}{F_2} = \dfrac{1.000 - 1.490}{-69.9796} = +7.0020^{-3}\,m$$

The radius of curvature of the back surface of the central distance portion is therefore **+7.0020 mm**.

Method 2

This alternative approach assumes that the lens is on the eye. We were told indirectly in the question that the radius of curvature of the back surface of the near peripheral portion is +7.8000 mm. The power of this back peripheral surface with the lens on-the-eye is:

$$F_2 = \dfrac{n_{tears} - n_{cl}}{r_2} = \dfrac{1.336 - 1.490}{+7.80^{-3}} = -19.7436\,D$$

If the on-eye-reading addition is +2.25 D, the back surface of the central distance portion must be 2.25 D more negative, i.e. −21.9936. The corresponding radius of curvature can be found using:

$$r_2 = \dfrac{n_{tears} - n_{cl}}{F_2} = \dfrac{1.336 - 1.490}{-21.9936} = +7.0020^{-3}\,m$$

The radius of curvature of the back surface of the central distance portion is therefore +7.0020 mm, which agrees with the result from method 1.

The question asks for all the surface radii, so:

- $r_1 = +7.0752\,mm$
- $r_2\,DV = +7.0020\,mm$
- $r_2\,NV = +7.8000\,mm$

Concluding points for Chapter 25

- Even with modern and sophisticated lens designs, monovision still has a higher success rate when compared with other methods of correcting presbyopia with contact lenses.
- It is essential to use diagnostic contact lenses that are as close as possible to the final prescription in order to obtain a good idea of potential success for any particular type of bifocal/multifocal contact lens.
- Bifocal/multifocal contact lens fitting often relies on a flexible and creative approach, including the combination of different lenses and/or methods.
- Always use a trial frame for subjective examination because a phoropter may influence the pupil size and also make it more difficult to obtain the reading addition under normal conditions for near vision.
- Once again the mathematical material in this chapter relies on an understanding of basic optical principles and the ability to perform step-along and step-back ray-traces.

Further recommended reading

De Carle J I (1997) Bifocal and multifocal contact lenses. In: Phillips A J, Speedwell L (eds), *Contact Lenses*, 4th edn. Butterworth-Heinemann, Oxford

Douthwaite W A (2006) *Contact Lens Optics and Lens Design*. Elsevier, Oxford

Efron N (2002) *Contact Lens Practice*. Butterworth-Heinemann, Oxford

GP Lens Institute (2006) *Correcting Presbyopia*. website www.gpli.info/correcting-presbyopia-04

Kerr C, Ruston D (2007) *The ACLM Contact Lens Year Book 2007*. Association of Contact Lens Manufacturers, London

Over-refraction Techniques in Contact Lens Practice

Caroline Christie

Introduction

This chapter discusses the necessity and techniques for the over-refraction of the contact lens patient. An over-refraction is often required when diagnostic or empirically ordered lenses are used for the initial fitting of a contact lens patient. An over-refraction is also an essential part of the after-care visit for every contact patient. The exact technique employed depends on the type of lens fitted and the required outcome. Methods for the over-refraction of spherical single vision, toric single vision and bifocal/ multifocal lenses are explained. The somewhat thorny issue of prescribing spectacles for contact lenses wearers is also addressed.

Chapter content

- Over-refraction techniques for diagnostic (trial) lenses
- Over-refraction at after-care visits
- Prescribing spectacles for contact lenses wearers

Over-refraction techniques for diagnostic (trial) lenses

A final contact lens prescription cannot be determined solely from the spectacle lens prescription. It must be determined from refracting over a contact lens of a known design with a specific radius (base curve) and back vertex power (BVP).

Rigid gas permeable lenses

Once the back optic zone radius (BOZR) has been chosen (in modern rigid gas permeable or RGP designs this is usually in alignment with flattest keratometry reading), a diagnostic contact lens of known radius and BVP is placed on the eye. A spherical over-refraction refraction is then performed in the same manner as determining the best vision sphere (BVS), outlined in detail in Chapter 10.

The contact lens over-refraction is also a means of checking the fit of an RGP lens. In an alignment fitting, the liquid lens power should be zero and the over-refraction (plus any power of the diagnostic lens) should be equal to the ocular refraction, after allowing for the vertex distance. If the over-refraction shows less negative or more positive power than the ocular refraction, a slightly flat fitting should be suspected. A steep fitting (positive liquid lens) usually gives a more negative or less positive over-refraction. If the refraction is performed through a diagnostic lens with a BOZR other than the one to be prescribed, the final contact lens power must be adjusted because the final fluid (liquid) lens prescription varies with BOZR.

> *Rule of thumb*: If the radius is in the region of 7.80 mm, a change in radius of 0.05 mm = 0.25 D change in power

Soft or silicone hydrogel lenses

Depending on the back surface design, centre thickness and water content, the optical performance and final power of a soft or silicone hydrogel lens can vary, even though the BVP, base curve and diameter are seemingly identical. It is important to note that apparently identical lenses can perform very differently on the same eye and a full assessment of lens fit and over-refraction is required to determine the correct lens specification and hence the optimum performance for the patient. It is important to note that lenses are not interchangeable from brand to brand, even when parameters appear to be identical. Small amounts of residual astigmatism that can be easily ignored when prescribing RGP lenses cannot always be so readily ignored with these lenses.

The limit for acceptable acuity with soft and silicone hydrogel lenses has historically been quoted as 1.00 DC, but critical observers may notice as little as 0.50 DC, especially if the cylinder is 'against the rule' or oblique in nature. By fully correcting the cylindrical component with a toric contact lens, a significant improvement in acuity results from the elimination of ghosting and distortion. Most toric soft and silicone hydrogel lenses now offer a 0.75 DC option and practitioners choosing not to offer this are doing their patients a disservice. They surely would not consider prescribing spectacle lenses in this rather reckless way!

Near vision should be assessed in *all* patients, because there can be many causes of near vision difficulty, especially with soft and silicone hydrogel lenses. These include borderline acuity for distance, reduced blink rate that causes drying of the lens surface, lid pressure that causes irregular buckling or flexure of the lens when the eyes are in the near vision position, and even decentration of the actual lens. As a result of differences in lens materials and manufacturing techniques, lenses from one laboratory cannot necessarily be duplicated by another merely by ordering the same nominal specification. This is especially true for silicone hydrogel products. It is common to find that binocular visual acuity is frequently better and out of all proportion to monocular results, especially with simultaneous multi-focal lenses.

Over-refraction at after-care visits

Distance and near visual acuity (binocular and monocular) must be measured and recorded during every after-care appointment. If the expected acuity is not achieved and an over-refraction does not give any improvement, retinoscopy may assist in indicating residual astigmatism, distortion, poor optical quality, excess surface deposits or even poor surface wetting. Where one eye requires more minus or plus and the other less by the same amount, this is often an indication that the right and left lenses have been reversed. Variable vision that is worse after a blink suggests a loose fitting lens or lens flexure. In such cases consider improving the fit, specifying increased centre thickness or use a material with increased modulus. Blurred vision as a result of residual astigmatism should be predictable at the initial fitting, but is sometimes not observed until lenses have fully settled, when patients typically complain of ghosting.

Flare can be experienced with small RGP lenses, especially if they decentre superiorly. In this case, consider a lens with a larger BOZD and larger total diameter; specify less edge lift (especially in the aspheric form) or change to a soft lens. Flare is uncommon with soft lenses, but can occur if the lens is badly decentred and of a high power, or with some toric designs. Where the patient cannot adapt, the lens should be re-made with a larger front optic zone diameter (FOZD) or re-fitted with a different design. In soft lenses, reports of a deterioration in vision during the course of the day may be the result of lens dehydration. In soft and RGP lenses, foggy vision as a result of greasy lenses tends to be worse immediately on insertion because of poor wetting or later in the day because of surface drying. Ensuring that care products are being used correctly, or changing to a different system or changing the lens material can help.

Remember it is *always* important to consider the possibility of ocular pathology being the cause of reduced acuity at after-care visits.

If the expected visual acuity and binocular balance are achieved the practitioner need determine only whether any additional plus is accepted. A hand-held +0.50 DS lens should blur distance acuity slightly (generally one line) if the contact lens power is correct.

Single vision lenses

First, look for any spherical over-refraction using the standard refraction technique for the BVS outlined in detail in Chapter 10. If the expected acuity is not obtained with a simple spherical over-refraction, residual astigmatism is suspected and this can best be confirmed by retinoscopy. After retinoscopy, the degree and axis of astigmatism can be refined by subjective refractions using the techniques outlined in Chapter 11. However, allowance should be made for the variation in subjective response that occurs with blinking. Longer viewing time of the alternatives should be allowed so that blinking can clear vision and stabilise the lens position. Even with these precautions, retinoscopy is often the 'main player' during an over-refraction.

It is worth considering the inclusion of a correction for residual astigmatism in any over-reading spectacles because, although reduced vision at distance is often readily accepted, poor near vision can be a major problem for prolonged near tasks.

Toric soft and toric silicone hydrogel contact lenses

Modern soft and silicone hydrogel lenses are generally fitted by the empirical method, outlined in Chapter 23. Key points for successful initial lens ordering involve performing an accurate initial refraction, correcting for back vertex distance in *all* meridians, and selecting the lower cylinder power when the cylinder falls between two available powers.

At the issue appointment, the lenses should be allowed to settle for about 5–10 minutes. If the fit is considered acceptable the lens orientation should be assessed using the markings on the lens. Where the lens orients with the markings aligned (no rotation) a simple ±0.25 DS over-refraction and duochrome test, to obtain the final binocular balance, should be performed and the lenses issued. However, provided that the lens is stable and re-orients to the same position each time, the lens may be re-ordered making the appropriate allowance for the mislocation, using the LARS (left add, right subtract) or the CAAS (clockwise add, anti-clockwise subtract) rule outlined in detail in Chapter 23. There is no benefit in carrying out an over-refraction at this stage because it will result only in a complicated Stoke's construction calculation (see Chapter 23).

If the lens is more than 20–30 degrees away from the expected axis, this suggests that there is inadequate stabilisation for the lid and corneal features of the patient and an alternative design should be considered.

Bifocal and multifocal lenses

The use of a phoropter is not advised when over-refracting a patient wearing bifocal or multifocal lenses because this may alter the level of light reaching the eye, therefore affecting the pupil size. It can also hinder quick assessment of power changes between distance and near and obscure facial expressions, which can often be quite informative. A ±0.25 DS twirl is a simple, versatile and quick method; alternatively hand-held trial lenses may be used.

In simultaneous vision lenses, the patient is making a mental selection between the near or distance vision image in each eye at any given time. Binocular over-refraction allows the practitioner to assess the patient's ability to combine the images from both eyes, and gives a better idea of their functional vision. This is a departure from the single vision lens over-refraction technique because neither eye is occluded or fogged during over-refraction.

Other than to record acuities, it may be preferable to use a 'real-life' scenario to judge power changes, rather than a conventional chart. Asking the patient to look out of a window often works really well. A disappointing trial after a 'successful' fitting can often be pre-empted in this way. Often a ±0.25 DS change has minimal or no effect in the consulting room but is more noticeable when 'true' distance is viewed. When over-refracting, use a large target such as the 6/9 line on a letter chart. This helps prevent over-minusing in the final lens power, but does not represent best acuity, which can be recorded later. A minimum settling time of 15–20 minutes is recommended because there appears to be some neurological adaptation in that short time and many patients give a much more positive response after 20 minutes.

When dealing with multifocal contact lenses, it is extremely important that after each demonstrated power change, the effect of that change should be checked at both distance and near. A change should be made only where there is significant improvement in distance vision without compromise at near, and vice versa. The objective is to achieve the best balance of distance and near vision, according to the patient's priorities.

Most final prescriptions will be within 0.25 DS of the initial trial lens selection. This minimum change can have a significant effect on acuity and often considerably more than would be expected with a single vision lens. Making large changes early in the routine may well 'over-shoot' the optimum lens power. The author recommends that two changes should be sufficient if this modality is going to be successful. Further changes are unlikely to succeed and not only waste chair time but are also disheartening to both patient and practitioner.

Prescribing spectacles for contact lenses wearers

Prescribing spectacle lenses to contact lens wearers can often be both time-consuming and problematic. Potential problems experienced include the following:

- A difference in retinal image size when spectacles are worn
- Different spatial perspective
- Visual distortion (sloping floors and bowed door frames)
- Restricted field of view
- Reflections, especially at night
- Intolerance to full correction, especially cylinders

Other causes include cases where no spectacles have been worn for several years, especially where there is a marked change in myopia, astigmatism or very different cylinder power and axis compared with the previous correction. Long-term contact lens monovision (deliberate or accidental) may give rise to difficulty with bifocal and even separate distance and near spectacles, because the visual system has become accustomed to working in a very different way.

In comparison with PMMA (polymethyl methacrylate) most modern RGP lenses cause few problems, although some degree of corneal moulding may be encountered. This is seen particularly with astigmatic eyes fitted with compromised spherical and even aspheric lenses, and in some instances back surface, simultaneous vision multifocals. Pronounced distortion occurs in cases of lens adhesion; in such cases refraction should be postponed until the patient's lenses are re-fitted.

Satisfactory refraction is usually achieved immediately on removal of the lenses. If the result does not correlate with the contact lens specification, keratometry and the previous spectacle refraction, the refraction should be repeated. Where a PMMA wearer has been refitted with RGP lenses, it is advisable to wait about 8–12 weeks before refracting for spectacles, by which time most corneal changes will have resolved. This is confirmed by monitoring the keratometry readings or, better still, by using topography maps.

Corneal curvature and refractive changes are far less common with soft lenses because there is little mechanical moulding. When oedema is present with soft lenses, it generally extends from limbus to limbus without localised changes in curvature or physical distortion. Refraction on removal of soft lenses generally gives a good result, but where there is any doubt it should be repeated the next day and in the morning.

Tips

- Explain potential problems to the patient before actually dispensing spectacles.
- Review any pre-contact lens spectacles, however old!
- Give maximum possible binocular addition.
- Under-correct or omit cylinders, especially if oblique.
- Repeat refraction on another occasion if it does not correlate with the contact lenses, keratometry and previous correction.

Concluding points for Chapter 26

- Distance acuity should be assessed separately from near and intermediate vision, because different and independent problems can arise.
- Binocular acuity is often significantly better than that expected from monocular results, particularly with multifocal contact lenses.

- The retinoscope is an important diagnostic tool to ascertain reasons for suboptimal acuity such as residual astigmatism.
- Spectacle blur should not be present with RGP, soft and silicone hydrogel lenses. It is, however, to be expected with PMMA lenses (now rarely fitted).
- Remember to consider the possibility of ocular pathology as the cause of reduced acuity at after-care visits.

Accommodation and Convergence: Spectacle versus Contact Lenses

Andrew Keirl

Introduction

The topics of accommodation and convergence were discussed in Chapters 13 and 15 respectively. This chapter does not introduce any new material and its objectives are quite straightforward. The reader should be able to compare, with the aid of appropriate calculations, the accommodative and convergence demands for ametropic patients corrected by both spectacles and contact lenses, and discuss the optometric significance of such calculations. These objectives are achieved using Examples 27.1 and 27.2. The reader is advised to revise the explanations of the following terms that were introduced in Chapter 13:

- Accommodation
- Amplitude of accommodation
- Spectacle accommodation
- Ocular accommodation
- Convergence

Chapter content

- Revision of basic terminology
- Calculations of ocular accommodation and convergence

Revision of basic terminology

Accommodation

The eye's ability to adjust in power in order to focus on objects at different distances.

The amplitude of accommodation

The maximum amount by which an eye can change in power.

Spectacle accommodation

The accommodation required to neutralise negative vergence from a near object measured in the plane of the spectacle lens.

Ocular accommodation

The accommodation required to neutralise negative vergence from a near object measured in the plane of the eye.

Convergence

The movement (rotation) required from the primary position, for the eyes to fixate an object point on the midline.

Calculation of ocular accommodation and convergence

Examples 27.1 and 27.2 use the same methods as discussed in Chapters 13 and 15.

Example 27.1

A patient is corrected using a pair of −5.00 D spectacle lenses positioned at a vertex distance of 15 mm and centred for distance vision. The distance interpupillary distance is 64 mm and the centres of rotation of the eyes lie 27 mm behind the spectacle plane. An object is placed on the

midline and at a distance of 20 cm from the spectacle plane.

Calculate the ocular accommodation and convergence required to view the near object when the spectacles are worn.

Calculate the ocular accommodation and convergence required to view the same near object if the patient is corrected with contact lenses.

Ocular accommodation when spectacles are worn

The determination of the ocular accommodation involves a step-along ray-trace from the near object to the eye. We need to find the vergence arriving at the eye L_2 and then compare this with the ocular refraction K. The difference between the two is the ocular accommodation.

The step-along is performed using two columns, one for vergences (in dioptres) and one for distances (in metres). The vergences are illustrated in **Figure 27.1**.

Vergences (D)   Distances (m)

$$L_1 = \frac{1}{l} \qquad \leftarrow \qquad l_1 = -0.20\,\text{m}$$

$$L_1 = \frac{1}{-0.20} = -5.00\,\text{D}$$

$$L_1 = -5.00\,\text{D}$$

$$F_1 = -5.00\,\text{D}$$

$$L_1' = L_1 + F_1$$

$$L_1' = -5.00 + (-5.00) \qquad \rightarrow \qquad l_1' = \frac{n}{L_1'} = \frac{1}{-10.00}$$

$$= -10.00\,\text{D} \qquad\qquad = -0.10\,\text{m}$$

$$l_2 = l_1' - d$$

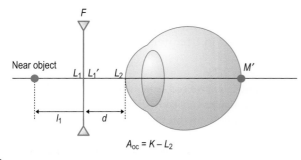

$$A_{oc} = K - L_2$$

288 **Figure 27.1** Ocular accommodation.

$$L_2 = \frac{n}{l_2} = \frac{1}{-0.115} \qquad \leftarrow \qquad l_2 = -0.10 - 0.015$$

$$= -8.69\,\text{D} \qquad\qquad = -0.115\,\text{m}$$

The vergence arriving at the eye from the near object is −8.69 D. We must now compare this with the ocular refraction to determine the ocular accommodation. So, using the distance spectacle correction and vertex distance:

$$K = \frac{F_{sp}}{1 - (dF_{sp})} \qquad K = \frac{-5.00}{1 - (0.015 \times -5.00)} = -4.65\,\text{D}$$

Now:

$$A_{oc} = K - L_2$$

$$A_{oc} = -4.65 - (-8.69) = +4.04\,\text{D}$$

Essentially, when viewing the near object, the eye receives more negative vergence than it needs (L_2 is what it is receiving and K is what it needs). The eye therefore needs to accommodate to reduce L_2 to the magnitude of K.

Convergence when spectacles are worn

From inspection of **Figure 27.2**, the following expression is used to calculate the convergence:

$$\text{Tan}\,\theta = \frac{h'}{l' + s}$$

Note that when finding the angle θ, the sign of l' is ignored in order to produce a positive value θ. The given data is summarised as follows:

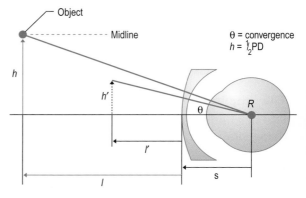

Figure 27.2 Convergence for a myopic patient corrected with a spectacle lens.

- $F = -5.00\,\text{D}$
- Interpupillary distance is 64 mm, so the ½ PD and therefore $h = 32\,\text{mm}$
- $s = 27\,\text{mm}$
- $l = -0.20\,\text{m}$ from the spectacle plane

$$L = \frac{1}{l} \qquad L = \frac{1}{-0.20} = -5.00\,\text{D}$$

$$L' = L + F \qquad L' = -5.00 + (-5.00) = -10.00\,\text{D}$$

$$l' = \frac{1}{L'} \qquad l' = \frac{1}{-10.00} = -0.1\,\text{m}$$

$$h' = h \times \frac{l'}{l} \qquad h' = 0.032 \times \frac{-0.100}{-0.200} = +0.016\,\text{m}$$

All values in the above equation are in metres.

And finally:

$$\text{Tan}\,\theta = \frac{h'}{l' + s} = \frac{+0.016}{|0.100| + 0.027} = 0.1260$$

In order to give a positive angle, the minus sign for l' has been ignored:

$$\text{Tan}^{-1} = 7.18°$$

The rotation or convergence required to view the near object placed 0.20 m from the lens is therefore 7.18°. In prism dioptres the convergence is given by:

$$p = 100 \tan \theta$$

which, in this example is:

$$p = 100 \tan 7.18 = 12.60\,\Delta$$

Ocular accommodation and convergence when contact lenses are worn

The object position for the contact lens wearer is the stated object position plus the vertex distance. Hence, l is 0.20 m + 0.015 = −0.215 m.

$$L = \frac{1}{l}$$

$$L = \frac{1}{-0.215} = -4.65\,\text{D}$$

As a contact lens wearer is effectively someone with artificial emmetropia, the eye must accommodate to neutralise the above negative vergence. The ocular accommodation A_{oc} is therefore +4.65 D.

The calculation of the convergence for the contact lens wearer is exactly the same as the calculation of the convergence for someone with emmetropia. The situation is illustrated in Figure 27.3. In the case of someone with emmetropia (or a contact lens wearer), there is of course no image formed because there is no spectacle lens. The eye can simply rotate to view the object directly. From Figure 27.3 the distance between the centre of rotation of the eye R and the object is given by:

$$l + s$$

and the convergence or rotation required to view the object is given by:

$$\text{Tan}\,\theta = \frac{h}{l + s}$$

Once again, h is equivalent to the ½ PD. The data used in this example is:

- $l = -20\,\text{m}$
- $s = 27\,\text{mm}$
- $h = 32\,\text{mm}$.

All values are entered in metres and, to give a positive angle, the minus sign for l' is ignored.

$$\text{Tan}\,\theta = \frac{h}{l + s} = \frac{0.032}{|0.200| + 0.027} = 0.1410$$

$$\text{Tan}^{-1} = 8.02°$$

The rotation or convergence required to view the near object is therefore 8.02°. The convergence in prism dioptres is:

$$p = 100 \tan 8.02 = 14.10\,\Delta$$

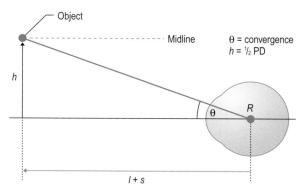

Figure 27.3 Convergence for an emmetropic patient or contact lens wearer.

Table 27.1 Summary of the results of Example 27.1

	Ocular accommodation (D)	Convergence (°)
Spectacles	+4.04	7.18
Contact lenses	+4.65	8.02

The patient therefore needs to accommodate *and* converge *more* when wearing contact lenses than when wearing spectacles.

Example 27.2

A patient is corrected using a pair of +6.00 D spectacle lenses positioned at a vertex distance of 12 mm and centred for distance vision. The distance interpupillary distance is 66 mm and the centres of rotation of the eyes lie 25 mm behind the spectacle plane. An object is placed on the midline and at a distance of 40 cm from the spectacle plane.

Calculate the ocular accommodation and convergence required to view the near object when the spectacles are worn.

Calculate the ocular accommodation and convergence required to view the same near object if the patient is corrected with contact lenses.

Ocular accommodation when spectacles are worn

Vergences (D)   Distances (m)

$$L_1 = \frac{1}{l} \qquad \leftarrow \qquad l_1 = -0.40\,\text{m}$$

$$L_1 = \frac{1}{-0.40} = -2.50\,\text{D}$$

$$L_1 = -2.50\,\text{D}$$

$$F_1 = +6.00\,\text{D}$$

$$L_1' = L_1 + F_1$$

$$L_1' = -2.50 + (+6.00) \qquad \rightarrow \qquad l_1' = \frac{n}{L_1'} = \frac{1}{+3.50}$$

$$= +3.50\,\text{D} \qquad\qquad\qquad = 0.2857\,\text{m}$$

$$l_2 = l_1' - d$$

$$L_2 = \frac{n}{l_2} = \frac{1}{+0.2737} \qquad \leftarrow \qquad l_2 = +0.2857 - 0.012$$

$$= +3.65\,\text{D} \qquad\qquad\qquad = +0.2737\,\text{m}$$

The vergence arriving at the eye from the near object is +3.65 D. We must now compare this with the ocular refraction to determine the ocular accommodation.

So, using the distance spectacle correction and vertex distance:

$$K = \frac{F_{sp}}{1 - (dF_{sp})} \qquad K = \frac{+6.00}{1 - (0.012 \times +6.00)}$$

Now: $\qquad\qquad\qquad\qquad = +6.46\,\text{D}$

$$A_{oc} = K - L_2$$

$$A_{oc} = +6.46 - (+3.65) = +2.81\,\text{D}$$

Essentially, when viewing the near object, the eye is not receiving enough positive vergence (L_2 is what it is receiving and K is what it needs). The eye therefore needs to accommodate to increase L_2 to the magnitude of K.

Ocular accommodation when contact lenses are worn

The object position for the contact lens wearer is the stated object position plus the vertex distance. Hence, l is 0.40 m + 0.012 = −0.412 m.

$$L = \frac{1}{l}$$

$$L = \frac{1}{-0.412} = -2.43\,\text{D}$$

As a contact lens wearer effectively has artificial emmetropia, the eye must accommodate to neutralise the above negative vergence. The ocular accommodation A_{oc} is therefore +2.43 D.

Convergence when wearing spectacles

Using **Figure 27.4**, the expression for finding the convergence is:

$$\text{Tan}\,\theta = \frac{h'}{l' - s}$$

When finding the angle θ the sign of h' is ignored in order to produce a positive value for θ. The data given in the example is:

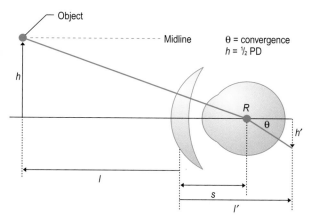

Object

Midline   θ = convergence
           h = ½ PD

R

θ

h'

l

s

l'

Figure 27.4 Convergence for a hypermetropic patient corrected using a spectacle lens.

- $F = +6.00\,D$
- Interpupillary distance is 66 mm, so the ½ PD and $h = 33$ mm
- $s = 25$ mm
- $l = -0.40$ m from the spectacle plane

$$L' = \frac{1}{l} = \frac{1}{-0.40} = -2.50\,D$$

$$L' = L + F = -2.50 + (+6.00) = +3.50\,D$$

$$l' = \frac{1}{L'} = \frac{1}{+3.50} = +0.2857\,m$$

$$h' = h \times \frac{l'}{l} = 0.033 \times \frac{+0.2857}{-0.4000} = -0.0236\,m$$

All values in the above equation are in metres.

$$\text{Tan}\,\theta = \frac{h'}{l' - s} = \frac{|0.0236|}{0.2857 - 0.025} = 0.0904$$

To give a positive angle, the minus sign for h' has been ignored:

$$\text{Tan}^{-1} = 5.17°$$

The rotation or convergence required to view the near object is therefore 6.53°
The convergence in prism dioptres is:

$$p = 100\tan 5.17 = 9.04\,\Delta$$

Convergence when contact lenses are worn

The convergence or rotation required to view the object is given by:

Table 27.2 Summary of the results of Example 27.2

	Ocular accommodation (D)	Convergence (°)
Spectacles	+2.81	7.18
Contact lenses	+2.43	5.17

$$\text{Tan}\,\theta = \frac{h}{l + s}$$

Once again, h is equivalent to the ½ PD. The data used in this example is:

- $l = -40$ m
- $s = 27$ mm
- $h = 33$ mm

All values are entered in metres and, to give a positive angle, the minus sign for l' is ignored:

$$\text{Tan}\,\theta = \frac{h}{l + s} = \frac{0.033}{|0.400| + 0.025} = 0.0776$$

$$\text{Tan}^{-1} = 4.44°$$

The rotation or convergence required to view the near object is therefore 4.44°. The convergence in prism dioptres is:

$$p = 100\tan 4.44 = 7.76\,\Delta$$

The patient therefore needs to accommodate *and* converge *less* when wearing contact lenses than when wearing spectacles.

Concluding points for Chapter 27

- If myopic, more accommodation *and* convergence is needed with contact lenses than with spectacles.
- If hypermetropic, less accommodation *and* convergence is needed with contact lenses than with spectacles.
- So, when changing from contact lenses to spectacles (and vice versa) the accommodation convergence ratio is only minimally disturbed.

Further recommended reading

Douthwaite W A (2006) *Contact Lens Optics and Lens Design*. Elsevier, Oxford

Efron N (2002) *Contact Lens Practice*. Butterworth-Heinemann, Oxford

Rabbetts R B (1998) *Bennett & Rabbetts' Clinical Visual Optics*. Butterworth-Heinemann, Oxford

Tunnacliffe A H (1993) *Introduction to Visual Optics*. Association of the British Dispensing Opticians, London

The Axial and Radial Edge Thicknesses of a Contact Lens

Andrew Keirl

Introduction

The radial edge thickness, t_r, at a given distance from the edge of a contact lens of positive power is shown in **Figure 28.1**, where it is compared with the axial edge thickness, t_a. The axial edge thickness is measured from the back surface to the front surface in a direction parallel to the optical axis of the contact lens at a specified diameter. The radial edge thickness is measured normal to the front surface of the contact lens, along a line connecting the point in question to the centre of curvature of the front surface C_1. This line is of course, the radius of curvature of the front surface r_1, hence the term 'radial edge thickness'.

Chapter content

- Calculation of the radial edge thickness of a monocurve contact lens
- Calculation of the radial edge thickness of a bicurve contact lens

Calculation of the radial edge thickness of a monocurve contact lens

Why is it necessary to calculate the radial edge thickness of a contact lens? The answer to this question lies in the assessment of the oxygen transmission through rigid gas permeable (RGP) and soft lenses, which can be made only if the thickness of the lens is calculated. The axial edge thickness of the lens is not used for this purpose because it is usually assumed that the gas flow is via the shortest route, and the radial edge thickness is less than the axial edge thickness. Both ISO 8320 and BS 3521 recommend that the radial edge thickness of a contact lens be measured normal to its front surface.

The radial edge thickness of a contact lens (t_r) is a component of the front surface radius of curvature, r_1. Figure 28.2(a) shows a positive monocurve lens, and illustrates the centre of curvature of the front surface (C_1), the centre of curvature of the back surface (C_2), the radius of curvature of the front surface (r_1) and the radius of curvature of the back surface (r_2). The radial edge thickness, (t_r), is to be calculated at some given point from the edge of the contact lens. The distance from this point to the optical axis of the lens is given the familiar symbol y (the semi-aperture). It should be noted that y is always measured on the *front surface* because the radial edge thickness of a contact lens is measured normal to its front surface. The centre thickness of the contact lens is labelled t_c. Inspection of Figure 28.2 shows that a triangle has been completed. The three angles have been labelled A, B and C, and the sides of the triangle *opposite* each angle have been labelled a, b and c.

Examination of this triangle shows that side c is equal to the radius of curvature of the front surface minus the radial edge thickness:

$$c = r_1 - t_r$$

The length of side b is equal to the radius of curvature of the second surface (the BOZR or r_2):

$$b = r_2$$

The length of the third side of the triangle is given by:

$$a = r_2 + t_c - r_1$$

If the surface radii r_1 and r_2 and the centre thickness t_c are known or can be calculated from the information given, the two side lengths, a and b, can be determined immediately. To obtain the radial edge thickness t_r we need to deduce the length of side c. One way of determining c is to make use of the sine rule:

$$\frac{a}{\sin A} = \frac{b}{\sin B} = \frac{c}{\sin C}$$

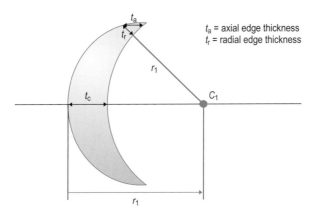

t_a = axial edge thickness
t_r = radial edge thickness

Figure 28.1 The axial and radial edge thicknesses of a monocurve contact lens. t_a = axial edge thickness; t_r = radial edge thickness. Note that as r_1 is measured from the front surface of the lens to c_1, t_r is a part of r_1.

From Figure 28.2 the angle B is given by:

$$B = 180 - \theta$$

and as angle θ is governed by the values of y and r_1

$$\sin \theta = \frac{y}{r_1}$$

so:

$$\theta = \sin^{-1}\left(\frac{y}{r_1}\right)$$

Having determined the angle B, we can use the sine rule to determine the angle A:

$$\frac{a}{\sin A} = \frac{b}{\sin B}$$

so:

$$\sin A = \frac{a}{b}\sin B$$

Angle C can be deduced from the geometry of the triangle:

$$C = 180 - (A + B)$$

The length of side c can now be determined by a second application of the sine rule:

$$\frac{c}{\sin C} = \frac{a}{\sin A}$$

so:

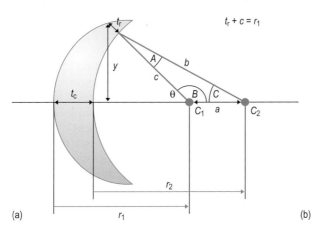

(a)

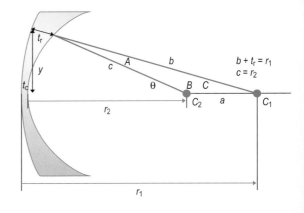

(b)

Figure 28.2 The radial edge thickness of (a) a positive monocurve contact lens and (b) a negative monocurve contact lens.

$$c = a\left(\frac{\sin C}{\sin A}\right)$$

Finally, the radial edge thickness can be obtained by rearranging the equation:

$$c = r_1 - t_r$$

to give:

$$t_r = r_1 - c$$

Example 28.1

Calculate the radial edge thickness 0.25 mm in from the edge of the lens:

C1: 7.75:10.00

where the radius of curvature of the front surface (r_1) is 7.45 mm and the centre thickness (t_c) = 0.25 mm.

Refer to Figure 28.2a during all stages of the calculation:

$$a = (t_c + r_2) - r_1$$

$$r_2 = \text{BOZR}$$

$$a = (0.25 + 7.75) - 7.45 = 0.55 \,\text{mm}$$

$$\sin\theta = \frac{y}{r_1}$$

$$y = \left(\frac{10}{2}\right) - 0.25 = 4.75 \,\text{mm}$$

$$\sin\theta = \frac{4.75}{7.45} = 0.6376$$

$$\theta = 39.6119°$$

$$B = 180 - \theta$$

$$B = 180 - 39.6119 = 140.3881°$$

Using the sine rule to find angle A:

$$\frac{a}{\sin A} = \frac{b}{\sin B}$$

and:

$$\sin A = \frac{a}{b}\sin B$$

$$b = r_2 = 7.75 \,\text{mm}$$

$$\sin A = \left(\frac{0.55}{7.75}\right)\sin 140.3881 = 0.04525$$

$$A = 2.5934°$$

$$C = 180 - (A + B)$$

$$C = 180 - (2.5934 + 140.3881) = 37.0185°$$

Using the sine rule to find side c:

$$\frac{c}{\sin C} = \frac{a}{\sin A}$$

and:

$$c = a\left(\frac{\sin C}{\sin A}\right)$$

$$c = 0.55\left(\frac{\sin 37.0185}{\sin 2.5934}\right) = 7.3183 \,\text{mm}$$

The radial edge thickness (t_r) is, therefore:

$$t_r = r_1 - c$$

$$t_r = 7.4500 - 7.3183 = 0.1316 \,\text{mm}$$

If the axial edge thickness were required, we would need to calculate the sag of each surface (using the accurate sag formula) at the given diameter. The accurate sag formula is:

$$s = r - \sqrt{r^2 - y^2}$$

$$r_1 = 7.45 \,\text{mm}$$

$$r_2 = 7.75 \,\text{mm}$$

$$y = \left(\frac{10}{2}\right) - 0.25 = 4.75 \,\text{mm}$$

The sag of the front surface is therefore:

$$s_1 = r_1 - \sqrt{r_1^2 - y^2}$$

$$s_1 = 7.45 - \sqrt{7.45^2 - 4.75^2} = 1.7107 \,\text{mm}$$

The sag of the back surface is:

$$s_2 = r_2 - \sqrt{r_2^2 - y^2}$$

$$s_2 = 7.75 - \sqrt{7.75^2 - 4.75^2} = 1.6263 \,\text{mm}$$

The axial edge thickness, t_a, is:

$$t_a = (t_c - s_1) + s_2$$

$$t_a = (0.2500 - 1.7107) + 1.6263 = 0.1656 \,\text{mm}$$

As mentioned at the start of this chapter, the radial edge thickness is shorter than the axial edge thickness.

In Example 28.1, $t_r = 0.1316$ mm and $t_a = 0.1656$ mm. Also all the surface radii are provided. However, the question may not give all the required surface radii, e.g. the back optic zone radius (BOZR) of a lens (r_2) and the back vertex power (BVP) may be given. In this case, r_1 would need to be found using step-back ray-tracing. Also, Example 28.1 used a contact lens of positive power as an example. If the lens were negative in power, the front surface would be less curved than the back surface. This means that the positions of the centres of curvature, c_1 and c_2, would reverse, i.e. c_2 would be *closer* to the lens than c_1. The calculation for side a of the triangle would therefore become:

$$a = r_1 - (t_c + r_2)$$

An example using a negative lens follows.

Example 28.2

Calculate the radial edge thickness 0.25 mm in from the edge of the lens:

C1: 7.50:9.00 BVP −4.25 D

where the centre thickness $(t_c) = 0.50$ mm and the refractive index of the lens material is 1.49. Refer to Figure 28.2b during all stages of the calculation.

In the example, the radius of curvature of the front surface of the contact lens has *not* been given. It must therefore be determined using a step-along ray-trace.

The radius of curvature of the front surface can be calculated because the refractive index, centre thickness and BOZR are known. As r_2 (the BOZR) has been given, the back surface power, F_2, can be calculated using:

$$F_2 = \frac{1 - n_{cl}}{r_2}$$

where the refractive index of air is taken to be 1, n_{cl} is the refractive index of the contact lens material and r_2 is the BOZR:

$$F_2 = \frac{1.00 - 1.49}{+7.50^{-3}} = -65.3333\,D$$

The equivalent air distance (EAD) for the *contact lens* (with t_{cl} substituted in metres) is given by:

$$EAD = \frac{t_{cl}}{n_{cl}} = \frac{5^{-4}}{1.49} = 3.3557^{-4}\,m$$

Now that the EAD has been determined. the refractive index between the two surfaces of the contact lens can be assumed to be **equal to that of air** $(n = 1)$. This value is used in the following ray-trace. A step-back ray-trace is needed to find the front surface power. As always, the computation takes place in two columns, one for vergences and the other for distances. The BVP of the contact lens in air is −4.25 D. This value is equal to L'_2 and is the starting point for the ray-trace.

Vergences (D)   Distances (m)

$$L'_2 = -4.2500\,D$$

$$L_2 = L'_2 - F_2$$

$$L_2 = -4.2500 - (-65.3333) = +61.0833\,D$$

$$L_2 = +61.0833\,D \quad \rightarrow \quad l_2 = \frac{1}{+61.0833} = +0.01637\,m$$

$$l'_1 = l_2 + \left(\frac{t_{cl}}{n_{cl}}\right) \text{ (step-back)}$$

$$L'_1 = \frac{1}{+0.01671} \qquad \leftarrow \qquad l'_1 = +0.01637 + 3.3557^{-4}$$
$$= +59.8564\,D \qquad\qquad = 0.01671\,m$$

As $L_1 = 0.00$, $L_1 = F_1$

$$F_1 = +59.8564\,D$$

The FOZR can be found using:

$$r_1 = \frac{n' - n}{F_1}$$

$$r_1 = \frac{1.49 - 1.00}{+59.8564} = +8.1863^{-3}\,m.$$

The front optic zone radius (FOZR) of the contact lens is therefore +8.1863 mm

As we now know the value of r_1, we can apply the sine rule as described in Example 28.1 to find side b and hence the radial edge thickness. It is most important that reference be made to Figure 28.2b during all stages of the calculation, which shows that:

$$a = r_1 - (t_c + r_2)$$

$$r_2 = BOZR = 7.50\,mm$$

$$t_c = 0.50\,mm$$

$$a = 8.1863 - (0.5000 + 7.5000) = 0.1863\,mm$$

$$\sin C = \frac{y}{r_1}$$

$$y = \left(\frac{9.00}{2}\right) - 0.25 = 4.25\,\text{mm}$$

$$\sin C = \frac{4.25}{8.1863} = 0.5192$$

$$C = 31.2759°.$$

Using the sine rule to find angle A:

$$\frac{a}{\sin A} = \frac{c}{\sin C}$$

and:

$$\sin A = \frac{a}{c}\sin C$$

$$c = r_2 = 7.50\,\text{mm}$$

$$\sin A = \left(\frac{0.1863}{7.5000}\right)\sin 31.2759 = 0.0129$$

$$A = 0.7389°$$

From the geometry of Figure 28.2b:

$$B = 180 - (A + C)$$

$$B = 180 - (0.7389 + 31.2759) = 147.9852°$$

Using the sine rule to find side b:

$$\frac{B}{\sin b} = \frac{a}{\sin A}$$

and:

$$b = a\left(\frac{\sin B}{\sin A}\right)$$

$$b = 0.1863\left(\frac{\sin 147.9852}{\sin 0.7389}\right) = 7.6586\,\text{mm}$$

The radial edge thickness (t_r) is therefore:

$$t_r = r_1 - b$$

$$t_r = 8.1863 - 7.6586 = 0.5277\,\text{mm}$$

Note that the calculation of the angle θ is not required when the lens is negative.

Calculation of the radial edge thickness of a bicurve contact lens

The calculation of the radial edge thickness of a contact lens now needs to be extended to include

calculations involving a bicurve lens which is illustrated in Example 28.3.

Example 28.3
Calculate the radial edge thickness 0.75 mm inward from the periphery of the lens:

C2: 7.70:6.50/9.00:9.50

where the BVP is +4.50 DS, the refractive index of the lens material 1.50 and the centre thickness 0.50 mm.

A bicurve (C_2) contact lens consists of back surface with a central radius and one flatter peripheral curve with a sharp transition between the two curves. This is illustrated in the somewhat exaggerated **Figure 28.3**, in which a monocurve (C_1) lens is compared with a bicurve (C_2) lens. An explanation of the numbers given above is as follows: **7.70** is the BOZR (radius of curvature of the central zone); **6.50** is the back optic zone diameter (BOZD or the diameter of the central zone); **9.00** is the radius of the peripheral curve; and **9.50** is the total diameter of the lens. All measurements are in millimetres.

There are three steps to solving this problem:

1. The radius of curvature of the front surface of the contact lens has not been given in the question. This needs to be calculated.
2. Sags need to be applied to the central zone of the lens in order to determine side a of the triangle ABC shown in Figure 28.5.

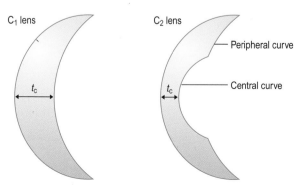

Figure 28.3 Monocurve versus Bicurve.

3. The radial edge thickness can now be calculated using the trigonometric method previously discussed.

Step 1

The radius of curvature of the front surface can be calculated because the refractive index, centre thickness and BOZR are known. We will call the power of the back optic zone F_2 and the BOZR r_2. Using the familiar equation, we have:

$$F_2 = \frac{1 - n_{cl}}{r_2}$$

where the refractive index of air is taken to be 1, n_{cl} is the refractive index of the contact lens material and r_2 the radius of curvature of the back optic (central) zone:

$$F_2 = \frac{1.00 - 1.50}{+7.70^{-3}} = -64.9351 \, D$$

The EAD for the *contact lens* (with t_{cl} substituted in metres) is given by:

$$EAD = \frac{t_{cl}}{n_{cl}} = \frac{5^{-4}}{1.50} = 3.3333^{-4} \, m$$

Now that the EAD has been determined, the refractive index between the two surfaces of the contact lens can be assumed to be **equal to that of air** ($n = 1$). This value is used in the following ray-trace. A step-back ray-trace is needed find the front surface power. As always, the computation takes place in two columns, one for vergences and the other for distances. The BVP of the contact lens is +4.50 D. This value is equal to L_2' and is the starting point for the ray-trace.

Vergences (D)   Distances (m)

$L_2' = +4.5000 \, D$

$L_2 = L_2' - F_2$

$L_2 = +4.5000 - (-64.9351) = +69.4351 \, D$

$L_2 = +69.4351 \, D \qquad \rightarrow \qquad l_2 = \dfrac{1}{+69.4351}$

$$= +0.01440 \, m$$

$$l_1' = l_2 + \left(\frac{t_{cl}}{n_{cl}} \right) \text{(step-back)}$$

$L_1' = \dfrac{1}{+0.01473} \qquad \leftarrow \qquad l_1' = +0.01440 + 3.3333^{-4}$

$$= +67.8643 \, D \qquad\qquad\qquad = 0.01473 \, m$$

As $L_1 = 0.0000$, $L_1 = F_1$

$$F_1 = +67.8643 \, D$$

The FOZR can be found using:

$$r_1 = \frac{n' - n}{F_1}$$

$$r_1 = \frac{1.50 - 1.00}{+67.8643} = +7.3676^{-3} \, m$$

The FOZR of the contact lens is therefore **+7.3676 mm.**

Step 2

The next step of the problem involves the central zone (the BOZR and BOZD). Inspection of **Figure 28.4** shows the sags of two curves, one formed by the central curve (s_1) and one by the peripheral curve (s_2). The sags of these curves must be used in order to determine side a of the triangle ABC shown in **Figure 28.5**. Side a is an unknown value. Side $b = r_2$, which in this case is the back peripheral radius (9.00 mm). Side c is a component of the front surface radius of curvature r_1 and is determined in stage 3 by making use of the sine rule. You are advised to inspect Figures 28.4 and 28.5 and locate the distance p. If the lens were monocurve (C1) in design, p would be equivalent to the centre thickness of this monocurve lens. If p is known, then:

$$a = (r_2 + p) - r_1$$

The radial edge thickness can be determined using the trigonometrical method outlined in

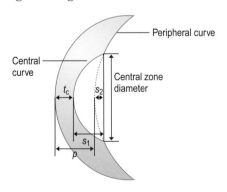

Figure 28.4 The bicurve contact lens: s_1 is the sag of the central curve at the central zone diameter; s_2 is the sag of the peripheral curve at the central zone diameter; $p = (t_c + s_1) - s_2$ and is equivalent to the centre thickness of a C1 lens.

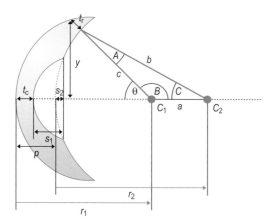

Figure 28.5 Diagram for the calculation of the radial edge thickness of a bicurve contact lens. The radial edge thickness of a bicurve contact lens $a = (r_2 + p) - r_1$.

Example 28.1. We need therefore to find p. With the aid of the Figure 28.4:

$$p = (t_c + s_1) - s_2$$

where t_c is the centre thickness of the contact lens, s_1 the sag of the central curve (7.70 mm) at an aperture of 6.50 mm (the BOZD) and s_2 is the sag of 9.00 mm (the back peripheral radius) at an aperture of 6.50 mm (the BPZD).

Using the familiar sag formula

$$s = r - \sqrt{r^2 - y^2}$$

and with reference to Figure 28.4, the values of the sags of the two curves s_1 and s_2 can be calculated. Care needs to be taken when inputting values into the sag formula (refer to Figure 28.4 at all times). Using

$$s_1 = r_c - \sqrt{r_c^2 - y^2}$$

where $r_c = 7.70$ mm (the radius of curvature of the back central curve) and $y = 3.25$ mm, y in this case being the semi-aperture of the central zone and is given by 6.50/2:

$$s_1 = 7.70 - \sqrt{7.70^2 - 3.25^2} = 0.7195 \text{ mm}$$

$$s_2 = r_p - \sqrt{r_p^2 - y^2}$$

where $r_p = 9.00$ mm (the radius of curvature of the back peripheral curve) and $y = 3.25$ mm (the semi-aperture of the central zone).

$$s_2 = 9.00 - \sqrt{9.00^2 - 3.25^2} = 0.6073 \text{ mm}$$

From Figure 28.4:

$$p = (t_c + s_1) - s_2$$

$$p = (0.5000 + 0.7195) - 0.6073 = 0.6122 \text{ mm}.$$

And from Figure 28.5:

$$a = (r_2 + p) - r_1$$

$$a = (9.0000 + 0.6122) - 7.3676 = 2.2446 \text{ mm}$$

Step 3

As side a of triangle ABC (Figure 28.5) is now known, we can now apply the sine rule exactly as described in Example 28.1 for the monocurve lens, to find side c and hence the radial edge thickness. It is most important that reference be made to Figure 28.5 during all stages of the calculation:

$$a = 2.2446 \text{ mm}$$

$$\sin \theta = \frac{y}{r_1}$$

$$y = \left(\frac{9.50}{2}\right) - 0.75 = 4.00 \text{ mm}$$

(9.50 mm is the total lens diameter; 0.75 mm is the point at which the radial edge thickness is to be found).

$$\sin \theta = \frac{4.0000}{7.3676} = 0.5429$$

$$\theta = 32.8825°$$

$$B = 180 - \theta$$

$$B = 180 - 32.8825 = 147.1175°$$

$$\frac{a}{\sin A} = \frac{b}{\sin B}$$

and:

$$\sin A = \frac{a}{b} \sin B$$

$b = r_2 = 9.00$ mm (the back peripheral radius).

$$\sin A = \frac{2.2446}{9.000} \sin 147.1175 = 0.1354$$

$$A = 7.7820°$$

$$C = 180 - (A + B)$$

$$C = 180 - (7.7820 + 147.1175) = 25.1005°$$

299

$$\frac{c}{\sin C} = \frac{a}{\sin A}$$

and:

$$c = a\left(\frac{\sin C}{\sin A}\right)$$

$$c = 2.2446\left(\frac{\sin 25.1005}{\sin 7.7820}\right) = 7.0321\,\text{mm}$$

The radial edge thickness (t_r) is, therefore:

$$t_r = r_1 - c$$

$$t_r = 7.3676 - 7.0321 = 0.3355\,\text{mm}$$

Concluding points for Chapter 28

In addition to introducing new material, the examples in this chapter demonstrate how other aspects of optics, such as ray-tracing and sags, can form integral and essential parts of the solution to a problem.

The calculation of the radial edge thickness in itself is not difficult, but the keys to understanding this topic are the construction of the diagrams and the geometry contained within the diagrams.

For a more detailed discussion of the thicknesses of contact lenses, the reader should consult Douthwaite (2006).

Further recommended reading

Douthwaite W A (2006) *Contact Lens Optics and Lens Design*. Elsevier, Oxford

Efron N (2002) *Contact Lens Practice*. Butterworth-Heinemann, Oxford

Contact Lens Verification

Caroline Christie

Introduction

Responsibility for the quality and subsequent clinical performance of contact lenses rests with the practitioner who supplies them. It is therefore essential that every contact lens is measured and/or inspected carefully to determine both its quantitative and its qualitative accuracy. All lenses should be checked:

- To verify the accuracy of lenses being supplied to the patient:
  - in the case of disposable lenses this would simply involve checking the details on individual blister packs or cartons
- To establish the specification of patient's existing lenses:
  - to confirm that lenses are being worn in the correct eyes (if no engravings)
  - to confirm that current and old lenses have not been mixed up
- If lens parameters are thought to have altered or distorted through use
- To verify the continued accuracy of trial (diagnostic) lenses being used.

Various instruments, although designed for a specific purpose, may also be adapted to check certain contact lens parameters and are outlined in this chapter.

Remember that any instrument used for accurate measurement of lens parameters requires regular calibration and checking against its own test standards.

Chapter content

- Rigid lens verification
- Soft and silicone hydrogel lens verification
- Dimensional tolerances

Rigid lens verification

The back optic zone radius (BOZR) and back vertex power (BVP) are arguably the most important dimensions in respect of a rigid contact lens. The former is vulnerable to change and distortion over a period of time and measurement is an essential feature of a routine after-care examination where problems with the lenses have been reported or observed. Verification also resolves the problem of lenses of similar power being transposed right for left by the patient, or indeed by the clinician, and also the wearing by the patient of an older 'spare' lens. It is not uncommon to be asked to provide after-care for lenses of unknown specification fitted elsewhere. Verification of the key parameters can help in the clinical decision-making process if a duplicate lens is the answer for a particular patient rather than a complete re-fit.

Both manufacturers and practitioners alike must accept certain tolerance limits on dimensional and optical properties and these should realistically reflect the accuracy of lens measurement and clinical limits of accuracy in fitting. The standards employed in the UK are those of the British Standards Institution and the International Standards Organization and are summarised at the end of this chapter.

Measurement of power

Instrument: focimeter

Method

1. Clean and dry the lens.
2. Set the focimeter into the vertical position or employ a special lens holder.
3. Place the lens as close as possible to the focimeter using a very small stop or the lens holder.
4. To measure the BVP, place the convex surface of the lens towards the focimeter: the reading may still give more plus or less minus than the true BVP, especially if the radius of the lens is very steep.
5. Note the image quality; distortion indicates a poor optic.
6. A good image does not, however, guarantee a distortion-free optic because a small stop is used and therefore only the centre of the lens is actually being measured.

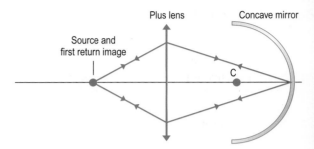

Figure 29.1

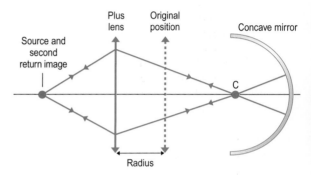

Figure 29.2

Measurement of radii

Instrument: radiuscope

The radiuscope (also called a microspherometer) can be used to measure the surface radii of a contact lens. The optical principles of the radiuscope are based on Drysdale's method for determining the radii of convex and concave reflecting surfaces. Drysdale's method is illustrated in Figures 29.1–29.4. In Figure 29.1, light is focused on to the pole of a concave reflecting surface by a positive lens. As light is focused at the pole of the reflecting surface, it reflects back along its own path and forms an image (the return image) at the source. If the lens (and light source) is now moved away from the reflecting surface (Figure 29.1), a second position is found where a return image can be seen at the source. This occurs when the light emerging from the lens is directed towards the centre of curvature (C) of the reflecting light, so the distance moved is equal to the radius of curvature of the reflecting surface. Exactly the same argument can be applied to convex mirrors (Figures 29.1 and 29.4), in which case the lens (and the source)

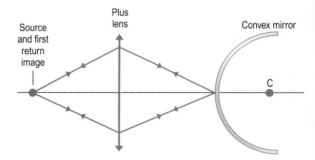

Figure 29.3

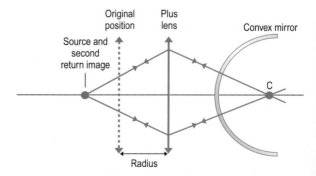

Figure 29.4

302

would have to be moved towards the convex reflecting surface.

So that Drysdale's method can be employed in the practical measurement of the radii of curvature of reflecting surfaces, a compound microscope is required in which a target is projected along the axis of the instrument. An image of this target is seen through the microscope when it is focused on a reflecting surface such as a contact lens. An image is also seen when the microscope is focused at the centre of curvature of the reflecting surface (the aerial image). The distance between the two positions where an in-focus image of the target is seen is equal to the radius of curvature of that surface (**Figure 29.5a**). By attaching a measuring device, such as a dial gauge, to the side of the microscope, a measurement of this distance may be obtained. A radiuscope that employs a digital gauge for the determination of the surface radius is shown in Figure 29.5b.

The radiuscope may be used to measure back optic zone radii, front optic zone radii, peripheral radii, lens thickness and axial edge lift. Soft lens radii can be recorded if the radiuscope is used together with a wet cell. In the case of a soft lens, the measured radius must be multiplied by the refractive index of the saline within the wet cell to obtain the actual radius. To obtain a good image in the case of a rigid lens, it is important to ensure that the lens is well dried before being placed on a drop of saline in the lens holder. It is also important to ensure that the instrument light is placed at the centre of the lens because this helps to locate the images. A toric lens produces one image when focused at the lens surface and two line images (at 90 degrees to each other) when focused at the centre of curvature. In the case of aspheric rigid contact lenses, it is the radius at the vertex of the aspheric surface that is required and, to assist in the taking of this measurement, the diameter of this area can be reduced by stopping down the aperture of the illuminator. Most instruments are provided with reduced apertures for this purpose.

Method

1. Place a few drops of saline into the concave recess of the lens support to

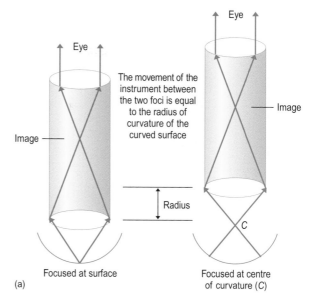

(a)

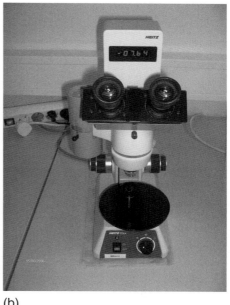

(b)

Figure 29.5 A radiuscope employing a digital gauge.

reduce reflection from the front surface of the lens.

2. Clean and dry the lens before positioning its front surface in contact with the saline on the support.

303

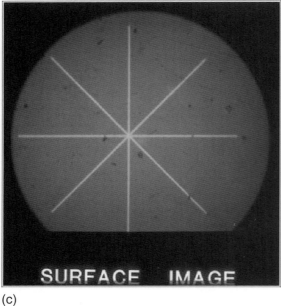

(c)

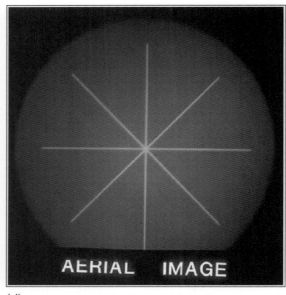

(d)

(e)

Figure 29.5 *Continued*

3. Handle the lens with care to avoid flexing, which may give rise to a distorted or even a false toric reading.

4. Focus the microscope on the aerial image at the centre of the back surface of the lens and adjust the stage until the image of the spoke-patterned target appears exactly centred.

5. Record the reading on the dial gauge (or set the dial gauge to zero).

6. Next, focus the microscope using the fine adjustment control on to the surface of the lens. Particles of debris and

scratches facilitate identification of this image.

7. The second reading is then recorded and the difference between the two readings gives the radius of curvature.

8. Three sets of readings should be taken and then averaged.

9. Remember that each reading must be obtained by focusing first on the aerial image and then on the surface image, or vice versa.

10. When measuring a toric surface, measurements in both principal meridians are required.

Tips

- Travelling from zero to the second position the image of the bulb filament is seen. The actual quality of this image gives an indication of any lens distortion.
- The image at the lens surface is usually much brighter and larger than at the centre of curvature and shows up any surface imperfections.

Figures 29.5c–e show the image formed by reflection from the lens surface, the aerial image and the image of the lamp filament.

Example 29.1

The saline in a wet cell has a refractive index of 1.336. When measuring the BOZR of a soft lens the readings at the lens surface and aerial image are 2.15 mm and 8.20 mm, respectively. Calculate the BOZR of this soft lens.

If the lens were rigid and measured without a wet cell, the BOZR would simply be 8.20 − 2.15 = 6.05 mm. To find the BOZR of a soft lens, the measured radius has to be multiplied by the refractive index of the saline in the wet cell. In this example, the BOZR of the soft lens would be 6.05 × 1.336 = 8.08 mm.

Instrument: keratometer

Several companies produce devices that can be clamped to the head-rest of a keratometer in order to measure the BOZR. These devices have a special contact lens holder that uses a front surface-silvered mirror and a lens support. The lens rests in saline within a small depression on a horizontal support. The mirror is set at 45° to the optical axis of the instrument and reflects light from the instrument on to the surface being measured. When attempting to measure the BOZR of a rigid contact lens with a keratometer, it is necessary to use a specially modified scale or conversion table because keratometers are primarily intended for the measurement of a convex surface (the cornea) and not a concave lens surface. If such tables are not available then 0.03 mm is added to correct for the concave surface.

Peripheral radii

It is possible to measure peripheral radii only with a radiuscope. The lens needs to be tilted and the actual bandwidth needs to be at least 1 mm and fairly well demarcated. In practice, both the narrow width of the peripheral curves and the presence of blending often render this measurement impossible.

Measurement of diameters

Total diameter

Instrument: band magnifier or V-gauge

Method for band magnifier

A band magnifier employs an engraved graticule with an adjustable eyepiece of ×7 magnification. The contact lens is repositioned on the scale to allow for the different zones of the lens and measurement is possible only if the transitions are sharp and not well blended.

Method for V-gauge

This method can measure only the total diameter (TD). The device consists of a V-shaped channel graduated between 6.0 and 12.5 mm.

Tip

Ensure that the lens is dry, otherwise it may prove difficult to remove from the smooth glass surface or the V-gauge without damaging it.

Other diameters (BOZD and BPZD)

The BOZD and back peripheral zone diameter (BPZD) can be measured using a band

305

magnifier or projection system in a similar manner to TD. In practice, location of these diameters may be difficult as a result of the blending of transitions and the increasing number of aspheric lenses now on the market. Measurements should be repeated in different meridians in order to verify that the back optic zone is circular. If the lens has a spherical BOZR and a toric periphery the back optic zone is oval in shape.

Measurement of thickness

Centre and edge thickness (t_c and t_e)

Instrument: thickness gauge

This instrument consists of a spring-loaded, ball-ended probe geared to a direct reading scale and is shown in **Figure 29.6**.

Method

1. When measuring the edge thickness, take several readings around the lens edge, because thickness may vary around the circumference.
2. Take care not to damage a thin edge.

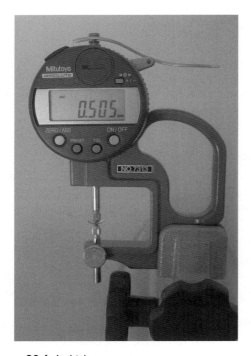

Figure 29.6 A thickness gauge.

Inspection of edge profile

The physical shape of the contact lens edge can have a significant influence on its comfort in wear. A simple and relatively quick method is to inspect the lens using a hand loupe, stereo-microscope or slitlamp. At high magnifications it is useful to support the lens on a holder.

Surface quality

Instrument: slitlamp, band magnifier or radiuscope

A magnification of ×20 or more is necessary to detect surface defects such as scratches, lathe marks, incomplete polishing or burning and mottling that have been caused by poor manufacture. Surface burning or mottling results from excessive heat during manufacture and is likely to cause poor in vivo wetting.

Special features

Special features to note and record include:

- engravings
- laboratory codes
- tints
- carrier design (assessed by edge measurement)
- prism ballast – increased edge thickness at base
- fenestrations (number, position, size and finish)
- truncation

Material

Confirmation of lens material is difficult, although comparison of specific gravity measurements can give an approximate guide. The simplest and most reliable indication can sometimes be colour, because certain gas permeable lenses are available in only one distinctive tint.

Dimensional tolerances

The object of specifying dimensional tolerances is to ensure that the lens is sufficiently accurate to fulfil its clinical function on the eye. The common tolerances for rigid gas permeable (RGP) contact lenses are listed in Table 29.1. Measurements should be taken at a temperature of $20 \pm 5°C$.

Dimensional tolerances for rigid contact lenses

BS 7208: Part 1: 1992 'Contact lenses – Part 1: Specification for rigid corneal and scleral contact lenses'. ISO 8321-1: 1991 'Optic and optical instruments – Contact lenses – Part 1: Specification for rigid corneal and scleral contact lenses'.

Table 29.1 Dimensional tolerances for rigid gas permeable contact lenses

Dimension	Tolerance
BOZR (mm)	±0.05
BPZR (where measurable) (mm)	±0.10
BOZD (mm)	±0.20
FOZD (mm)	±0.20
TD (mm)	±0.10
t_c (mm)	±0.02
BVP (in weaker meridian) (D)	
Up to ±5.00 D	±0.12
±5.00 D to ±10.00 D	±0.18
±10.00 D to ±15.00 D	±0.25
±15.00 D to ±20.00 D	±0.37
>20.00 D	±0.50
Cylinder power (D)	
Up to 2.00 D	±0.25
>2.00–4.00 D	±0.37
>4.00 D	±0.50
Cylinder axis (°)	±5

BOZD, back optic zone diameter; BOZR, back optic zone radius; BVP, back vertex power; FOZD, front optic zone diameter; t_c, centre thickness; TD, total diameter.

Soft and silicone hydrogel lens verification

Soft and silicone hydrogel lenses require verification for the same reasons as gas permeable ones, although checking is considerably more difficult because parameters vary depending on the:

- degree of hydration
- pH and tonicity of storage solution

- air or indeed saline temperature
- time taken if measurement taken in air
- method of supporting the lens

BOZR (base curve)

Instrument: optical spherometer

This is simply an instrument that measures the sag over a given chord and from this parameter the radius of curvature of the surface being measured can be calculated. The moving micrometer probe is raised until it just touches the back surface of the contact lens. The sag is measured, which is then converted to the more familiar radius of curvature. One well-known optical spherometer containing a wet cell is the Optimec, which includes a projection magnifier to assist with other key soft lens measurements and observations.

Method

1. Rinse and place the lens in a wet cell containing 0.9% saline solution.
2. Ensure that the lens is well centred on a supporting pillar.
3. The probe is raised towards the back surface of the lens until it just touches (established by viewing lens movement); the radius is then read directly off the instrument scale in 0.1 mm steps.
4. When contact has been made any further raising of the probe causes visible lens movement to be seen.
5. This technique should be repeated a minimum of three times and the average recorded.

Instrument: Keratometer

Although this is a possible method, a wet cell device would be required to keep the lens fully supported, fully hydrated and at a controlled temperature. Once a measurement had been obtained this would need to be multiplied by the refractive index of the saline (average value 1.336) and also the correction factor of +0.03 mm for the fact that a concave not a convex surface is being measured. It is not surprising that the accuracy of the result obtained by this method is suspect and unlikely to be of any clinical value.

Total diameter

Instrument: projection magnifiers

An optical system projects a magnified image of the lens on to a calibrated screen, from which linear measurements can be directly taken.

Back vertex power

The most accurate method of measuring power is by power profile mapping; such devices include the Visionix (**Figure 29.7**). These are generally only employed by manufacturers or in research facilities because of the very high cost of the instrumentation.

The most usual method used in the practice setting is the focimeter.

Focimeter: air measurement

Method

1. Pre-set the focimeter power scale to approximately the expected power.

2. Rinse the lens with saline.
3. Shake off surplus saline and dry carefully with a lint-free tissue.
4. Quickly but carefully place the lens concave surface down on a reduced aperture focimeter stop.
5. Read power directly from focimeter scale.

Tips

- Power is more easily and more accurately measured in air.
- Power in air is reasonably constant for up to about 1–4 minutes, depending on the water content.
- Any error in wet cell measurement is magnified fourfold: an error of 0.25 D = 1.00 D in air.
- The best clinical method for power measurement is simply the focimeter using the in-air method outlined above.

Thickness

Instrument: optical spherometer or projection magnifier

Many of the wet cell devices allow for measurement of lens centre thickness when observing the sagittal projection, which can be read directly off a millimetre scale pre-calibrated for the projection magnifier.

Instrument: radiuscope

A standard radiuscope can be used to determine lens thickness. However, this is more of a theoretical than a practical means of measurement.

Method

1. Place the lens, concave surface down, on a convex spherical trial lens with a curvature steeper than the lens to be measured to ensure contact at the centre.
2. Focus the target on the surface of the sphere with the lens in place.
3. Set the scale to zero.
4. Focus the target on the front surface of the lens.

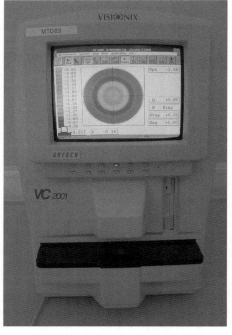

Figure 29.7 The Visionix for power profile mapping.

5. Read the distance travelled from the scale.
6. Multiply by the refractive index of the material to give the actual centre thickness: assumes knowledge of the actual lens material.

Edge form and surface quality

These are best observed with projection magnification however, the slitlamp is another useful and readily available tool allowing observation at high levels of magnification.

Deposits

Projection magnifiers reveal discrete areas of deposit such as white spots. Films such as protein are more easily observed by looking into the wet cell without magnification. The best method of observing deposits is by dark-field illumination. High magnification can be achieved with the slitlamp by using an oblique beam and selecting a dark background.

Water content and material

Water content can be estimated with a refractometer because of its inverse relationship with refractive index. One such instrument developed for contact lens measurement is the Atago CL-1 Refractometer.

Method

1. Rinse the hydrated lens with saline, shake off surplus and blot with a lint-free tissue.
2. Flatten the lens against the prism face, taking care not to cause damage.
3. Use an external light source and read directly from the internal scale, the water content at the junction between light and dark areas.

Assessing the lens material from the water content alone is far from certain, so identification may depend on other clues such as style of engraving, handling tint or type of edge bevel.

General tips for soft and silicone hydrogel lenses

- A useful instrument to have in the consulting room is a projection magnifier. It is used not only for lens inspection, but also for demonstrating to the patient surface and edge flaws and also deposits.
- Remember that wet cells are a potential source of possible cross-infection and require regular cleaning and disinfection.
- Particular care is required when flattening high water content lenses against glass surfaces because of the difficulty of removal and the risk of damage.

Dimensional tolerances

Soft lens specification is currently based on BS EN ISO 8321-2:2000 (BS 7208-24:2000) Ophthalmic optics – Specifications for materials, optical and dimensional properties of contact lenses Part 2: Single vision hydrogel contact lenses.

Table 29.2 Dimensional and optical tolerances of soft lenses

Dimension	Tolerance
BOZR (mm)	±0.20
Sag at specified diameter (mm)	±0.05
Total diameter (mm)	±0.20
Central optic zone diameter (mm)	±0.20
Centre thickness (mm):	
Up to 0.1 mm	±0.01 + 10%
>0.1 mm	±0.015 + 5%
BVP (in weaker meridian) (D):	
Plano to ±10.00 D	±0.25
±10.00 D to ±20.00 D	±0.50
>±20.00 D	±1.00
Cylinder power (D):	
Plano to 2.00 D	±0.25
2.25–4.00 D	±0.37
>4.00 D	±0.50
Cylinder axis (°)	±5

BOZR, back optic zone radius; BVP, back vertex power.

Concluding points for Chapter 29

This chapter has discussed the parameters, tolerances and instruments necessary for the verification of rigid and soft/silicone hydrogel contact lenses. Practitioners who regularly fit RGP lenses should at least have access to a focimeter and radiuscope. However, if one takes a real world approach to the verification of soft and silicone hydrogel lenses, most of which are now supplied in blister packs, the key responsibility lies in the comparison of the printed lens parameters with the written order.

Further recommending reading

Douthwaite W A (2006) *Contact Lens Optics and Lens Design*. Elsevier, Oxford

Efron N (2002) *Contact Lens Practice*. Butterworth-Heinemann: Oxford

Hough D A (2000) *A Guide to Contact Lens Standards*. British Contact Lens Association, London

Loran D F C (1989) The verification of hydrogel contact lenses. In: Phillips AJ, Stone J (eds), *Contact Lenses*, 3rd edn. Butterworths, London: 463–504

Watts R (1997) Rigid lens verification procedures. In: Phillips AJ, Speedwell L (eds), *Contact Lenses*, 4th edn. Butterworth-Heinemann, Oxford: 407–25

Binocular Vision: The Basics

Bruce Evans

Introduction

The editors of this text feel that previous works on the subject of contact lenses have given little or often no attention to the subject of binocular vision. This is surprising because most contact lens patients do in fact have two eyes! As many contact lens practitioners do not include a basic binocular vision assessment as part of routine contact lens after-care, it was felt that a discussion of binocular vision considerations in contact lens practice (see Chapter 31) was long overdue. As a precursor to this discussion, Chapter 30 attempts to give an insight into the basics of binocular vision and its assessment. As a stand-alone subject, binocular vision is discussed in much greater detail in the texts mentioned at the end of the chapter. However, all the material discussed in this chapter should be considered to be essential reading for all optometrists, dispensing opticians and contact lens opticians.

Chapter content

Normal binocular vision

Normal binocular vision occurs when the two eyes work together in a comfortable and coordinated way, so that they are both accurately aligned (motor fusion) with the object of regard and the monocular images are combined (sensory fusion) to give a clear, single, and stereoscopic view. This is illustrated schematically in Figure 30.1.

If one eye is covered, or if the eyes are dissociated in some other way, most people develop a **heterophoria**: the eyes move out of perfect alignment. If people have normal binocular vision then they must, under normal conditions, be overcoming the heterophoria and they achieve this through **motor fusion**. The adequacy of motor fusion can be assessed by measurement of the fusional reserves, the most relevant of which is the one that opposes the heterophoria, e.g. patients with a large exophoria at near exert their convergent fusional reserve to overcome the exophoria. The convergent fusional reserve is measured by using base-out prisms to force the eyes to converge.

In addition to the motor fusion that maintains physical alignment of the visual axes, the sensory visual system also needs to match the two monocular images and to fuse them together to obtain a single perception. This is the process of **sensory fusion**. Good sensory fusion requires a **fusion lock**: monocular images that are similar, ideally differing only by virtue of the different positions of the two eyes. If the monocular images differ in other ways (e.g. as a result of macular changes or anisometropia) fusion is impaired

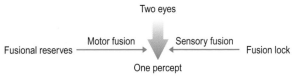

Figure 30.1 Simple model of binocular function (see text for explanation). (Adapted from Evans (2007).)

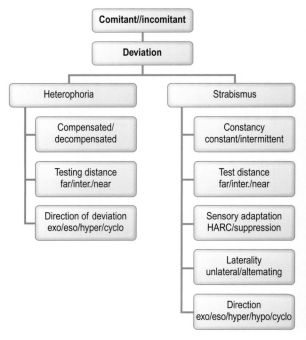

Figure 30.2 Classification of binocular vision anomalies. (Reproduced with permission from Evans (2005).)

and the patient has an increased risk of binocular vision anomalies.

Binocular vision anomalies

It follows from Figure 30.1 and the discussion above that three types of problem are likely to cause binocular vision anomalies: first, if the visual axes are grossly misaligned, to a degree that cannot be overcome by the fusional reserves. For example if a child has high uncorrected hypermetropia then he has to accommodate to obtain clear images and this induces convergence. This may cause such a large convergent deviation that the person cannot overcome it with divergent fusional reserves, and an esotropia develops. Alternatively, such as person may be barely able to overcome the deviation and may develop a decompensated esophoria and/or an intermittent esotropia.

A second type of problem causing binocular vision anomalies is a general health problem that causes the fusional reserves to worsen. When we have a febrile illness (e.g. influenza) we feel weak: our motor systems are less capable than usual. This reduces the fusional reserves. To take the example above, a child with high hypermetropia and associated esophoria may be just managing to overcome this with his divergent fusional reserves. If he develops a febrile illness his fusional reserves become impaired, and he may develop an esotropia.

A third type of problem is if there is a degradation of one eye's image, such as in uncorrected anisometropia, asymmetrical cataract or macular disease. This can interfere with sensory fusion.

Classification of binocular vision anomalies

A classification of binocular vision anomalies is reproduced in **Figure 30.2**. The classifications of comitancy (comitant/incomitant) and deviation (heterophoria/strabismus) are independent, i.e. a patient with heterophoria may be comitant or incomitant, as may one who has strabismus.

Decompensated heterophoria

If a person has a binocular vision anomaly that is very marked (e.g. a very large deviation, grossly inadequate fusional reserves or severely impaired sensory fusion), the ocular alignment may break down completely and that person develops a strabismus. Decompensated heterophoria is a less severe binocular vision anomaly, where the person is able to maintain fairly accurate ocular alignment but has to strain excessively to do this. This can produce symptoms

Table 30.1 Symptoms of decompensated heterophoria

Number	Type of symptom	Symptom
1		Blurred vision
2	Visual	Double vision
3		Distorted vision
4		Difficulty with stereopsis
5	Binocular	Monocular comfort
6		Difficulty changing focus
7		Headache
8	Asthenopic	Aching eyes
9		Sore eyes
10	Referred	General irritation

Reproduced with permission from Evans (2005).

(Table 30.1) and may, if untreated, lead to the development of a strabismus.

Investigation and diagnosis

The symptoms of decompensated heterophoria (Table 30.1) could result from a variety of conditions, and clinical tests are necessary to determine whether they do result from a decompensated heterophoria.

There is no one test that reliably diagnoses decompensated heterophoria. It is necessary to combine different test results to obtain a diagnosis, with the most important tests being cover test, Mallett's fixation disparity test and fusional reserves. The relevant tests have been combined to produce a diagnostic algorithm that can be a useful clinical guide; it is reproduced in **Figure 30.3** and includes the related condition of binocular instability, which is characterised by an unstable heterophoria and low fusional reserves. Binocular instability is a correlate of dyslexia (Evans 2001).

The Cover test

The cover test is used to differentially diagnose strabismus (Figure 30.4) from heterophoria (Figure 30.5). The test is carried out using as a target a letter from the line above the worst eye's acuity. If the visual acuity is worse than 6/60, a spotlight can be used. If the eye examination results in a significantly different prescription to that previously being worn, the cover test should be repeated with the proposed new prescription in place.

Often, the patient's history or previous records lead the practitioner to suspect that a strabismus may be found in one eye, which is likely to be the eye with the worse acuity. If so, the other eye should be covered first. This first cover is the 'purest' orthoptic test of all, because the moment before the cover is applied the patient has normal binocular vision and views the target in a completely natural way. The eyes should be watched as the cover approaches because, if a dissociated vertical deviation is present, a movement of an eye is often seen before the cover actually reaches the eye.

The use of the cover/uncover test to detect strabismus is described in Table 30.2 and Figure 30.4, and its use to detect heterophoria is described in Table 30.3 and Figure 30.5. It is useful to estimate the angle of any eye movements, and a method for this is described in Table 30.4. In heterophoria, the quality of recovery movement should also be quantified, and a grading system for this is described in Table 30.5.

The description above is of one form of the cover test: the cover/uncover test. This is useful for detecting strabismus, estimating the magnitude of the deviation under normal viewing conditions (Table 30.4) and evaluating the recovery movement in heterophoria (Table 30.5). But it is also useful to know how much the angle increases ('builds') as the patient is dissociated to greater degrees by alternate covering. So, after the cover/uncover test, it is advisable to alternate the cover from one eye to the other for about six further covers to see how the angle changes (Table 30.4). At the end of this alternating cover test as the cover is removed, the recovery movement can be observed again to estimate the effect of alternate covering on the recovery (Table 30.5). A cover/uncover test is then performed once more on the other eye to assess any change in the recovery of this eye. An example of the recording of cover test results for a patient is given in Table 30.6.

A great deal of information can be gleaned from the cover test. Table 30.7 includes some

DISTANCE/NEAR (delete)	Score
1. Does the patient have one or more of the symptoms of decompensated heterophoria (headache, aching eyes, diplopia, blurred vision, distortions, reduced stereopsis, monocular comfort, sore eyes, general irritation)? *If so, score +3 (+2 or +1 if borderline)* Are the symptoms at D ☐ or N ☐ *All the following questions apply to D or N, as ticked (if both ticked, complete 2 worksheets)*	
2. Is the patient orthophoric on cover testing? Yes ☐ or No ☐ *If no, score +1*	
3. Is the cover test recovery rapid and smooth? Yes ☐ or No ☐ *If no, score +2 (+1 if borderline)*	
4. Is the Mallett Hz aligning prism: <1Δ for patients under 40, or >2Δ for patients over 40? Yes ☐ or No ☐ *If no, score +2*	
5. Is the Mallett aligning prism stable (Nonius strips stationary with any required prism)? Yes ☐ or No ☐ *If no, score +1*	
6. Using the polarised letters binocular status test, is any foveal suppression < one line? Yes ☐ or No ☐ *If no, score +2*	
Add up score so far and enter in right hand column *if score: <4 diagnose normal, >5 treat, 4–5 continue down table adding to score so far*	
7. Sheard's criterion: (a) measure the dissociated phoria (e.g. Maddox wing, prism cover test): record size & stability (b) measure the fusional reserve opposing the heterophoria (i.e., convergent, or base out, in exophoria). Record as blur/break/recovery in Δ. Is the blur point, or if no blur point the break point, [in (b)] at least twice the phoria [in (a)]? Yes ☐ or No ☐ *If no, score +2*	
8. Percival's criterion: measure the other fusional reserve and compare the two break points. Is the smaller break point more than half the larger break point? Yes ☐ or No ☐ *If no, score +1*	
9. When you measured the dissociated heterophoria, was the result stable, or unstable (varying over a range of ±2Δ or more). (e.g., during Maddox wing test, if the Hz phoria was 4Δ XOP and the arrow was moving from 2 to 6, then result unstable) Stable ☐ or Unstable ☐ *If unstable, score +1*	
10. Using the fusional reserve measurements, add the divergent break point to the convergent break point. Is the total (= fusional amplitude) at least 20Δ? Yes ☐ of No ☐ *If no, score +1*	
Add up total score (from both sections of table) and enter in right hand column. If total score: <6 then diagnose compensated heterophoria, if >5 diagnose decompensated heterophoria or binocular instability.	

Figure 30.3 Scoring system for diagnosing decompensated horizontal heterophoria and binocular instability. This scoring algorithm is designed for horizontal heterophoria. If a vertical aligning prism of 0.5Δ or more is detected then, after checking trial frame alignment, measure the vertical dissociated phoria. If this is more than the aligning prism and there are symptoms then diagnose decompensated heterophoria; but still complete the worksheet for any horizontal phoria. Reproduced with permission from Evans (2007).

additional comments. Other methods of assessing ocular alignment are available, including the Hirschberg and Krimsky tests which are based on an observation of corneal reflexes. These tests are inaccurate, however. With practice, cover testing is nearly always possible, even with infants.

Further investigation of heterophoria

Mallett's fixation disparity test (**Figure 30.6**) is a useful tool in the diagnosis of decompensated heterophoria at near. The fixation disparity is not measured, but rather the prism (or sphere) that eliminates the fixation disparity is detected. This prism is much less than the heterophoria

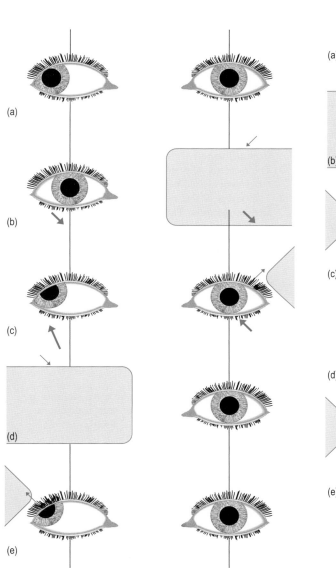

Figure 30.4 Cover test in right divergent strabismus with right hypertropia (movement of the eye is signified by the solid arrow, movement of the cover by the dotted arrow): (a) right eye deviated out and up; (b) left eye covered both eyes move left and downwards so that the right eye takes up fixation; (c) left eye uncovered, both eyes move right and upwards so that the left eye again takes up fixation; (d) and (e) no movement of either eye as the strabismic right eye is covered and uncovered. (Adapted from Evans (2007) with permission of Elsevier Ltd.)

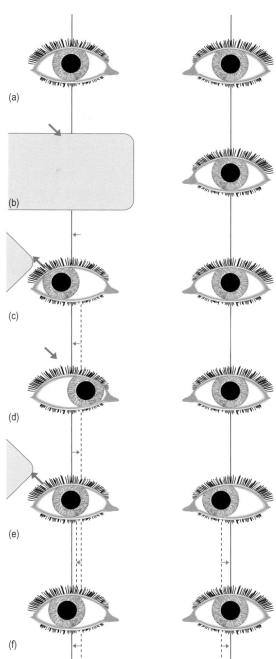

Figure 30.5 The cover test in esophoria (movement of the eye is signified by the solid arrow, movement of the cover by the dotted arrow). (a)–(c) From the 'straight' active position, the right eye moves inwards when dissociated by covering (b); it moves smoothly outwards to resume fixation with the other eye when the cover is removed (c). Note that the left (uncovered) eye does not move during the simple pattern of movements. (d)–(f): The 'versional pattern': the right eye moves inwards under the cover, as in the simple pattern (d); on removing the cover, both eyes move to the right by the same amount (about half the degree of the esophoria) (e); both eyes then diverge to the straight position (f). (Adapted from Evans (2007) with permission of Elsevier Ltd.)

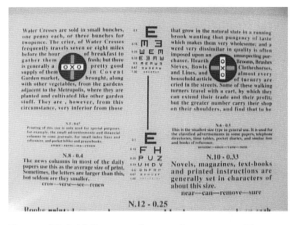

Figure 30.6 Mallett unit fixation disparity test. The patient wears cross-polarised filters so that the page is seen by both eyes except the green lines, one of which (for each OXO test) is seen by either eye.

Table 30.2 Detection of strabismus with the cover/uncover test

1. As the cover moves over one eye then the practitioner should watch the uncovered eye. It is the behaviour of the uncovered eye that reveals whether the patient has a strabismus
2. For example, as the left eye is covered the practitioner should watch the right eye. If the right eye moves then this suggests that there is a strabismus in this eye (Figure 30.4, the behaviour of the right eye from (a) to (b))
3. The direction and amplitude of the movement should be estimated (see Table 30.4)
4. The cover is then slowly removed from the eye that has been covered and this eye is observed to see if a movement occurs, signifying heterophoria instead of strabismus (see Table 30.3)

and usually reflects the portion of the heterophoria that is uncompensated.

Other tests for investigating heterophoria are listed in Figure 30.3.

Management

The management of decompensated heterophoria starts with determining the cause of decompensation and attempting to eliminate this cause. Figure 30.1 can help with this, as detailed below.

The first question to ask is whether there is an impairment to sensory fusion, e.g. look for

Table 30.3 Detection of heterophoria with the cover/uncover test

1. As the cover is slowly removed from the eye that has been covered, this eye is observed to see if a movement occurs. If a movement occurs as an eye regains fixation after being covered then this indicates a heterophoria (e.g. in Figure 30.5 the right eye moves out as the cover is removed from (b) to (c))
2. The direction and amplitude (Table 30.5) of the movement should be estimated
3. The quality of this recovery movement should also be recorded. This gives an objective indication of how well the patient is able to compensate for the heterophoria (Table 30.5). A higher grading in Table 30.5 is more likely to be associated with decompensation and thus more likely to require treatment

Table 30.4 Estimating the amplitude of movement on cover testing

1. The amplitude of movement should always be estimated (in Δ) and recorded during cover testing
2. It is easy to train yourself to be quite accurate at this, and regularly to 'calibrate' your estimations. On a typical Snellen chart, the distance from a letter on one end of the 6/12 line to a letter on the other end is about 12 cm (measure this on your chart to check). If the distance is 12 cm, this means that when patients changes their fixation between these two letters the eyes make a saccade of 2Δ (1Δ is equivalent to 1 cm at 1 m)
3. If you place two markings on the wall near the letter chart that are 24 cm apart, when patients change their fixation between these two marks then the eyes are moving by 4Δ
4. After you have done the cover test and estimated the amplitude (in Δ) of the strabismic or heterophoric movement, remove the cover and have the patient look between these two marks, or between the two letters on the 6/12 line while you watch the eyes. Compare the amplitude of this eye movement with the amplitude of movement that you saw during cover testing, to check the accuracy of your estimate
5. A similar method can be used for larger amplitudes
6. At near, this task is even easier. Use as your fixation target the numbers on a centimetre ruler which you hold at $\frac{1}{3}$ m. If the patient looks from the 1 to the 2, then the patient's eyes are moving by 1 cm which, at $\frac{1}{3}$ m, equates to 3Δ

Table 30.5 A grading system that can be used to gauge cover test recovery in heterophoria

Grade	Description
1	Rapid and smooth
2	Slightly slow/jerky
3	Definitely slow/jerky but not breaking down to a strabismus
4	Slow/jerky and breaks down with repeat covering, or only recovers after a blink
5	Breaks down readily after one to three covers

Table 30.6 Example of a recording of cover test results

Distance 2Δ XOP G1 → 2Δ XOP G2
Near 8Δ XOP G1 → 12Δ XOP G3

Key: at distance, the cover/uncover test reveals 2Δ exophoria with good (grade 1) recovery. After the alternating cover test the angle does not change but the recovery becomes a little slower (grade 2 recovery)

At near, the cover/uncover test reveals 8Δ exophoria with good recovery. After the alternating cover test the angle builds to 12Δ exophoria with quite poor recovery, but not quite breaking down into a strabismus (grade 3 recovery)

Table 30.7 Additional comments on cover testing

- With the alternating cover test, as the angle builds towards the total angle, it becomes easier to detect vertical deviations which are typically smaller than horizontal deviations
- Vertical deviations also can sometimes be spotted by watching for movement of the eyelids
- In some patients, an eye is deviated before the test (strabismus) or becomes deviated during the test and is very slow to take up fixation. So, when the dominant eye is covered there may be no apparent movement, even though the uncovered eye is not fixating the target. A movement of the deviated eye can sometimes be elicited by asking the patient to 'look directly' at the target, or by moving the fixation target a little
- The magnitude of the deviation can also be measured using a prism bar or loose prisms, typically during the alternating cover test. This is especially useful in large angles where accurate estimations of angular movements becomes difficult

anisometropia (see Chapter 31), visual field loss, unilateral cataract or macular problems. If these problems can be found and corrected, no further treatment may be required.

The second question is whether the magnitude of the heterophoria is atypical. Norms for heterophoria vary with different types of dissociation test, but it is unusual to find a distance heterophoria of more than about 2 Δ esophoria, 3 Δ exophoria and 1 Δ hyperphoria, or a near heterophoria of more than about 2 Δ esophoria, 8 Δ exophoria and 1 Δ hyperphoria. Significant degrees of esophoria in children should always raise the suspicion of uncorrected hypermetropia and a cycloplegic refraction is likely to be necessary. An esophoria that is greater at distance than at near might indicate a lateral rectus palsy. A vertical heterophoria is unusual and often indicates a cyclo-vertical incomitancy. A heterophoria that changes markedly can be a sign of pathology, in which case referral is indicated.

If the heterophoria is not atypical and has not changed, has it become decompensated because the fusional reserves are inadequate? If so, these might be improved with eye exercises. If the fusional reserves are inadequate because the general health is poor, or if there are marked symptoms, eye exercises are likely to be an uphill struggle.

Exophoric conditions can respond well to eye exercises in patients of any age. There are many types of exercises that can work for decompensated exophoria or convergence insufficiency, including the Institute free-space stereograms (Figure 30.7).

A simpler approach to correcting many forms of decompensated heterophoria is to reduce the degree of heterophoria to a level that can be comfortably compensated by the patient. If there is a hyperphoria in the primary position, a vertical prism may alleviate symptoms. Similarly, a decompensated exophoria at near in an older patient may be compensated by a base-in prism in reading spectacles. The Mallett unit can be used to determine the prism power for eliminating any fixation disparity.

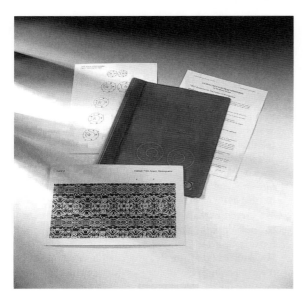

Figure 30.7 Institute free-space stereogram (IFS) exercises. (Reproduced with the permission of the Institute of Optometry (www.ioosales.co.uk).)

In children, horizontal heterophoria can be reduced by refractive modification, e.g. a child with emmetropia and a decompensated exophoria could be given minus lenses. These cause accommodation that will induce convergence, reducing the exophoria. Multifocals can be used for patients with a decompensated esophoria at near. The Mallett unit can be used to determine the minimum sphere that eliminates any fixation disparity, and this can be checked with a cover test. The goal is to reduce the refractive modification over time.

Strabismus

In strabismus the visual axes are misaligned to an extent that makes binocular single vision impossible. The most difficult cases of strabismus to detect are microtropia, when there may be no movement seen on cover testing.

If adults develop strabismus, they are likely to experience diplopia (seeing two images of each object) and confusion (seeing two dissimilar images superimposed). Young children who develop strabismus usually develop sensory adaptations that allow them to avoid diplopia and confusion. In small angle deviations they tend to develop harmonious retinal correspondence and in larger deviations they are more likely to develop suppression.

Diagnosis, investigation and management

Usually, strabismus is detected with the cover test, although this is not always the case. Microtropia is often, but not always, associated with anisometropia and it is important to know whether a person with anisometropia also has microtropia, because the management of any amblyopia may be different (see below). Other tests that help to detect microtropia are eccentric fixation, stereoacuity and the 4 Δ base-out test (Evans 2005).

The investigation and management of strabismus depend on the details of the case. Most commonly, optometrists see adults with long-standing strabismus that was investigated and treated in childhood. Usually, there are no symptoms and no treatment specific to the strabismus is required.

If an optometrist examines an adult with recent-onset strabismus, it is most likely that there is a pathological cause and the patient requires referral. The rapidity of onset of symptoms is a good indicator of the urgency for the referral.

Misalignments of the visual axes are quite common in the first month of life, but should be reducing by the end of the second month (Horwood 2003). A marked esotropia in the first 6 months is quite likely to be infantile esotropia syndrome, which requires referral.

If an optometrist examines an older child who has recently developed a strabismus, the first task is to rule out pathology (Evans 2005). A related task is to look for a cause of the strabismus. If a refractive cause is found (e.g. hypermetropia causing esotropia), this may be the only cause, but pathology should still be excluded by the primary eye-care examination. Next, determine whether the strabismus is amenable to optometric management. In children with an esotropia, before inserting a cycloplegic determine the maximum 'dry' hypermetropia and carry out a cover test with this correction to see if it corrects the esotropia. Then carry out a cycloplegic refraction to determine the full hypermetropia. If correction of the hypermetropia is likely to eliminate the esotropia prescribe spectacles. The child needs to be monitored closely (e.g. in 3 months) and, if the strabismus does not respond to optometric management, referral is required. Always warn the parents to return immediately if the situation worsens, or if any symptoms (e.g. diplopia) are reported.

If the esotropia is at near, more plus can be given in the form of multifocals. Multifocals for children need to have the segment height situated higher than usual, typically to bisect the lower edge of the pupil. The children need regular visits to the dispensing optician to ensure that the fit of the spectacles remains optimal. This is also the case for patients with hypermetropic accommodative esotropia who will need regular spectacle adjustments to make sure that they are not looking over the top of the spectacles.

If there is a long-standing exophoria that is breaking down into an exotropia (e.g. with increased studying), consider a negative add or eye exercises.

If there is a new deviation or marked change in the deviation and this is not caused by a refractive error, the patient requires referral for neuro-ophthalmological investigation.

This chapter can provide only an overview of this complex area. A significant number of optometrists specialise in orthoptics and these practitioners are likely to use more sophisticated interventions than those outlined here, including treating strabismus with eye exercises (Evans, 2007). For optometrists who have not specialised in this area, it is safest to limit the strabismic cases that they manage to children with a strabismus of recent onset that responds to refractive management. Other cases need to be referred to a more specialist optometrist or ophthalmologist/orthoptist team.

Amblyopia

Amblyopia is defined as a visual loss resulting from an impediment or disturbance to the normal development of vision. It usually occurs when an early interruption to the development of vision causes a visual deficit that, in later life, cannot be immediately corrected refractively. Early detection and treatment are important because the plasticity of the visual system, and hence its response to treatment, decline throughout life. A working definition of amblyopia is visual acuity worse than 6/9 and/or two lines or more of acuity difference in the two eyes. Functional amblyopia is classified in Figure 30.8.

Investigation and diagnosis

The acuity deficit in strabismic amblyopia is worsened with crowded stimuli, so it is best to use a visual acuity test that employs crowding, i.e. the classic Sheridan–Gardiner test, with just one optotype presented at a time, is not a good test. The computerised Test Chart 2000 (available from IOO Sales – 020 7378 0330) is ideal because the optotypes can be randomised, and lower case letters, numbers and pictures can be used with younger children.

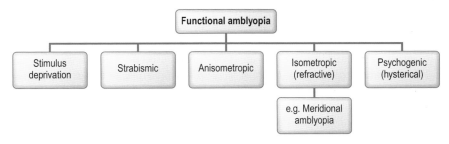

Figure 30.8 Classification of functional amblyopia. (Reproduced with permission from Evans (2005).)

The differential diagnosis of amblyopia involves two approaches: detecting a **negative sign** and a **positive sign**. The negative sign is to rule out pathological causes of poor vision. The positive sign is the presence of an amblyogenic factor, most commonly anisometropia and/or strabismus. The detection of anisometropia often requires a cycloplegic refraction and the detection of strabismus has been outlined above. It is important to detect any microtropia, so it is useful to test for eccentric fixation using the fixation graticule of the direct ophthalmoscope.

Management

The management of amblyopia varies with the age of the patient and its type. Recent research has demonstrated that many cases of amblyopia in children achieve an improvement of visual acuity simply with refractive correction alone (Stewart et al. 2004). Surprisingly, this improvement was found in strabismic as well as anisometropic amblyopia, although researchers studied only the effect of spectacle correction. The evidence discussed in Chapter 31 indicates that these individuals, many of whom are anisometropic, would be likely to do better with contact lenses. The current recommendation is that all patients with amblyopia who are to be treated should try refractive correction first, and only progress to other treatments if the amblyopia persists after 18 weeks of refractive correction.

If amblyopia remains after refractive correction for 18 weeks, the further treatment for amblyopia is occlusion of the dominant eye. Recent research suggests that patching for 2 hours a day, including 1 hour of detailed near visual activity, is as effective as patching for 6 hours (Pediatric Eye Disease Investigator Group 2003)

In strabismic amblyopia (and in mixed anisometropic and strabismic amblyopia) there is quite strong evidence to suggest that treatment becomes less effective by about the age of 7–12 years. There is also the risk of breaking down sensory adaptations and causing diplopia. So, patching is not recommended over the age of 12

and should be used cautiously over the age of 7 (Evans 2002).

In pure anisometropic amblyopia, the research evidence is more equivocal, but several studies suggest that treatment can be attempted at any age. Refractive correction is important, ideally with contact lenses to eliminate problems from aniseikonia and differential prismatic effects (see Chapter 31). If acuities still do not equalise, occlusion can be tried. Prolonged occlusion is not recommended for older children or adults but, if there is no strabismus, incomitancy or decompensated heterophoria, occlusion for up to 2 hours a day is unlikely to cause problems, although it should still be monitored closely.

The purpose of occlusion is to encourage use of the amblyopic eye, so it is helpful to have the patient carry out detailed tasks. Computer games often increase motivation.

Incomitant deviations

An incomitant deviation is a deviation in which the angle varies in different positions of gaze, depending on which eye is fixing. Incomitant deviations can be congenital or acquired. A new or changing incomitancy requires urgent referral.

Knowledge of the position and actions of the extraocular muscles is essential to understand incomitant deviations. The most complex muscles to understand are the cyclotorsional muscles, which are illustrated in **Figure 30.9**. Only the right eye's superior rectus (left panel) and superior oblique (right panel) are shown, but the inferior rectus and inferior oblique muscles follow similar lines of action underneath the globe.

A consideration of the anatomy of the extraocular muscles can be used to arrive at the familiar motility diagram, illustrated in **Figure 30.10**. The positions of the muscles will, of course, change as the eye moves, so the actions of the muscles change depending on the position of the eye. This is why there are some differences between the actions of the muscles in the primary position (Table 30.8) and the positions of gaze in which the

Table 30.8 Actions of the extraocular muscles in the primary position

Muscle	Primary action	Secondary action	Tertiary action
Medial rectus	Adduction	None	None
Lateral rectus	Abduction	None	None
Superior rectus	Elevation	Intorsion	Adduction
Inferior rectus	Depression	Extorsion	Adduction
Superior oblique	Intorsion	Depression	Abduction
Inferior oblique	Extorsion	Elevation	Abduction

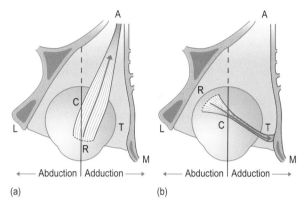

(a)   (b)

Figure 30.9 Schematic diagram showing approximate positions of (a) right superior rectus and (b) right superior oblique. (Adapted from Evans (2007) with permission of Elsevier Ltd.)

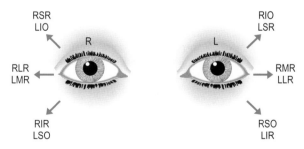

Figure 30.10 Cardinal positions of gaze in which the extraocular muscles have their maximum actions. RSR, right superior rectus.

muscles have their maximum action (Figure 30.10).

The lateral rectus is innervated by its own nerve, the abducens, and the superior oblique also has its own nerve, the trochlear. The rest of the extraocular muscles are innervated by the third nerve. The nerves that supply the superior oblique and the lateral rectus are relatively thin nerves with long pathways, and so are prone to damage.

Investigation and diagnosis

Symptoms and history usually reveal whether the incomitant deviation is of recent onset or long standing. Ocular motility testing is the basic diagnostic tool for the primary care optometrist, using the information in **Figure 30.10**. The basic motility test (Table 30.9) can be expanded to obtain more information by cover testing in peripheral gaze and by asking patients about any diplopia that they perceive. The motility test can work well for diagnosing recent onset incomitancies, although long-standing incomitant deviations often develop secondary sequelae, which make it harder to determine the original underacting muscle (Evans, 2007). Other methods, including the computerised Hess screen (available from IOO Sales – 020 7378 0330) and diagnostic algorithms (Evans, 2005) are helpful for more complex cases.

These additional tests are especially useful for diagnosing cyclo-vertical incomitancies (e.g. a superior oblique palsy), which can be difficult to detect on motility testing. The double Maddox rod test is an extremely valuable test for such cases. Two identical Maddox rods are placed in the cylinder section of a trial frame, with the grooves on the filters exactly at 90°. The patient views a spotlight and if two red lines are perceived this indicates a hyperdeviation. If one of the lines is tipped this eye has a cyclodeviation, the angle of which can be

Table 30.9 Procedure for carrying out the basic objective ocular motility test

1. The best target is a point light source, which should be bright enough to allow the corneal reflexes to be clearly seen, but not so bright as to cause blepharospasm. For infants, any target that catches their attention is recommended
2. The light should be held at about 50 cm from the patient. It should be moved quite slowly, so that it takes about 5 seconds to move from extreme gaze on one side to extreme gaze on the other
3. The light should be moved in an arc, as if using an imaginary perimeter bowl
4. It is easier to observe the eye movements in extreme gaze if no spectacles are worn, but if there is an accommodative strabismus then spectacles should be worn, or the test repeated with and without spectacles
5. Various authors have preferences for different patterns of movement of the target. The star pattern is often used, but other patterns have been recommended and each has its own merits
6. The straight up and down (in the midline) positions are often also tested to look for an A or V syndrome
7. In young children, the head may need to be gently held. With infants, it is best not to hold the head but to move the target very far around the patient so that the eyes are forced to move when the head cannot turn any more
8. The motility test is performed while watching the reflection of the light in the corneas to detect any marked under-actions or over-actions
9. Any over- or under-actions can be graded from grade 1 (just detectable) to grade 5 (extremely noticeable)
10. Throughout the test, the corneal reflexes of the light are observed. If one disappears, then either the light is misaligned or the patient's view of the light has been obscured (e.g. by the nose). This means that the light has been moved too far: the test should be carried out within the binocular field
11. In downgaze, patients often need to be asked to keep their eyes as wide open as possible. The lids should be physically held open if this is the only way that the reflexes can be seen
12. If any abnormality is observed on this binocular motility test, then each eye is occluded and a monocular motility test carried out in turn

From Evans (2005).

measured by rotating the relevant Maddox rod until the two red lines are parallel.

Management

For primary care optometrists and opticians, most of the patients whom they see with incomitant deviations have a long-standing condition, typically congenital. The history reveals this and the symptom of diplopia is usually absent. It is possible for a long-standing incomitancy to decompensate, and this can be triggered by forcing the patient to look in the field of action of the affected muscle for prolonged periods (see box). Decompensation can occur spontaneously, and these patients are likely to report recent-onset diplopia and will require referral to rule out a pathological cause of decompensation.

> **Top tips for spectacle and contact lens dispensing**
>
> - Some patients have orthoptic conditions that make it difficult for them to look down when reading (e.g. superior oblique palsies, A syndromes)
> - These cases can often be identified by a chin-down position, to avoid looking down when they read.
>
> In these cases, multi-focal spectacles, or alternating-vision RGP contact lenses are contraindicated

Incomitant deviations that are long standing or have been medically investigated sometimes benefit from a prism, particularly those where there is already binocular single vision in some positions of gaze. As the deviation varies in different directions of gaze there is no single prism power that is a perfect correction. However, incomitant deviations usually become less incomitant over time and often a prism can be found that extends the field of binocular comfortable single vision. The tests outlined in the section on decompensated heterophoria can be used.

For new or changing incomitancy the management for the primary care practitioner is referral. The speed of onset of symptoms is a good guide to the urgency of referral.

Other conditions

Nystagmus

Nystagmus is a regular, repetitive, involuntary movement of the eyes, the direction, amplitude and frequency of which are variable. Nystagmus can be classified into three main categories: early onset nystagmus (synonyms: congenital nystagmus, infantile nystagmus syndrome); latent nystagmus; and acquired nystagmus. Congenital nystagmus occurs in the first 6 months of life and may be associated with a sensory defect (e.g. a cause of reduced vision such as albinism) or not, in which case it is sometimes called motor defect nystagmus. Latent nystagmus is associated with an early (in first year) interruption to binocularity, typically in infantile esotropia syndrome. Latent nystagmus has the characteristic feature of occurring or being much worse only when one eye is occluded. Acquired nystagmus can be a sign of pathology and can occur at any age.

The history is usually strongly indicative of the type of nystagmus. If an optometrist detects nystagmus in a child who has not been medically investigated, the child should be referred. If the nystagmus has been investigated and is not changing, referral is not required. These patients often have a high refractive error and careful refraction is important. Patients with early onset nystagmus often do better in contact lenses than in spectacles, particularly RGP contact lenses. Patients may have a null position: a direction of gaze in which the nystagmus is reduced. These patients may have a compensatory head posture and should not be dispensed with spectacles that make it difficult for them to use their null position.

Nystagmus that is new or changing can be a sign of pathology and requires early referral.

Accommodative anomalies

Accommodation is commonly assessed by the push-up test, which measures the amplitude of accommodation. Accommodative accuracy (lag) can be measured objectively using dynamic retinoscopy, and this can be useful to confirm a diagnosis of accommodative insufficiency (low amplitude of accommodation) (Evans 2005). Accommodative infacility (an inability to change accommodation rapidly) can be detected and treated with flippers (Evans 2005). Flippers can also be used to treat accommodative insufficiency, as can push-up exercises similar to those sometimes used to treat convergence insufficiency.

Concluding points

This chapter has attempted to give an insight into the basis of binocular vision and its assessment. As a stand-alone subject, binocular vision is discussed in much greater detail in the texts given below. However, the material discussed in this chapter should be considered to be essential reading for all optometrists, dispensing opticians and contact lens opticians.

References

Evans B J W (2001) *Dyslexia and Vision*. Whurr, London

Evans B J W (2007) *Pickwell's Binocular Vision Anomalies*, 5th edn. Elsevier, Oxford

Evans B J W (2005) *Eye Essentials: Binocular Vision*. Elsevier, Oxford

Horwood A (2003) Neonatal ocular misalignments reflect vergence development but rarely become esotropia. *British Journal of Ophthalmology* **87**:1146–50

Pediatric Eye Disease Investigator Group (PEDIG) (2003) A randomized trial

of patching regimens for treatment of moderate amblyopia in children. *Archives of Ophthalmology* **121**:603–11

Stewart C E, Moseley M J, Fielder A R, Stephens D A, MOTAS cooperative (2004) Refractive adaptation in amblyopia: quantification of effect and implications for practice. *British Journal of Ophthalmology* **88**:1552–6

Binocular Vision Considerations in Contact Lens Practice

Bruce Evans

Introduction

Although spectacle lenses and contact lenses are both methods of correcting refractive errors, there are some optical differences between these two modes of correction. These differences are summarised in Part 2 of this book and they have some important effects on binocular vision anomalies. For some binocular vision anomalies, contact lenses bestow advantages compared with spectacle lenses and contact lenses are therefore clinically indicated, if other factors are also favourable. For other binocular vision anomalies, contact lenses are associated with disadvantages and are therefore contraindicated.

In practice, this means that knowledge of the orthoptic status of a patient is important before they are fitted with contact lenses. In cases where the same optometrist carries out the eye examination and the contact lens fitting, this is not likely to be a problem, especially if the patient's interest in contact lenses is made clear before the eye examination. In cases where the contact lens fitting is separated from the eye examination, as a precaution the contact lens practitioner may wish to carry out the 'essential investigations' in Table 31.1. This is especially important in patients with high myopia and in those who are fitted with monovision or multifocal contact lenses (see below). If any of these essential investigations reveal suspicious findings (see Chapter 30), the 'additional investigations' in Table 31.1 may also be appropriate (Evans 2005). If a patient requires a prismatic correction (or decentration to give a prismatic effect) in spectacles, and this cannot be replicated in contact lenses, contact lens wear is contraindicated.

Chapter content

- Orthoptic indications for contact lenses
- Orthoptic contraindications for contact lenses
- Monovision

Orthoptic indications for contact lenses

It was noted in Chapter 17 that there are optical problems associated with the correction of some refractive errors with spectacle lenses. In high ametropia, significant aberrations result when the patient looks away from the optical centre of a spectacle lens. This is minimised by wearing contact lenses because the lens moves with the eye. It was noted in the discussion of Figure 30.1 that improving the clarity of the optical image is likely to improve sensory fusion, which might therefore improve the orthoptic status.

The most commonly encountered refractive error, where there are marked orthoptic advantages to wearing contact lenses, is anisometropia.

Anisometropia

Anisometropia of 2.00 D or more affects 1.5% of white populations and 2.7% of Singapore adolescents (Logan and Gilmartin 2004). As contact lenses move with the eyes, the patient is likely to keep looking approximately through

Table 31.1 Binocular vision work-up for contact lens patients

Essential investigations
History: 'Have you ever had double vision, a turning eye, a lazy eye, or eye muscle surgery?'
Symptoms: 'Do you experience eyestrain/headaches/ blurring/diplopia associated with any visual task (e.g. reading, VDU use)?'
Accurate neutralisation of current spectacles to detect prisms or decentration
Cover test at distance and near
Ocular motility test (including asking for any reports of double vision)
Additional investigations (appropriate in some cases, e.g. if any of the above are abnormal)
Mallett's fixation disparity test at distance and near
Dissociation tests (particularly useful for detecting small vertical deviations)
Double Maddox rod test (particularly useful for detecting cyclo-vertical deviations) (Evans 2007)
Computerised Hess screen

These tests do not necessarily have to be carried out by the contact lens practitioner, because they may have been included in a recent eye examination. However, this should not be assumed and it is best to check this with the examining optometrist.

From Evans (2005).

the optical centres of both contact lenses, even when the position of gaze is eccentric. This means that differential prismatic effects are not likely to be problematic, whereas they may cause problems with spectacles.

There is another advantage to contact lens wear in anisometropia: they greatly reduce aniseikonia. It used to be thought that, compared with spectacle lenses, contact lenses reduce aniseikonia in refractive anisometropia (e.g. aphakia) but increase anisometropia in axial anisometropia. This theoretical prediction, known as Knapp's law, was disproved by research that revealed that contact lenses reduce aniseikonia in all forms of anisometropia (Winn et al. 1988). This finding was not a complete surprise (Edwards 1979), indicating that contact lenses are a superior method of optical correction for all cases of anisometropia and provide a more potent binocular stimulus to the visual system (Evans 2006b).

This means that non-strabismic patients with anisometropia are likely to achieve their best binocular visual acuity and stereoacuity when wearing contact lenses rather than spectacles. It is known that refractive correction with spectacles alone can improve the best corrected visual acuity in an amblyopic eye, even without patching (Gibson 1955, Moseley et al. 2007) and this therapeutic effect might be enhanced by the superior optical correction with contact lenses (Evans 2006b). It should be noted that, in patients with pure anisometropic amblyopia (no strabismus), amblyopia can respond to treatment at almost any age (Evans 2007), so contact lens correction for these individuals may have beneficial effects in some adults.

One difficulty with correcting anisometropic amblyopia with any form of refractive correction is that patients typically do not experience an immediate improvement in their visual performance. For comparison, when someone with bilateral myopia inserts contact lenses they can immediately look in the distance and experience a dramatic improvement in vision. Psychologically, this probably offsets the inconvenience of lens insertion each morning. Most individuals with anisometropic amblyopia wake up with clear vision and so there is no immediate benefit from wearing a contact lens. This makes these individuals more likely to drop out of contact lens wear, and continuous-wear contact lenses may make contact lens wear more achievable for them (Edwards 1979). The advent of silicone hydrogel lenses has made continuous wear of contact lenses safer and is likely to have a major impact on the management of anisometropic individuals, for both adults and children. Of course, the usual clinical indications for contact lenses need to be met, and a careful approach is required when considering continuous-wear contact lenses for children (Evans 2006b).

Correction of motor deviations with contact lenses

Some individuals with decompensated hetero-phoria or strabismus can be cured simply with a refractive correction. The most obvious example is fully accommodative esotropia: where there may be latent hypermetropia, causing an esotropia, and correction of the hypermetropia cures the strabismus. Often, any amblyopia resolves once the eyes have been straightened, especially if the strabismus is of recent onset. Clearly, these cases are relatively easy to treat in optometric practice and, if the patient is motivated and suitable, contact lens correction is an option.

In other individuals, the deviation may not be caused by a refractive error, although it may nevertheless be possible to correct the deviation with a refractive correction, e.g. some cases of decompensated exophoria can be treated by over-minusing the patient. A patient with emmetropia might be given −1.00 DS, which induces accommodation; this, in turn, induces accommodative convergence, reducing the exo-deviation. The goal in these cases is gradually to reduce the over-correction, and daily dispos-able contact lenses are an alternative to specta-cles that some patients find preferable.

It is also possible to prescribe prismatic cor-rections that can be helpful in some binocular vision anomalies, e.g. a small hyperphoria can be corrected by using a prism on one rigid gas permeable (RGP) contact lenses, but the effect of gravity limits this to base-down in one eye. Some soft toric designs have adequate stabilisa-tion to allow up to 2 Δ of horizontal prism in each lens (Evans 2006b).

Other orthoptic uses of contact lenses

Contact lenses of grossly inappropriate power or with opaque optic zones can be used as 'occluders'. These can be useful in amblyopia treatment or to alleviate intractable diplopia (Evans 2001).

Another use of contact lenses is for congeni-tal nystagmus (also called early onset nystag-mus or infantile nystagmus syndrome) (Evans 2005). Some evidence suggests that contact lenses can result in a reduction in nystagmus

amplitude and an improvement in visual acuity through providing tactile feedback about the eye movements (Evans 2006b).

Although some binocular vision anomalies are contraindications for monovision, occasion-ally monovision may be a solution to an orthoptic anomaly. Acquired incomitant strabis-mus in adults frequently results in diplopia and in some cases (e.g. small-angle cyclo-vertical deviations) monovision may eliminate the competitive non-fusible images and alleviate symptoms (London 1987). Acquired incomitant deviations should be medically investigated first (Evans 2005), and if monovision is to be tried the orthoptic status and symptoms should be carefully monitored.

Orthoptic contraindications for contact lenses

Contact lenses are worn on the cornea and move as the eye moves. They are therefore a more 'natural' method of correcting refractive errors than spectacles. Consequently, this section on orthoptic contraindications for contact lens wear is much briefer than the section on orthop-tic indications for contact lenses.

Contact lenses are particularly well suited to monovision because of the lack of differential prismatic effects. However, the monocular blur that occurs in monovision has a dissociating effect and monovision is therefore contraindi-cated in patients whose orthoptic status is likely to be easily compromised.

In some forms of contact lens wear, a visual compromise is deemed acceptable because of the cosmetic advantages of contact lenses, e.g. a high toric correction may not be available in a disposable lens design for a patient's most astigmatic eye, but the patient is happy to live with slight blur in this eye. If the patient has a decompensated heterophoria, strabismus or an incomitancy, this blur could conceivably cause the binocular vision to decompensate. So, atten-tion to the tests listed in Table 31.1 is important in these cases.

If a highly myopic patient wears specta-cle lenses that are correctly centred for dis-tance vision, when he reads his interpupillary distance is less than the centration distance of the lenses. This causes a base-in effect with spectacles that is helpful in cases of near exophoria. If the patient is fitted with contact lenses, he may miss this base-in prism, resulting in symptoms. A similar effect could occur in those with high hypermetropia who have a near esophoria. The investigations listed in Table 31.1 help to identify these cases who, depending on the severity of binocular vision anomaly, might need to be warned to look for symptoms or might require treatment of their binocular vision anomaly (see Chapter 30) before being fitted with contact lenses.

Superior oblique palsies can decompensate if the person is forced to fixate in the field of action of the weak muscle, i.e. to look down and in. Alternating vision multifocals are therefore contraindicated in these cases, and this applies to both spectacle and contact lens designs (Evans 2005).

Monovision

Monovision is a method of correcting presbyopia where one eye is focused for distance vision and the other for near (Evans 2006a). Although the earliest form of prescribed monovision may have been the use of monocles, monovision is not a popular method of correcting presbyopia with spectacles. This may be partly explained by the availability of very successful progressive addition lens designs for spectacles, but is no doubt also attributable to the optical problems associated with anisometropic spectacle corrections, as noted above. It was also noted that contact lenses avoid or minimise these problems, and this in part explains why monovision has proved much more popular in contact lens wearers than in spectacle wearers. Another reason is probably the compromises that are often associated with multifocal contact lenses, which are generally less successful than multifocal spectacle lenses.

Indeed, the compromises associated with multifocal contact lenses and monovision frequently lead practitioners to try a form of modi-

fied monovision. In **modified monovision,** multifocal lenses are fitted to each eye, but one is biased towards distance vision and the other towards near vision. In **enhanced monovision**, a multifocal lens is worn in one eye and a single vision lens in the other. There are some limitations of monovision contact lenses with respect to orthoptic function and these are discussed below. These limitations also often apply to modified monovision.

Effects of monovision on binocularity

A consequence of monovision is that one eye is blurred at each distance. Figure 30.1 provides a simple model of binocular vision, which stresses that a clear image in each eye is necessary to provide a good fusion lock for sensory fusion.

Monovision therefore weakens fusion and there is some evidence to suggest that tests of orthoptic function are sometimes worse in monovision wearers (Evans 2006a). For some patients this can cause a barely compensated heterophoria to decompensate and such patients can be said to have orthoptic contraindications to monovision.

Even in patients with normal orthoptic status, monovision inevitably causes impaired stereoacuity (Evans 2006a). This in itself is a contraindication for monovision in patients whose activities require good stereoacuity, e.g. pilots should not be fitted with monovision and there is a reported case in which an aircraft accident was partly attributable to the pilot wearing monovision contact lenses (Nakagawara and Veronneau 2000).

The reduction in stereoacuity in monovision is not generally considered to be a contraindication to driving, although it is advisable to warn patients who are being fitted with monovision to drive only when they have adapted to it. Most patients seem to adapt to the reduced stereoacuity, just as they adapt to suppressing the blurred image from the out-of-focus eye. Occasionally, patients do not make these adaptations, in which case they are not suitable for monovision and the fitting should be abandoned. In other cases, the patient may be able to make these adaptations under most but not all conditions. The limiting case for monovision, when it is hardest to adapt, is usually during night driving. If this is a problem for patients, they can be prescribed spectacles to wear over the contact lenses, which provide a distance over-correction for the eye with the near vision contact lens. These spectacles may also need to be worn a few times for the patient to adapt. For these, as with monovision itself, it is sensible to advise the patient to wear the new correction as a passenger in a car at night, to ensure that vision and depth perception appear normal before driving.

Contraindications for monovision

The monocular blur that is inevitable in monovision can make binocular fusion impossible or difficult for some patients. As noted at the start of Chapter 30, three factors can make it difficult for a patient to compensate for a heterophoria: an unusually large heterophoria, poor sensory fusion and poor motor fusion. Monovision makes sensory fusion more difficult than normal, because the foveal receptive fields are suppressed in one eye. However, the larger receptive fields away from the fovea are relatively insensitive to blur and are probably not suppressed, allowing sensory fusion to occur. In other words, binocular fusion may be slightly compromised, but for most patients this is tolerable. For some patients, this may be intolerable, and this is particularly likely when one of the other two factors mentioned above is also compromised: when there is an unusually large heterophoria and/or impaired motor fusion (low fusional reserves).

Clinically, this means is that, if a practitioner is contemplating fitting a patient with monovision, it is advisable to check that the heterophoria is well compensated. It is usually enough to carry out the items listed as essential investigations in Table 31.1 and, if any of these are abnormal, to consider the need for undertaking the additional investigations in Table 31.1 (see Chapter 30). These procedures may also be required in any patient who is wearing monovi-

sion and develops symptoms that are suggestive of binocular vision anomalies (principally, diplopia or asthenopia). Monovision is certainly contraindicated in patients with intermittent strabismus (Kushner and Kowal 2003).

Monovision may force patients with long-standing constant non-alternating stabismus (e.g. microtropia) to fixate some of the time with their strabismic eye, which they habitually suppress. This can cause **fixation switch diplopia**, and monovision is therefore contraindicated in these cases (Kushner and Kowal 2003).

In Chapter 30, incomitant deviations were defined as deviations with an angle that varies in different positions of gaze and depending on which eye is fixating. The latter part of this definition relates to the fact that the angle of deviation characteristically increases when the patient is fixing with the paretic eye. As monovision forces the patient to fix with one eye for each distance, the angle of the deviation may be increased at one distance compared with the angle to which the patient has adapted; this might cause decompensation (Kushner and Kowal 2003). Decompensation could also result from the impairment to sensory fusion (as discussed above) and, in the case of a superior oblique palsy, if the paretic eye is forced to look down when reading. For all these reasons, it is advisable to avoid monovision in patients with incomitant deviations.

Ocular dominance

An interesting decision when fitting people with monovision is the determination of which eye should be used for which distance. A common belief is that the **dominant eye** should be fitted with the distance lens, but this is an unsatisfactory answer because the dominant eye varies with different tasks (Evans 2006a). Often contact lens wearers who are becoming presbyopic present at a contact lens check and the practitioner finds that the refractive error of one eye has changed to make this eye more myopic or less hyperopic (Evans 2006a), i.e. the patient has spontaneously started monovision without any help from the practitioner. If patients are asymptomatic, it is best to leave them as they are, and to modify the lens powers as required to continue with monovision. This, of course, assumes that the patient remains happy with the vision and the concept of monovision, which should be explained.

If the practitioner wishes to commence monovision, then if the patient is of an inquisitive nature he can be given trial contact lenses and instructed to try the distance correction in one eye for a few hours or days, and then the distance correction in the other eye for a similar period.

Some patients are eager to participate in clinical decisions in this way, but others are happier with a more didactic role for the practitioner. For these patients, the practitioner can use either of two plus lens blurring tests. For both tests, the patient starts by wearing the distance correction in both eyes and fixating a distance optotype that is close to the threshold for visual acuity. With one test, the distance eye is selected as the eye that requires the least amount of plus for the patient to detect blur at distance under binocular condition. The other test involves holding a +2.00 DS lens over each eye in turn and asking the patient when he feels most comfortable.

In all cases, patients should be advised not to carry out tasks that require stereoacuity (e.g. driving) until they have adapted to monovision, and care should be taken to ensure that they understand the limitations as well as the advantages of monovision.

Concluding points

Habitual contact lens wearers often present for an eye examination wearing their spectacles rather than their contact lenses. This is unfortunate because many of the tests that the optometrist performs during a routine eye examination do not relate to most of the patient's day. As part of an eye examination, an optometrist assesses the patient's binocular vision. If the patient wears spectacles only, say, 10% of the waking day, the results of such an assessment are relevant to only 10% of the patient's day. The patient

may have a binocular vision problem while wearing the contact lenses (for 90% of the day) but not when wearing spectacles, and the optometrist will never know! A basic binocular vision assessment should therefore be included as part of every contact lens after-care examination. If any anomalies are found by a contact lens practitioner, these should be discussed with the optometrist. Unless a practice fits contact lenses to one-eyed patients only, binocular vision cannot be ignored!

References

Edwards K H (1979) The management of ametropic and anisometropic amblyopia with contact lenses. *Ophthalmic Optician* December 8:925–9

Evans B J W (2001) Diplopia: when can intractable be treatable? In: Evans B, Doshi S (eds) *Binocular Vision and Orthoptics*. Butterworth-Heinemann, Oxford: 50–7

Evans B J W (2007) *Pickwell's Binocular Vision Anomalies*, 5th edn. Elsevier, Oxford

Evans B J W (2005) *Eye Essentials: Binocular Vision*. Elsevier, Oxford

Evans B J W (2006a) Monovision: a systematic review. *Ophthalmic and Physiological Optics* in press

Evans B J W (2006b) Orthoptic indications for contact lens wear. *Journal of the British Contact Lens Association* in press

Gibson H W (1955) Amblyopia. In: *Textbook of Orthoptics*. Hatton Press, London: 170–94

Kushner B J, Kowal L (2003) Diplopia after refractive surgery: occurrence and prevention. *Archives of Ophthalmology* 121:315–21

Logan N S, Gilmartin B (2004) School vision screening, ages 5–16 years: the evidence-base for content, provision and efficacy. *Ophthalmic and Physiological Optics* 24:481–92

London R (1987) Monovision correction for diplopia. *Journal of the American Optometry Association* 58:568–70

Moseley M J, Neufeld M, McCarry B, Charnock A, McNamara R, Rice M, Fielder A (2002) Remediation of refractive amblyopia by optical correction alone. *Ophthalmic and Physiological Optics* 22:296–9

Nakagawara V B, Veronneau S J H (2000). Monovision contact lens use in the aviation environment: a report of a contact lens-related aircraft accident. *Optometry* **71**:390–5

Winn B, Ackerley R G, Brown C A, Murray F K, Prais J, St John M F (1988). Reduced aniseikonia in axial anisometropia with contact lens correction. *Ophthalmic and Physiological Optics* **8**:341–4

Index

Only information relating to contact lenses in general is entered under that heading; information relating specifically to RGP or soft lenses is given under the more specific headings.

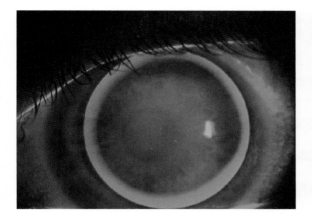

Figure 17.2 An RGP contact lens that has been fitted steep relative to the curvature of the cornea. This relationship will produce a positive tear (or liquid) lens.

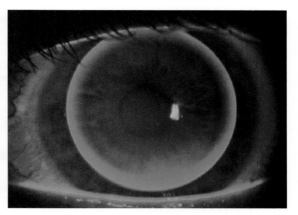

Figure 17.3 An RGP contact lens that has been fitted in alignment with the cornea. This relationship will produce an afocal or plano tear lens.

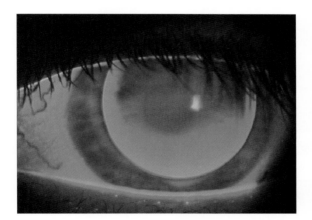

Figure 17.4 An RGP contact lens that has been fitted flat relative to the curvature of the cornea. This relationship will produce a negative tear lens.

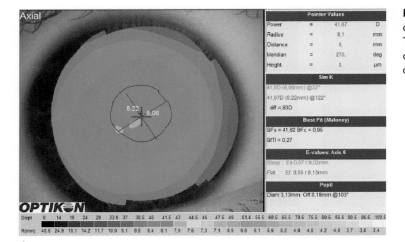

A

Figure 18.9 (a) A topographical contour map for a spherical cornea. Topographical contour maps for (b) a cornea with high astigmatism and (c) a cornea with keratoconus.

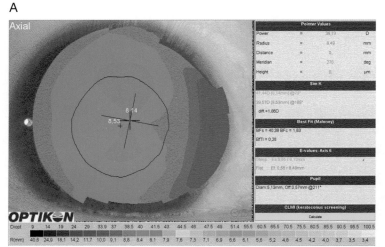

B

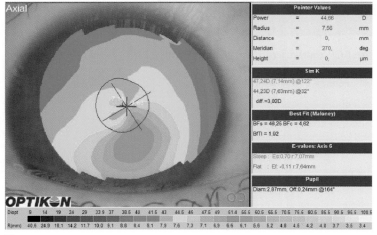

C

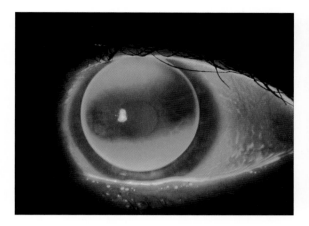

Figure 22.1 A spherical RGP contact lens fitted to a toroidal cornea.

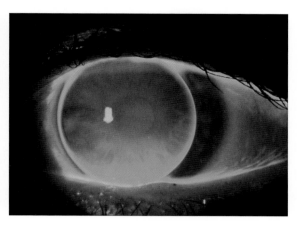

Figure 22.2 The same eye as shown in Figure 22.1 but fitted with a back surface toric RGP lens.

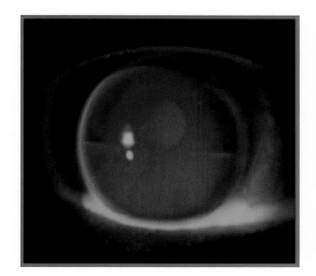

Figure 25.1 Fluorescein pattern of the Tangent Streak RGP bifocal (translating design).

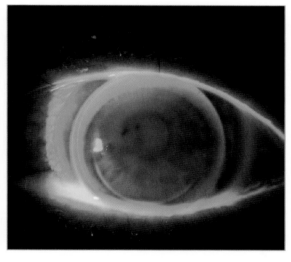

Figure 25.2 Fluorescein pattern of the PresbyLite RGP bifocal (translating design).

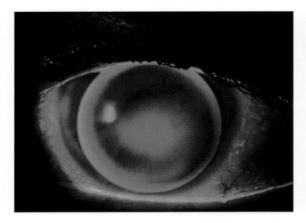

Figure 25.5 Fluorescein pattern of a well fitted RGP, back surface, aspheric multifocal (fitted typically 0.4 mm steeper than flattest *K*).

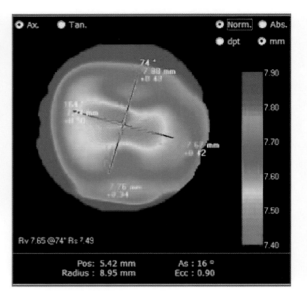

Figure 25.6 A topography map clearly demonstrating corneal moulding after wear of an RGP back surface aspheric multifocal.

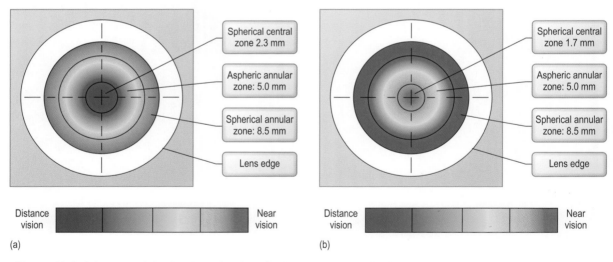

(a)   (b)

Figure 25.8 Schematic of the Proclear Multifocal (multi-zone multifocal). (a) D lens: dominant eye; (b) N lens: non-dominant eye.